Pesticides in Aquatic Environments

Pesticides in Aquatic Environments

Edited by:
Erick Soares Lins

Delve
PUBLISHING
www.delvepublishing.com

Pesticides in Aquatic Environments

Erick Soares Lins

Delve Publishing
2010 Winston Park Drive,
2nd Floor
Oakville, ON L6H 5R7
Canada
www.delvepublishing.com
Tel: 001-289-291-7705
001-905-616-2116
Fax: 001-289-291-7601
Email: orders@arclereducation.com

ISBN: 978-1-77361-033-7 (Hardcover)

DECLARATION

Some content or chapters in this book are open access copyright free published research work, which is published under Creative Commons License and are indicated with the citation. We are thankful to the publishers and authors of the content and chapters as without them this book wouldn't have been possible.

ABOUT THE EDITOR

Erick Soares Lins

Erick Lins was born and raised in Brazil, and attended the University of Brasília where he obtained his bachelor's degree in Chemistry. Right after graduating, he began working at the General Coordination of Laboratories of the Ministry of Agriculture, Livestock and Food Supply. In that position, Erick was part of the team in charge of setting the strategic goals to the National Agricultural Laboratory Network, in order to support the inspection branch with an advanced analytical basis. This work was directly focused on food production systems and the identification of pesticide residues in several products, and had state-of-the-art analytical instrumentation and quality assurance as fundamental parts of it. After obtaining his master's degree in Chemistry, working with the development of analytical systems based on flow injection, Erick went to pursue his PhD at the University of Saskatchewan, in Canada, with *in situ* spectroelectromicroscopy, along with his wife Verônica and their daughter Rebeca.

TABLE OF CONTENTS

SECTION 2

Latest Resarch

LIST OF CONTRIBUTORS

Erick Soares Lins

Francisco Sánchez-Bayo

Faculty of Agriculture and Environment, The University of Sydney, Eveleigh, NSW, Australia

Koichi Goka

National Institute for Environmental Studies, Tsukuba, Japan

Daisuke Hayasaka

Faculty of Agriculture, Kindai University, Nara, Japan

Francisco Antonio Guardiola

Department of Cell Biology and Histology, Faculty of Biology, University of Murcia, Murcia 30100, Spain

Alberto Cuesta, José Meseguer

Department of Cell Biology and Histology, Faculty of Biology, University of Murcia, Murcia 30100, Spain

Maria Angeles Esteban

Department of Cell Biology and Histology, Faculty of Biology, University of Murcia, Murcia 30100, Spain

Sajal Ray

Aquatic Toxicology Laboratory, Department of Zoology, University of Calcutta, Kolkata, West Bengal, India

Soumalya Mukherjee

Aquatic Toxicology Laboratory, Department of Zoology, University of Calcutta, Kolkata, West Bengal, India

Niladri Sekhar Bhunia

Aquatic Toxicology Laboratory, Department of Zoology, University of Calcutta, Kolkata, West Bengal, India

Anindya Sundar Bhunia

Aquatic Toxicology Laboratory, Department of Zoology, University of Calcutta, Kolkata, West Bengal, India

Mitali Ray

Aquatic Toxicology Laboratory, Department of Zoology, University of Calcutta, Kolkata, West Bengal, India

Paul K. Mensah

Institute for Water Research, Rhodes University, Grahamstown, South Africa

Carolyn G. Palmer

Institute for Water Research, Rhodes University, Grahamstown, South Africa

Wilhelmine J. Muller

Institute for Water Research, Rhodes University, Grahamstown, South Africa

PREFACE

The goal of this book is to bring together several factors regarding a threatening concern of our society today: how our production systems impact life on Earth. All ecosystems around us endure the effects of the way we manage resources in order to survive and thrive. All biomes are highly interconnected, and it is clear that we cannot live with unhealthy land, atmosphere and water.

The aquatic ecosystems are incredibly delicate, yet, they have been suffering a lot from man-made pollution in several forms, from industrial effluents to the presence of pesticides, mostly coming from agricultural practices. Humans have a special connection with water, not only because it is the origin of life itself, but also because it is a primary material for so much that we need, even if we do not see it clearly. Understanding the mechanisms by which these chemicals reach water bodies, how they behave in such systems, and the impact they have on organisms of all sizes, shapes and roles in the food web, is essential to better protect them. By doing this, directly and indirectly we are protecting ourselves.

This book is a collection of basic literature, which present findings that are foundational for the environmental research of the forthcoming decades, recent mechanistic discoveries at cellular levels, as well as a summary of useful strategies to assess risk and to execute monitoring with advanced laboratory techniques.

Editor

SECTION 1

1
CHAPTER

INTRODUCTION TO THE CLASSIFICATION OF PESTICIDES

Erick Soares Lins

THE FIRST OF THEIR KIND

The Era of chemical pesticides as we know today began in the 1940s, with the introduction of organo-chlorinated compounds, such as Aldrin, Dieldrin, Chlordane and others. The promise of those substances was to exterminate bugs and thus stop the spread of the diseases they carried and were vectors to. Their introduction in the market helped shape what would become an incredibly powerful chemical industry.

Among those, DDT (Dichlorodiphenyltrichloroethane) is probably the most widely known, and is considered by many to be the great initiator of the use of synthetic chemicals to protect crops against pests. DDT had a major role in World War II, when it was widely used to effectively control the incidence of diseases spread by insects, mainly malaria, yellow fever, and typhus, given its recently discovered broad-spectrum insecticide properties and inexpensive synthesis. This development granted the Swiss chemist Paul Hermann Müller the Nobel Prize of Medicine in 1948 for "his discovery of the high efficiency of DDT as a contact poison against several arthropods".

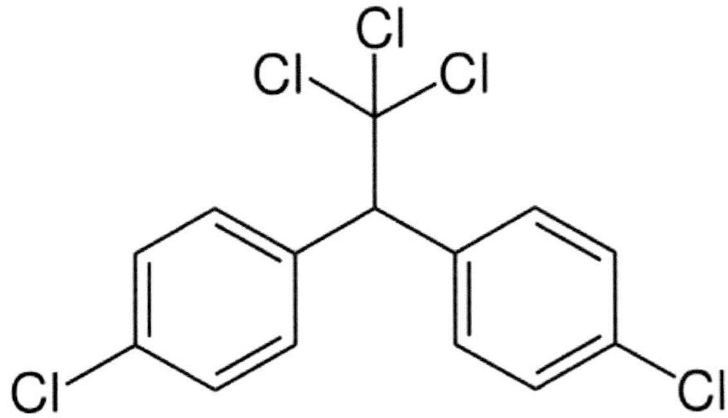

Figure 1: DDT molecule

For some time, little or no concern was raised about the safety of the application of DDT and the like, since no immediate adverse effects were ever observed. In fact, those new chemicals seemed safer to work with than their predecessors, as some of them were based on arsenic or mercury derivatives, and the few documented farm occurrences of DDT poisoning were attributed to misuse. The clear benefits and the unclear risks made DDT the weapon of choice not only to fight insects in urban areas, but largely in agriculture. In 1945, the year it was officially released commercially, production of DDT reached around 15,000 tonnes, nearly three times the amount produced the previous year, and it kept growing until 1963 when it peaked at 81,000 tonnes produced in the US.

During that time, however, some research started to focus on the properties of these chemicals, and the better known risks associated to DDT and similar substances ultimately led to a ban that was enacted on January 1st 1970 in Sweden. Nowadays, most countries have prohibited the use of DDT on crops, with one last production plant still operational in India. Although not extensively used in agriculture as in the past, DDT is still useful for vector control of malaria in underdeveloped and tropical countries, being deemed as the salvation of millions of lives that would otherwise succumb to insect-borne diseases.

Figure 2: An advertisement for Killing Salt Chemicals, from 1947, stating the many benefits of the use of DDT, not only in farms, but also in households.

It is notable that the issues surrounding pesticides are treated differently depending on the forces at play and their interests, or what it is that is judged to be most important. While some populations are able to choose safer alternatives instead of being chronically exposed to a chemical hazard, a cheap and effective insecticide can save populations from diseases and famine when nothing else is in place. Even the terminology in the regulatory basis is still a matter of debate. While some countries prefer to call such chemicals "plant protection products" and somehow emphasize the reason why they are being used and their inherent benefits, other choose to call them "agrotoxic" assuming that it will give the public the notion that these products are not safe or healthy. The FAO (Food and Agriculture Office) has a technical definition for the term **Pesticide** in its International Code of Conduct on the Distribution and Use of Pesticides (FAO, 2003):

Pesticide *means any substance or mixture of substances intended for preventing, destroying or controlling any pest, including vectors of human or animal disease, unwanted species of plants or animals causing harm during or otherwise interfering with the production, processing, storage, transport or marketing of food, agricultural commodities, wood* *and wood products or animal feedstuffs, or substances which may be administered to animals for the control of insects, arachnids or other pests in or on their bodies. The term includes substances intended for use as a plant growth regulator, defoliant, desiccant or agent for thinning fruit or preventing the premature fall of fruit, and substances applied to crops either before or after harvest to protect the commodity from deterioration during storage and transport.*

From a more economic standpoint, "Pests" are defined by some specialists as any *"animal or plant pathogen whose population density exceeds some unacceptable threshold level, resulting in economic damage"* (Horn, 1988).

Figure 3: Adult Mexican bean beetle (photo by omafra.gov.on.ca)

CLASSIFICATIONS OF PESTICIDES

It is common sense that there may be too many pesticides in the market for anyone to keep track of, and the number of registered substances for pest control worldwide is difficult to estimate. Different regulatory bodies have different approval criteria, and the simple reality of the crops in different regions demand different controls. That body of information is not always easy to retrieve, although it is somehow available to the public. Nonetheless, a good hint to start getting an idea of the size of the universe we are considering here is the scope of multi-residue analytical methods applied to food and environmental samples, usually at the service of governmental surveillance programs. In those cases, several techniques can come up to about 900 different molecules, being most of them of the active ingredient, although it includes also metabolites and degradation products that may also be toxicologically relevant, as well as forbidden pesticides.

The analytical approach to the presence of pesticides in environmental samples will be explored later cn, but for now it is enough to say that the analytes are addressed as groups with certain similarities in their chemical and physical properties, which will be more or less amenable to different analytical methods. Likewise, other categorizations of pesticides are in place, and they will take certain characteristics into account. A very important one considers their main uses and type of pest as established by the World Health Organization (WHO), although their recommendation stresses that some pesticides may be effective for more than one of those uses:

Table 1: WHO classification of pesticides considering their main intended utilization

AC -acaricide	L - larvicide
AP - aphicide	M - molluscicide
B - bacteriostat (soil)	MT - miticide
FM - fumigant	N - nematocide
F - fungicide, other than for seed	O - other use for plant pathogens treatment

FST - fungicide, for seed treatment	PGR - plant growth regulator
H - herbicide	R - rodenticide
I - insecticide	-S applied to soil: not used with herbicides or plant growth regulators
IGR - insect growth regulator	RP( ) - repellant (species)
Ix - ixodicide (for tick control)	SY - synergist

In terms of their intended use by applicant certification, the categories often used are shown in Table 2.

Table 2: Classification of pesticides regarding application authorization

Category
Agricultural Plant Pest Control
Agricultural Animal Pest Control
Antimicrobial Pest Control
Aquatic Pest Control
Forest Pest Control
Ornamental and turf Pest Control
Right-of-Way Pest Control
Public Health Pest Control
Seed Treatment
Wood Treatment

There are categories of pesticides whose primary objective is not necessarily direct pest elimination.

Table 3: Classes of pesticides for indirect pest elimination

Category
Defoliant

Desiccant
Insect Growth Regulator
Plant Growth Regulator

Also by mode of action:

Table 4: Pesticide categories by general mode of action

Group	Action	Description
Herbicide	Contact	Action on plant parts happens in the surfaces it covers.
	Systemic	Acts inside the plant system, after absorbed.
	Selective	Acts in certain plants and not others, even when applied to them alongside.
	Non- selective	Acts on all plants where it is applied.
	Residual	Proceeds treatment for longer periods by slowly releasing and breaking down the main product.
	Non-residual	Action lasts only the short period that follows application.
Fungicide	Protective	Acts preventively to an infection.
	Eradicant	Acts on an already settled disease.
Insecticide	Contact	Acts on insect by contact with the product.
	Inhalation	Acts on insect at inhalation of the product.
	Ingestion	Acts on insect at feeding.

And some categories based on their mechanism of action:

Table 5: Classification of insecticides based on their biological action. (Reproduced from IRAC)

Main Group	**Primary site of action**
Acetylcholinesterase (AChE) inhibitors	Nerve action
GABA-gated chloride channel blockers	Nerve action
Sodium channel modulators	Nerve action

Nicotinic acetylcholine receptor (nAChR) competitive modulators	Nerve action
Nicotinic acetylcholine receptor (nAChR) allosteric modulators	Nerve action
Glutamate-gated chloride channel (GluCl) allosteric modulators	Nerve and muscle action
Juvenile hormone mimics	Growth regulation
Miscellaneous nonspecific (multi-site) inhibitors	
Chordotonal organ TRPV channel modulators	Nerve action
Mite growth inhibitors	Growth regulation
Microbial disruptors of insect midgut membranes	
Inhibitors of mitochondrial ATP synthase	Energy metabolism
Uncouplers of oxidative phosphorylation via disruption of the proton gradient	Energy metabolism
Nicotinic acetylcholine receptor (nAChR) channel blockers	Nerve action
Inhibitors of chitin biosynthesis, type 0	Growth regulation
Inhibitors of chitin biosynthesis, type 1	Growth regulation
Moulting disruptors, Dipteran	Growth regulation
Octopamine receptor agonists	Nerve action
Mitochondrial complex III electron transport inhibitors	Energy metabolism
Mitochondrial complex I electron transport inhibitors	Energy metabolism
Voltage-dependent sodium channel blockers	Nerve action
Inhibitors of acetyl CoA carboxylase	Lipid synthesis, growth regulation
Mitochondrial complex IV electron transport inhibitors	Energy metabolism
Mitochondrial complex II electron transport inhibitors	Energy metabolism
Ryanodine receptor modulators	Nerve and muscle action

Chordotonal organ Modulators - undefined target site	Nerve action
Compounds of unknown or uncertain MoA	Target protein responsible for biological activity is un-known, or uncharacterized

And there are toxicity-based classifications, also established by the WHO and GHS ("The Globally Harmonized System of Classification and Labelling of Chemicals"), which considers LD50 values for rats and sets groups separated by toxicity ranges:

Table 6-A: Classification of pesticides based on toxicity established by the World Health Organization

WHO Class		LD50 for the rat (mg/kg body weight)	
Oral		Dermal	
Ia	Extremely Hazardous	< 5	< 50
Ib	Highly Hazardous	5 to 50	50 to 200
II	Moderately Hazardous	50 to 2000	200 to 2000
III	Slightly Hazardous	> 2000	> 2000
U	Unlikely to present acute hazard	5000 or higher	

Table 6-B: Classification of pesticides based on toxicity established by the Globally Harmonized System of Classification and Labelling of Chemicals.

GHS Category	Classification criteria			
	Oral		Dermal	
	LD50 (mg/kg bw)	Hazard Statement	LD50 (mg/kg bw)	Hazard Statement
Category 1	< 5	Fatal if swallowed	< 50	Fatal in contact with skin
Category 2	5 to 50	Fatal if swallowed	50 to 200	Fatal in contact with skin
Category 3	50 to 300	Toxic if swallowed	200 to 1000	Toxic in contact with skin
Category 4	300 to 2000	Harmful if swallowed	1000 to 2000	Harmful in contact with skin
Category 5	2000 to 5000	May be harmful if swal-lowed	2000 to 5000	May be harmful in contact with skin

All categorization serves a particular purpose, and varies in terms of amplitude and complexity. They carry in their nomenclatures information that is important for management, risk assessment, biochemical investigation and so on. A very important classification is the one that is based in the chemical families and molecular structures. The idea is that understanding what pesticides have in common at molecular level will help understand their chemical and physical properties, as well as their behavior in an organism. It will also increase understanding of its mechanisms by knowing the molecules' structural backbone and functional groups. This classification is also important for analytical purposes, as mentioned before.

Table 7: Classification of pesticides based on their chemical families, and most important functional groups represented (data from *Guide de Classement des Ingredients Actifs par Groupes Chimiques, Quebec,* 2007)

Group	General structures
Aryloxyphenoxypropionic acids and derivatives	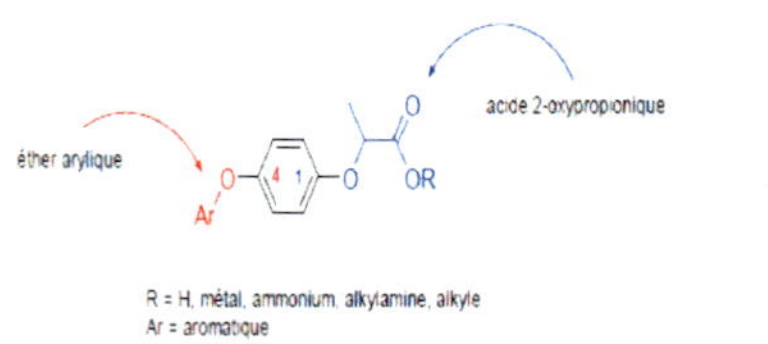
Aryloxycarboxylic acids and derivatives	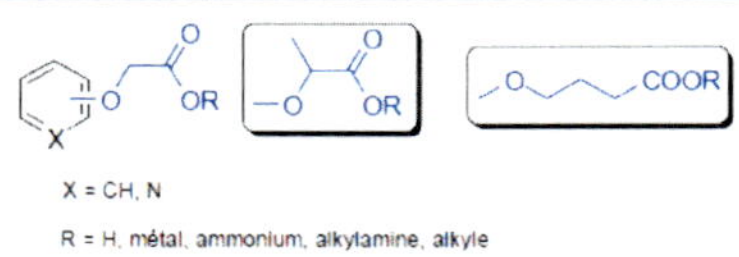
Benzoic acids and derivatives	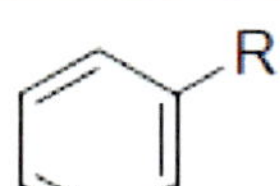 R=COOH, SO_3H etc
Phosphonic acids and derivatives	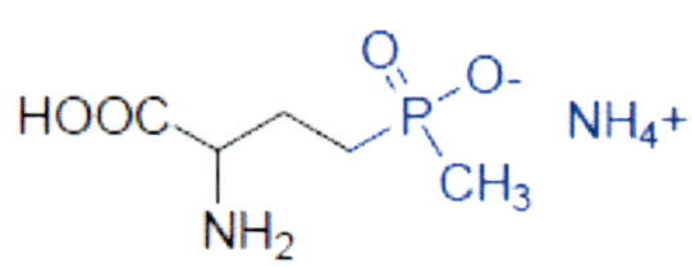
Phtalic acids and derivatives	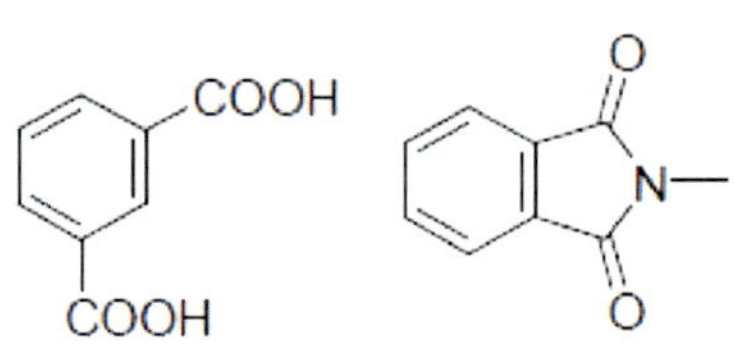

Fatty acids and surfactants	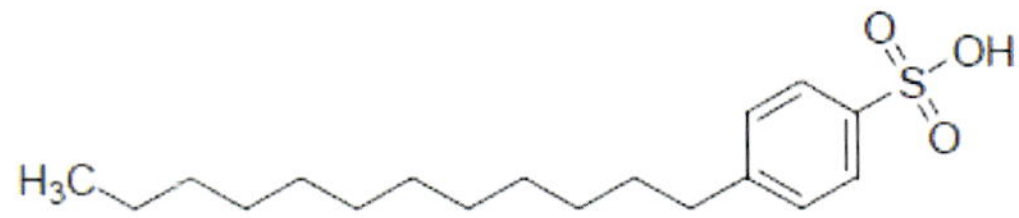
Halogenated Organic acids and derivatives	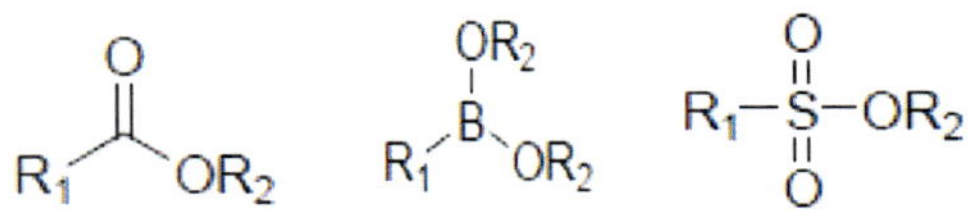
Acylureas	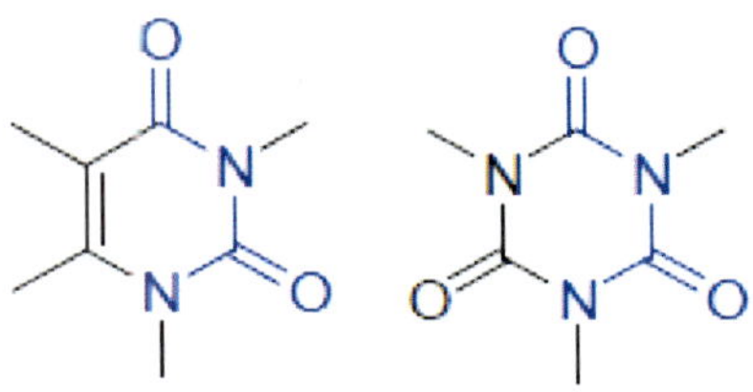
Alcohols	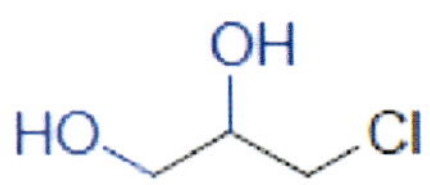
Aldehydes	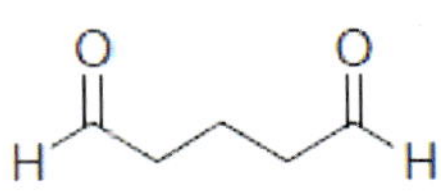
Amides	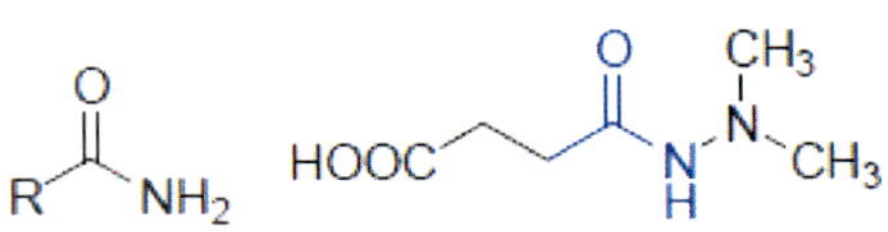
Amines	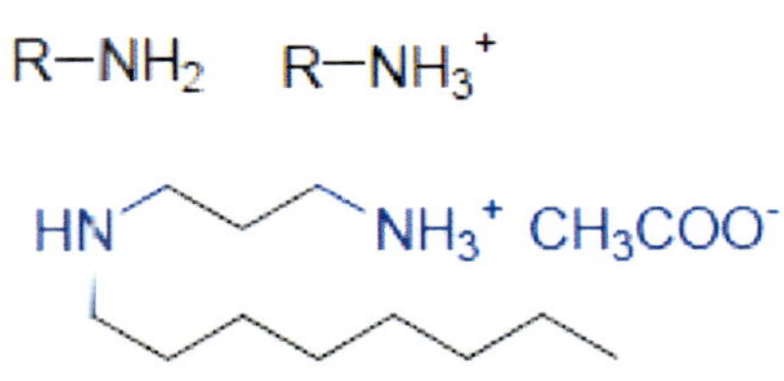
Quaternary ammonium	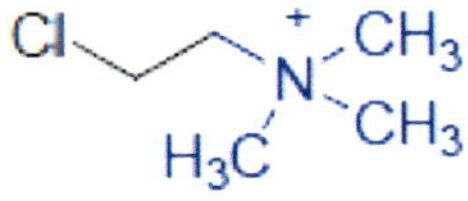

Anilides	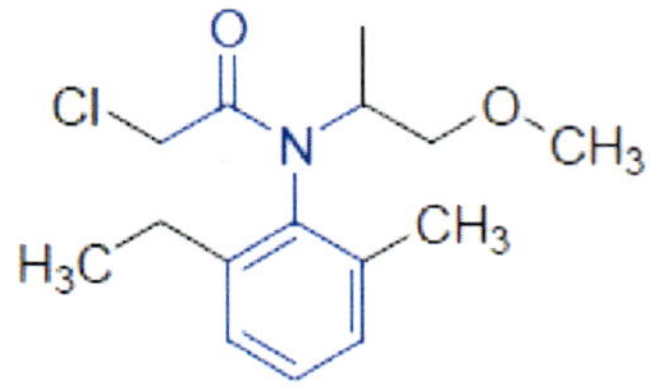
Anilines	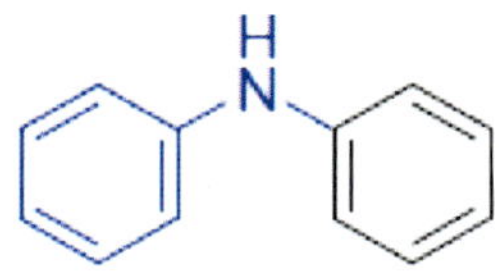
Azoles, oxazoles and thiazoles	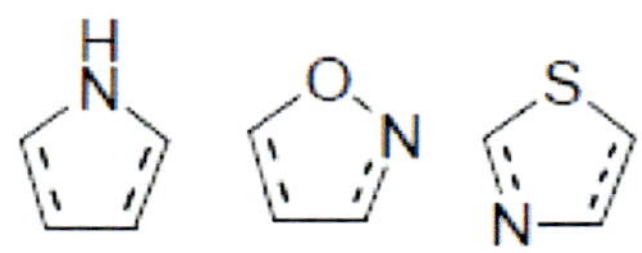
Benzamide	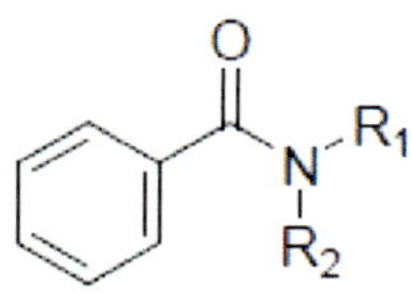
Benzonitriles	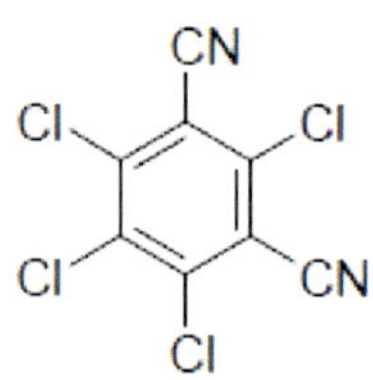
Biscarbamates	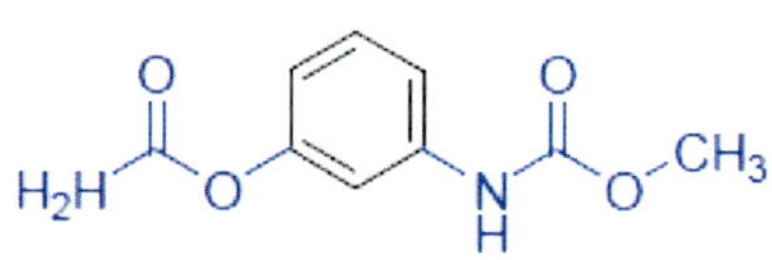
Carbamates	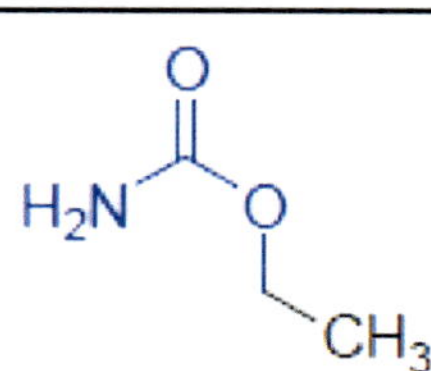

Chlorophenols	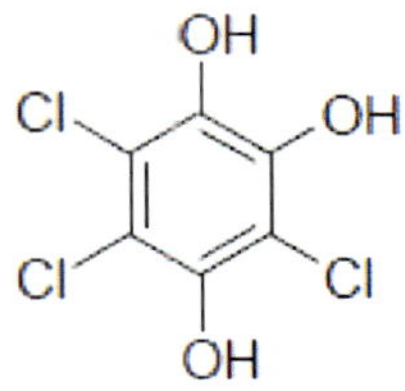
Chlorotriazines	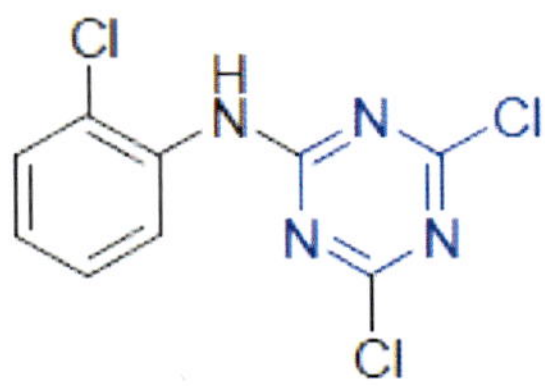
Chromenones	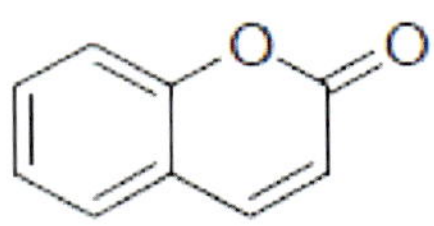
Cyclohexanedione-oximes	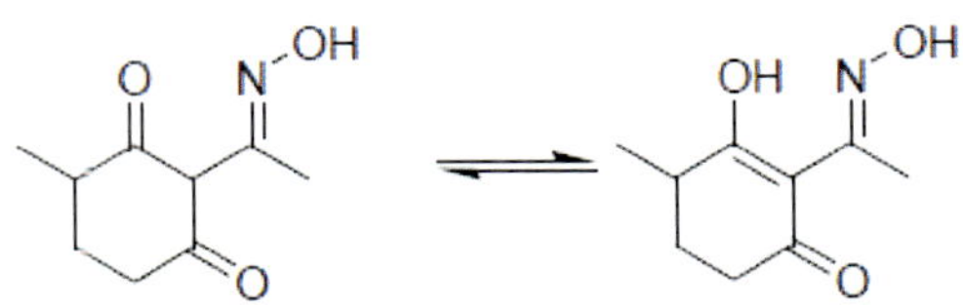
Diazines	
Dinitrobenzenes	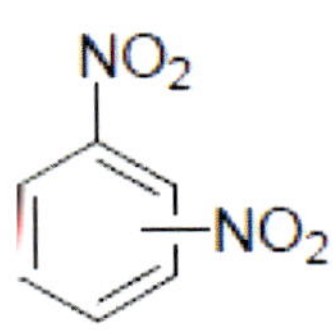
Dithiocarbamates	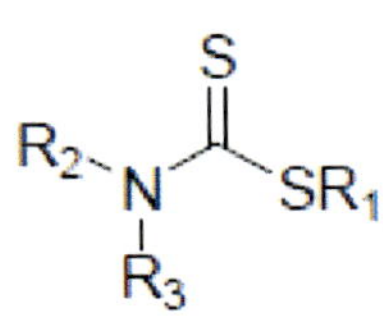

Dithiophosphates	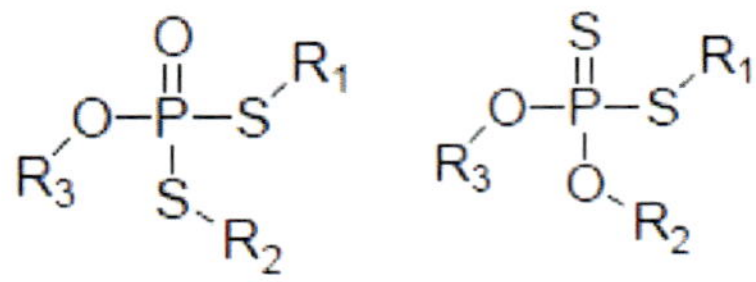
Guanidines	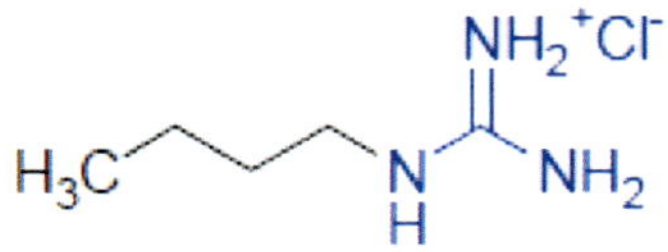
Mineral and plant oils	Long chain hydrocarbons
Hydrocarbons	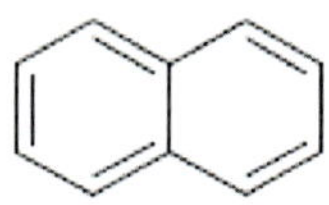 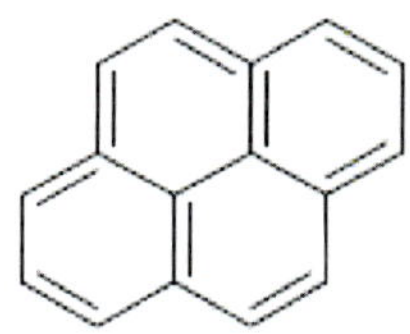
Imidazolinones	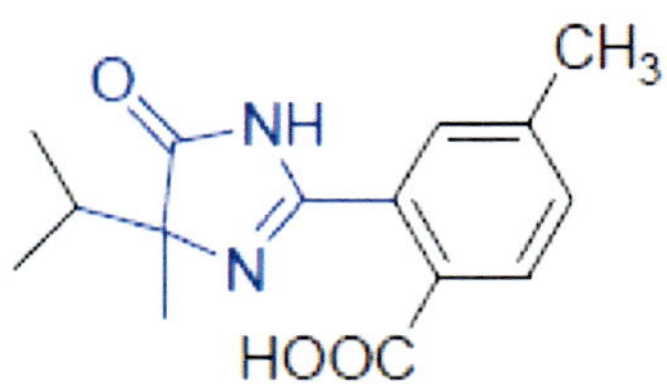
Indanediones	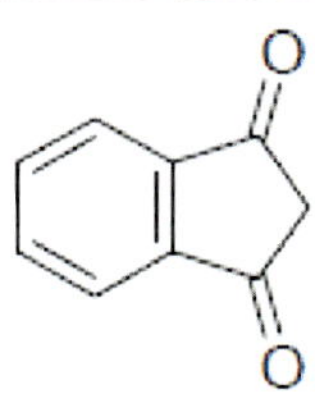

Inorganics	$Pb(CH_3CO_2)_2$ Na_2CO_3 S_8 AlP $CuSO_4$
β-Methoxyacrylates	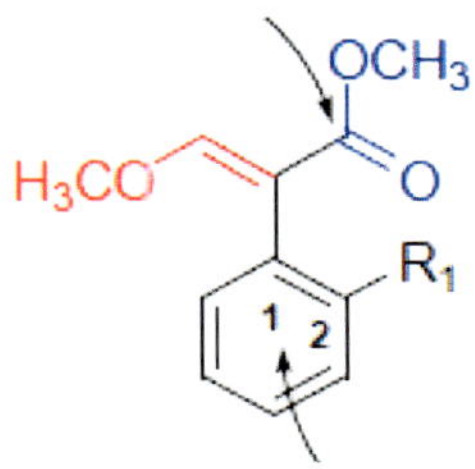
Morpholines and oxathi-ines	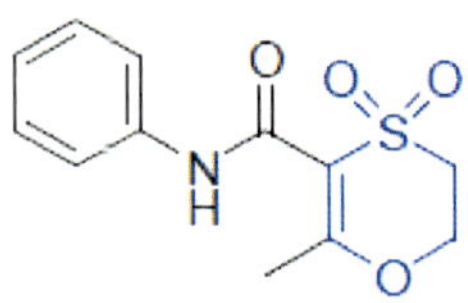
Nitrobenzenes	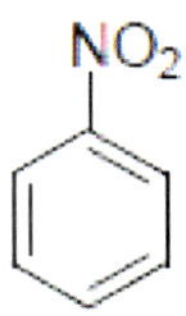
Organochlorinated	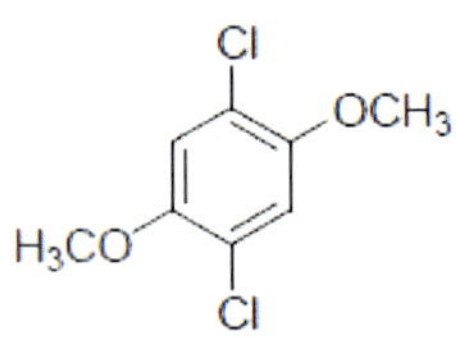
Organohalogenated	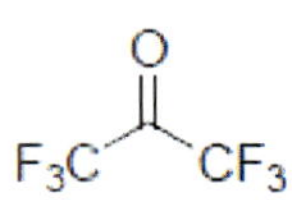

Oxime-carbamates	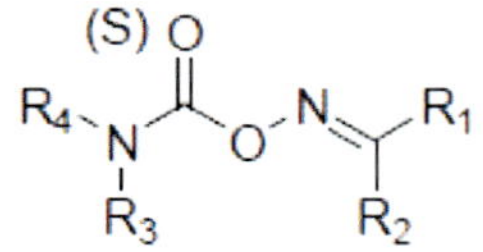
Phenols	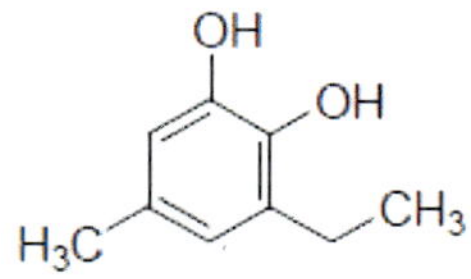
Pheromones	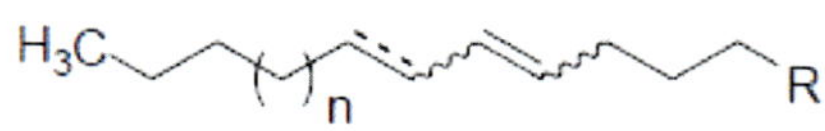 R = CH_3, CH_2OH, CH_2OAc, $COOR_1$ n = 1, 2, 3...
Phosphates	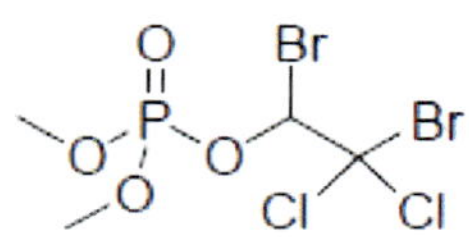
Phosphoramidothioates	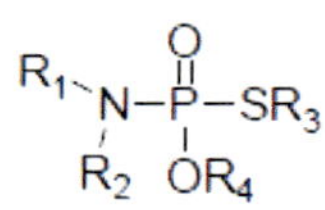
Pyrethrinoids	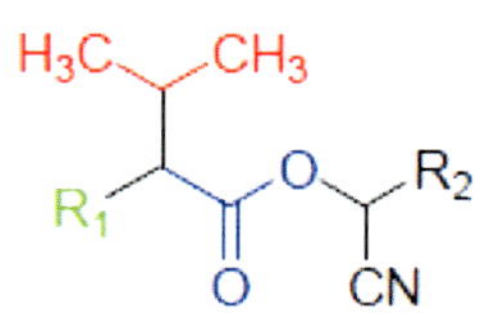
Pyridines	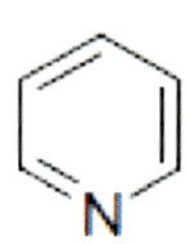

Sulfonylureas	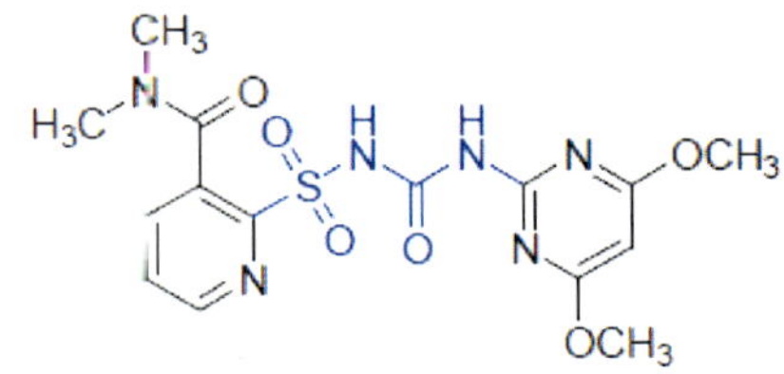
Thiocarbamates	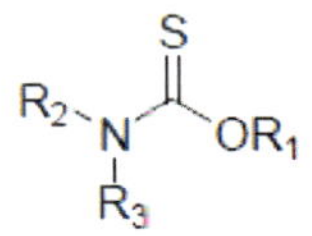 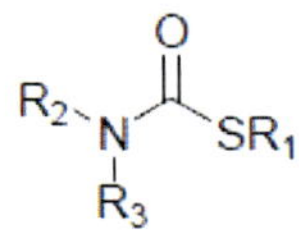
Thiophosphates	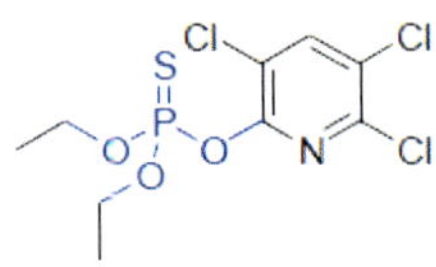
Triazines and tetrazines	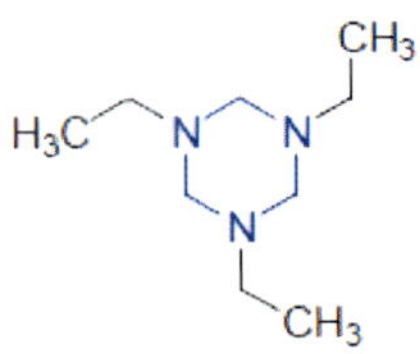
Triazoles	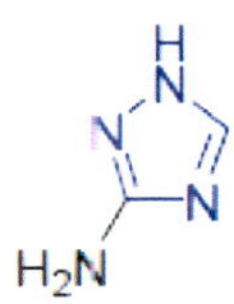
Ureas	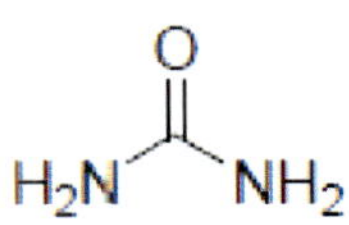 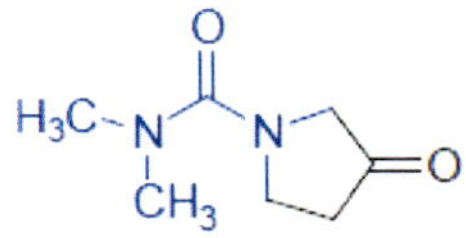

Following that same approach, allied to a regulatory viewpoint, the Regulation (EC) No 1185/2009 of the European Parliament and of the Council of 25 November 2009 lists a total of more than 500 active ingredients that have been identified at some point, and that may be subject to monitoring depending on their properties, toxicity and usage. It is a more detailed categorization and, therefore, longer. (See Annex I).

Pesticides may also be divided by the way their formulations are presented:

Table 8: Classification of pesticide based on the form the products are presented and made commercially available.

Presented form	Most common examples
Solid form	Bait, powder, wettable powder
Liquid form	Spray, emulsifiable concentrate, solution
Gaseous form	fumigant

CONSIDERATIONS ABOUT TOXICITY RISKS OF PESTICIDES

History shows that the advent of chemical pesticides in the first moment brought great benefits, such as: lowering incidence of diseases that public policies failed to cover; decreasing food prices, and providing better supply specially in post-war times; and generated a large industry that counts on technological innovation, creates jobs, and moves an enormous revenue. It all comes with a price, though. The example of DDT is remarkable because of its social impact, and it set the pace to how society perceived pesticides in the following decades. These are the main concerns that culminated in the ban of DDT and organochlorinated pesticides, and which are still present in this discussion:

Risks to human health - the safety of these chemicals are assessed in terms of acute and chronic effects upon exposure. As defined by IUPAC:

- Acute toxicity is the ability of a substance to cause adverse effects within a short time of exposure, which is equals to a period of 14 days from the administration of a single dose, or with multiple doses within 24 hours.
- Chronic toxicity means the adverse effects that persist over a long period of time from the exposure, even if those effects are delayed and not immediately manifest. Chronic exposure, which is a continuous or repetitive contact with the toxic substance, may lead to different adverse effect.

Pest resistance - eventually, pesticides become less effective because the pests themselves become more resistant. It is a micro-evolutionary process, in which a genetic adaptation occurs. If but a few individuals survive the effects of a pesticide due to any genetic characteristic, they have a chance to pass those to upcoming generations. After a while, a whole population shares that genetic expression. An additional problem is that, before replacing the pesticide when it is no longer showing the same performance, it is common to increase doses, and that means there is more material being released in the environment.

Environmental persistence - pesticides can be carried away in the air, as well as reach water streams, and adsorb to soil and sediments. Some might be resistant to degradation process such as hydrolysis, photolysis and biodegradation. Eventually, those molecules are absorbed by organisms and enter the food chain and, in many cases, they bioaccumulate possibly even reaching foodstuffs that humans consume.

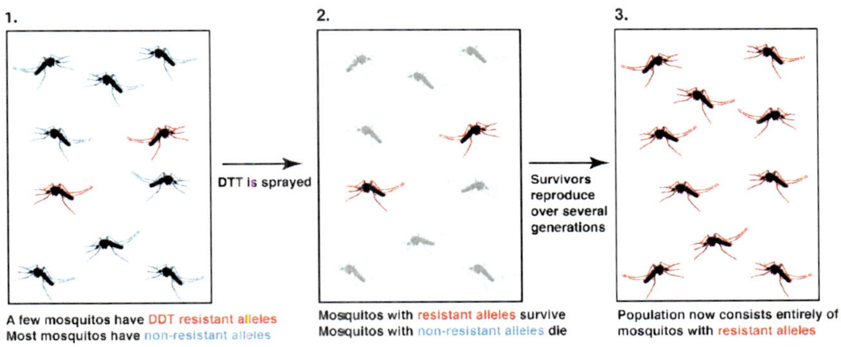

Figure 4: Schematic representation of pesticide resistance (adapted from khanacademy.org)

The properties of DDT, its effectiveness and usually silent consequences, brought up thorough research and discussion, and today's regulation for all pesticides registration and use is heavily based on the concerns mentioned above, sustained by a precautionary principle that dictates international regulations.

We shall see in later chapters more detailed aspects of the environmental impact of chemical pesticides, especially in water bodies and their ecosystems.

2
CHAPTER

THE ROLE OF PESTICIDES IN FOOD PRODUCTION

Erick Soares Lins

HISTORIC BACKGROUND OF AGRICULTURAL USE PESTICIDES

Before the discovery of the effects of DDT, there were some inorganic chemicals like ferrous sulphate, ammonium sulphate and sodium arsenate (used as herbicide), sodium chlorate and sulphuric acid (for pest control), and the Bordeaux mixture, a fungicide based on copper sulphate and lime. Some organic compounds derived from industrial processes such as nitrophenols, chlorophenols and naphthalene were also tried against fungal infestations and insects, but had collateral effects and required large doses. DDT came to solve those problems, being inexpensive, notably effective and at that moment apparently safe.

Since the beginning of that practice back in the decade of 1940, several products have been introduced into the market, their popularity varied through time, and some of them ended up being banned or had their uses restricted by regulation. Table 9 below lists some of the most important pesticides introduced in the market, and their current situation regarding registration with the USEPA.

Table 1: Situation of several commercialized products regarding their registration for use.

Year introduced	Pesticide	Type	Current situation (in the US – EPA summary report issued in 2016)
1939	DDT (commercialized 1944)	Insecticide	Banned
1942	2,4-D	Herbicide	Banned
1943	Zineb	Fungicide	Authorized
1944	HCH (lindane)	Insecticide	Banned
1946	Parathion	Insecticide	Banned
1948	Aldrin, dieldrin	Insecticide	Banned (both)
1949	Captan	Fungicide	Authorized
1952	Diazinon	Insecticide	Restricted use from certain manufacturers
1953	Mercoprop	Herbicide	Authorized
1955	Paraquat (commercialized 1962)	Herbicide	Restricted use from certain manufacturers
1956	Carbaryl	Insecticide	Authorized
1965	Aldicarb	Nematicide	Authorized
1968	Benomyl	Fungicide	Authorized
1971	Glyphosate	Herbicide	Authorized
1972	Diflubenzuron	Insecticide	Authorized
1973	Permethrin	Insecticide	Restricted use from certain manufacturers
1990	Imidacloprid	Insecticide	Restricted use from certain manufacturers
	Azoxystrobin	Fungicide	Authorized
	Spinosad	Insecticide	Authorized

Many of the inorganic pesticides are either banned or with restricted use, such as arsenic and cadmium compounds. It is the same case for most of the organochlorine pesticides, such as Chlordane and Heptachlor.

The agricultural use of pesticide is by far the most relevant in terms of volume and diversity, and probably the one that may bring greatest environmental impact, when compared to other more localized utilizations, e. g. gardening, or residential pest control. It is estimated that the agricultural sector is responsible for ca. 80% of pesticide usage in 2007, which corresponds to 684 million pounds of conventional pesticides used in the United State in that year. Historical data shows an increase in that percentage in the decade of 1970, after which it remains quite stable, even with fluctuations in total absolute amounts of pesticides in use per year.

QUANTITATIVE ASSESSMENT REGARDING PESTICIDE USAGE

A report released by the United States Environmental Protection Agency (USEPA) with estimated sales and usage of pesticides in the country and worldwide. The graphs and estimates come from proprietary data sources, i.e. they are approximate estimates with information gathered by research companies, and not an official survey. After a stupendous usage growth until the early 1980s peaking at 1,121 million tonnes only in the US, the trend started to decrease again although quite slowly during that decade and the following, arriving at ca. 857 million tonnes by the end of 2007, as seen in Figure 5.

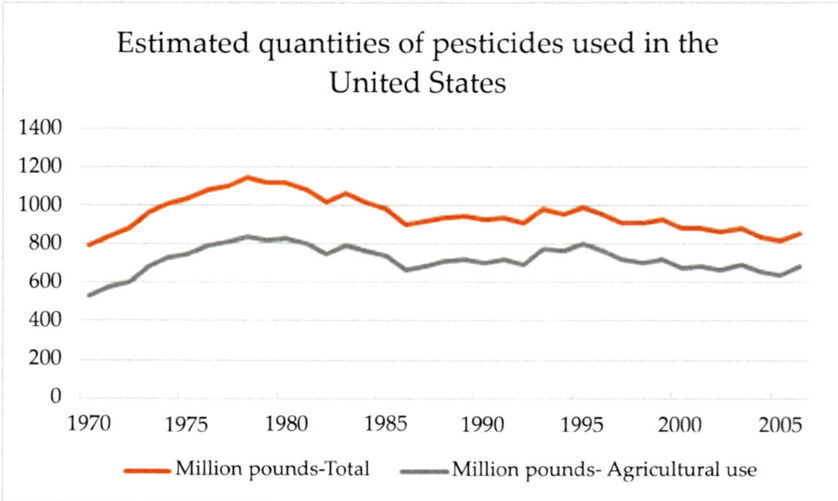

Figure 1: Estimated amounts of pesticide products at use in the US since the 1970s (Data from USDA)

The agricultural use is the most relevant, and contributes roughly to 80% of the total amount of pesticides. Other non-agricultural uses include: home & garden, right-of-way, railroads, sport fields.

Specialists of the US Department of Agriculture (Osteen and Padgitt, 2002) attribute that drop to at least two major causes, both of them inherent to technologic evolution: i) the development of new pesticides that required lower application rates, and ii) the advent of genetically modified crops, which would demand lower quantities of defensive products. Both of these assumptions are still a matter of debate especially because of the further consequences to human health and the environment. The public perception about dietary risks and exposure, as well as environmental impact, may really have had some role catalyzing certain changes in policy and the food production system as a whole, starting in the 1980s. A more detailed analysis by the USDA (2014) describes some other crucial factors at play, objectively measured:

- Pest infestation levels

It sounds like an intuitive direct relationship between the degree of infestation and the damage it causes, and the amount of pesticides required to fight it. It means that simply increasing the quantities may actually turn out to be not cost effective. Instead, knowledge of the infestation level and the acceptable damage range is key to keep production at a reasonable pace economically, both in short and long terms.

- Pesticides Technical Improvements

The developments in synthesis science and methods led to new active substances over the decades, which came to the market with higher efficiency or lower application rates.

- Integrated Pest Management

This is probably the state-of the-art in pest control, with a more holistic approach. Agricultural practices such as crop rotation, alternating pesticides, biological monitoring, weather predictive models, optimization of planting and harvesting dates, are applied together with the objective of diminishing pest resistance and making the spraying less subject to losses.

- Genetically engineered crops

Seeds can be genetically modified to be insect resistant or herbicide tolerant. Plants that can tolerate more powerful herbicides allow, however, for a higher use in fighting unwanted pervasive weeds, and that is only one of the current debates over this technique.

- Conservation practices

Aimed at reducing soil erosion by, for example, use of conservation tillage that is in turn associated with Glyphosate use. This is an example of a practice that does not necessarily results in less pesticide being used.

- Pesticide regulations

Pesticides on the market may be subject of reevaluation with the best scientific data available, and that takes into account toxicology studies as well as other impact research. As it happened with DDT and others, those findings may result in ban, which will force growers and producers to find alternatives.

- Changes in crop acreage

Increasing or decreasing planted acreage will result in variable crop production, thus demanding proportionally more or less aggregate pesticide use.

- Increase in crops acreage intended for biofuels

The case of corn acreage in the US is illustrative, since it was greatly enlarged for the production of ethanol. It increased the pesticide use and it changed the active ingredients for that crop. A notable increase in the use of Glyphosate was observed.

TEMPORAL TRENDS

Corn, Soybeans, Cotton, Wheat, and Potatoes - those are the top five crops in the US in terms of quantities of pesticide applied. They are also major crops in terms of overall production, reaching around 360 million tonnes of corn, 106 million tonnes of soybeans, 55 million tonnes of wheat and 20 million tonnes of potatoes only in the year 2014 according to FAOStat. The US production is only matched by other commodities superpowers like China and Russia producing 126 million tonnes and 59 million tonnes wheat respectively, and 95 million tonnes and 31 million tonnes potatoes in that same year.

Figure 2: Wheat harvest (photo from pixabay.com)

The study made by the USDA splits the pesticides timeline in two periods: from 1960 to 1982 and from 1982 to 2008. The latter is subdivided in two other periods, from 1982 to 1995 and from 1996 to 2008, because 1996 is when genetically engineered seeds were first made commercially available, and that was deemed to become a major influential factor for the amounts of pesticide use. Data presented here was retrieved from USDA/NASS, Agricultural Chemical Usage: Field Crop Summaries, various years.

1960-1981

Corn

Most used pesticide until the early 1960s was 2,4-D, an herbicide. Later in that decade, however, 2,4-D represented only 13% of pesticide use in corn, being mostly replaced by atrazine, which accounted for almost 60%. The trend for insecticides, especially the organochlorinated ones, can be illustrated by aldrin, which accounted for 15% pesticide use in 1968 and was not around the late 1970s anymore, being replaced by carbofuran, phorate, fonofos and terbufos. This had a lot to do with regulation over chlorinated hydrocarbons and their environmental impact.

Soybean

Aggregate pesticide use had a huge leap: from less than 3 million pounds in 1960 to 146 million pounds in 1982. The rate also followed that, climbing from 0.1 pound per acre to 2.1 in the same period. The first dominant herbicide was chloramben, which accounted for more than 30% of all pesticide use in 1968, being after that replaced by active ingredients such as alachlor, metribuzin and trifluralin.

Cotton

Insecticide use in cotton acres peaked at 105 million pounds in 1972, rising from 67 million pounds in 1960, and then severely declining to 19 million pounds by the end of the period studied. DDT and similar organochlorines were being substituted until ruled out of use in the mid-1970s. Methyl-parathion and synthetic pyrethroids became most widely used, as the application rates were much lower, Farmers also applied new agricultural practices to control insect infestation, such as crop rotation (with sorghum, for example). Herbicide use, however, increased from 0.1 pound per planted acre to 1.9, led by trifluralin and fluometuron.

Potato

Total pesticide use in potato crops had a slow increase in the period, starting around 9 million pounds and reaching over 15 million pounds in the early 1980s. DDT and malathion were the most used pesticides in the 1960s, and disulfoton after that. Fumigants and desiccants (often referred as 'Other Pesticides') were the major contributors for most of the period studied, reaching 11.8 pounds per planted acre in 1982. Herbicides are less used when compared to the other major crops, but they are still present in most planted acres.

Wheat

Until 1971, 2,4-D was the most common pesticide, having MCPA and dicamba new herbicides introduced starting 1972. By 1980, herbicide used increased to 0.32 pound per planted acre.

Figure 7 (a, b and c) presents a quantitative breakdown of the amounts of each of the most used pesticides in that period. We can see that for each of these major crops (corn, soybean and cotton), there is always one or two that are more common, such as atrazine, which accounted for more than half of the overall usage in corn, or the herbicides chloramben and trifluralin in soybean. It is noticeable that DDT was indeed the main insecticide used in that period.

Appendix figure 4.6
Pesticide use by active ingredient (a.i.) on corn in 1968, percent of total pounds a.i. applied[1]

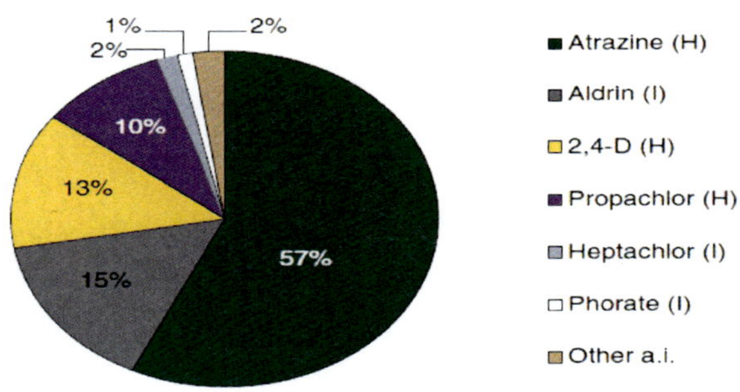

[1]This graph shows the top pesticide a.i. used in corn in 1968.
Sources: Economic Research Service with USDA and proprietary data. See Appendix 2.

Appendix figure 4.8
Pesticide use by active ingredient (a.i.) on soybeans in 1968, percent of total pounds a.i. applied[1]

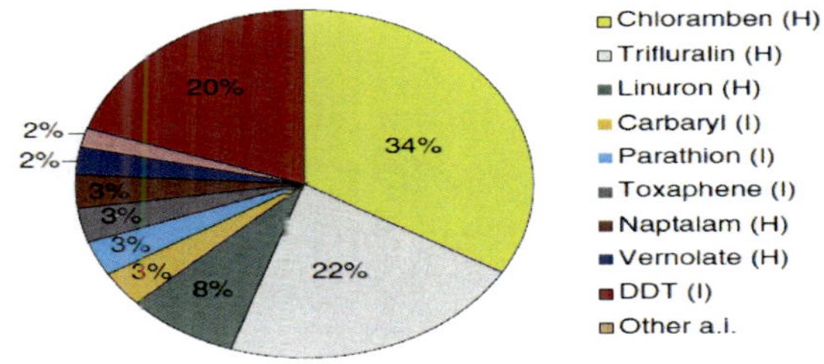

[1]This graph shows the top pesticide a.i. used in soybeans in 1968.
Sources: Economic Research Service with USDA and proprietary data. See Appendix 2.

Appendix figure 4.10
Pesticide use by active ingredient (a.i.) on cotton in 1968, percent of total pounds a.i. applied[1]

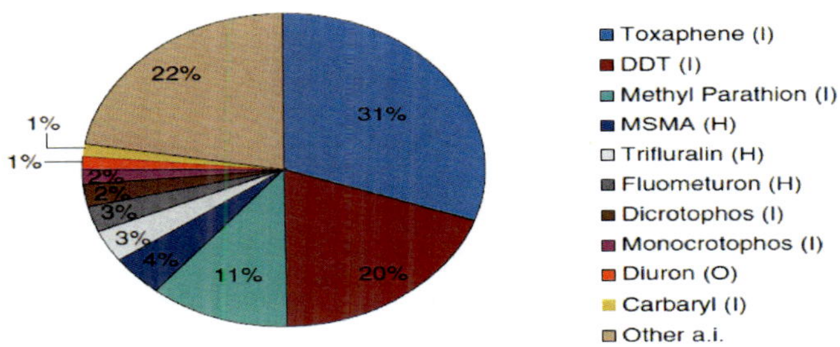

[1]This graph shows the top pesticide a.i. used in soybeans in 1968.
Sources: Economic Research Service with USDA and proprietary data. See Appendix 2.

Figure 7 - Relative percentages of active ingredients used on most important crops in 1968 - a) Corn; b) Soybean; c) Cotton. Source: USDA/NASS Agricultural Chemical Usage.

1982-1995 and 1996-2008

Corn

Atrazine became the dominant herbicide in the first period, accounting for more than 25% of herbicide use. For insecticides, use fluctuated through the 1980s and 1990s, decreasing to a level of 0.26 pounds per planted acre, and the most used ingredients were terbufos and chlorpyrifos.

Genetically modified corn seeds then made the use of both herbicides and insecticides drop from 2.67 pounds per planted acre to 2.3, led by nicosulfuron and then glyphosate.

Soybean

By 1995, the most used herbicides were pendimenthalin, trifluralin and glyphosate. In 2008, a total of 106 million pounds of herbicides were used, in contrast with 4.7 million pounds of insecticides and 1.04 million pounds of fungicides. In 1996, herbicide-tolerant genetically modified seeds were introduced, and their adoption rose from only 7% of soybean acres to over 90% in 2008. In the first moments, herbicide uses declined but rose again reaching 1.4 pounds per planted acre. Glyphosate represented more than 85% of the total herbicide use, and this raised concerns about weed resistance.

Cotton

Pest resistance to pyrethroids, namely the tobacco budworm, made farmers turn to azinphos-methyl and methyl parathion, organophosphates, as well as adopting new cropping strategies and integrated pest management. Introduction of genetically modified cotton in 1996 contributed to the reduction in insecticide use, since the seeds were made to be resistant to insects. Total insecticide use increased, however, to over 40 million pounds in 1999 and 2000, before finally declining to 5 million pounds in 2008. Likewise, herbicide-tolerant cotton seeds were in place in 1996, when trifluralin, monosodium methanearsonate (MSMA), pendimentalin, and fluometuron were the most used herbicides. When the modified cotton occupied 70% of acres, total herbicide use per planted acre decreased while glyphosate was the dominant active ingredient. In 2007 and 2008 there was an increase in the amount used per acre, possibly due to glyphosate resistant weed.

Potato

The early 1990s brought a sharp increase in the use of 'Other Pesticides' in potato production, while herbicides, fungicides and insecticides remained quite stable. The total amount applied reached 52 million pounds in 2008, of which 43.5 million pounds referred to the desiccants, fumigants and others. Herbicides are still present, and by 2005 metribuzin was used in 74% of acres, followed by pendimenthalin and rimsulfuron. In terms of insecticides, instead of DDT and malathion, metam sodium, dicholopropene and metam potassium, all fumigants, have been more used in the recent years.

Wheat

2,4-D, MCPA and dicamba were still common pesticides in the period, although now surpassed by glyphosate. Total herbicide use in wheat in 2008 was estimated to be 20 million pounds, followed by 1.26 and 1.68 million pounds of insecticides and fungicides, respectively.

Figure 8 (a, b and c) shows some relevant changes in the profile of pesticide usage. Glyphosate has established as the herbicide of choice, largely dominating soybean production, and surpassing atrazine in corn crops. Other non organochlorinated insecticides dominate the market.

Appendix figure 4.7
Pesticide use by active ingredient (a.i.) on corn in 2008, percent of total pounds a.i. applied[1]

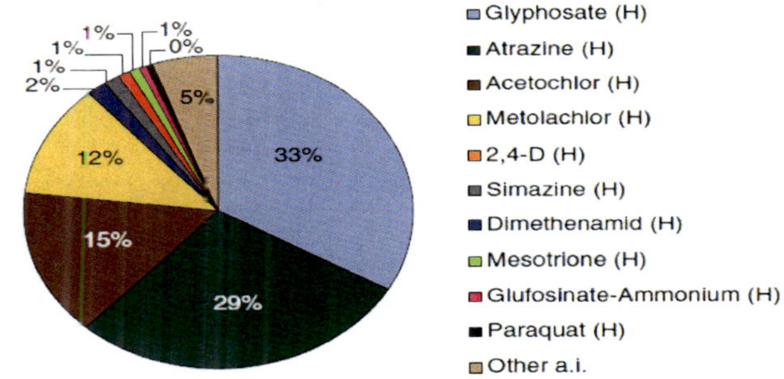

[1]This graph shows the top pesticide a.i. used in corn in 2008.
Sources: Economic Research Service with USDA and proprietary data. See Appendix 2.

Appendix figure 4.9
Pesticide use by active ingredient (a.i.) on soybeans in 2008, percent of total pounds a.i. applied[1]

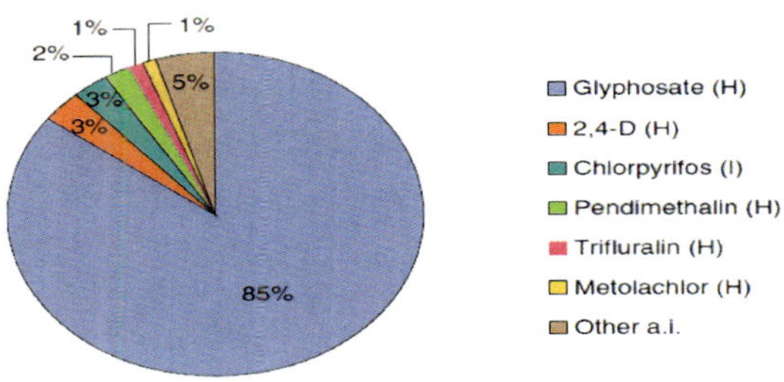

[1]This graph shows the top pesticide a.i. used in soybeans in 2008.
Sources: Economic Research Service with USDA and proprietary data. See Appendix 2.

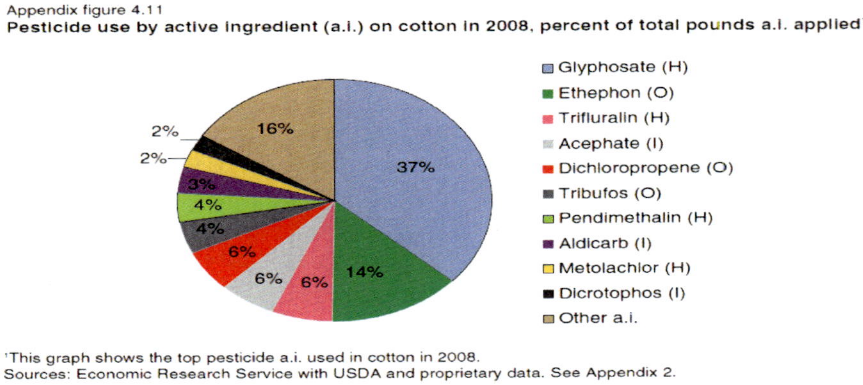

Figure 3: Relative percentages of active ingredients used on most important crops in 2008 - a) Corn; b) Soybean; c) Cotton. Source: USDA/NASS Agricultural Chemical Usage.

INSECTICIDES VERSUS HERBICIDES USE LEVELS AND GENETICALLY MODIFIED CROPS

It is important to assess the effectiveness of genetically engineered plants in leading to a decreased in pesticides use, as it was foreseen in the early years of the technology. In fact, we can see that since the introduction of herbicide-tolerant seeds and their spread in acres along the US, the use of glyphosate increased as well as the amount per planted acre, but until 2008 it was still lower than the sum of other herbicides, and it did decrease the amount of herbicide applied per acre for a while. However, a recent rise, possibly due to the development of resistant weed in the fields threatens to bring those levels back to what they were in 1996, as depicted in Figure 9.

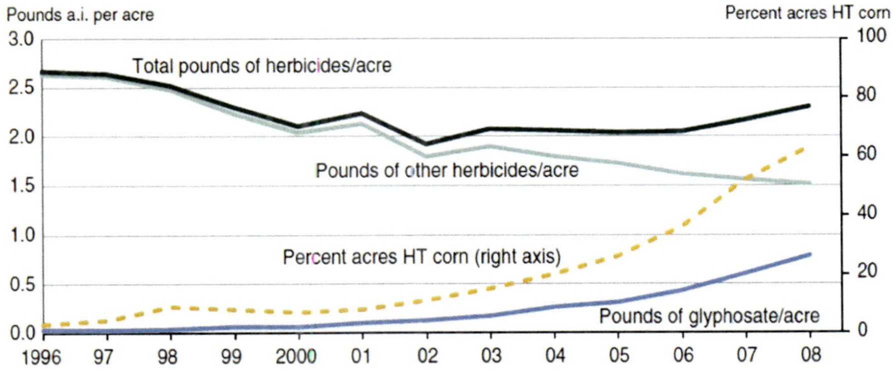

Figure 4: Herbicide use per planted acre in pounds and percent acres of herbicide tolerant corn since its introduction (reproduced from USDA/NASS)

Something similar can be seen with soybeans, except that in 2008 the total use of herbicides per acre was already higher than in 1996, given the wide adoption of herbicide-tolerant plants.

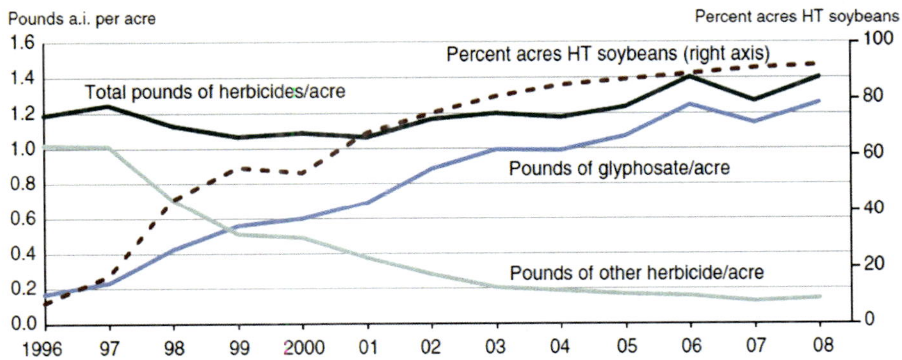

The results seem to be more effective in genetically modified cotton and corn, where an actual decrease in the number of products per acre resulted after adoption of these seeds, represented by the time graph in Figure 11 (a and b). It must be noted that in 1999 and 2000 a pest eradication program was in place for cotton, fostering the application of insecticides.

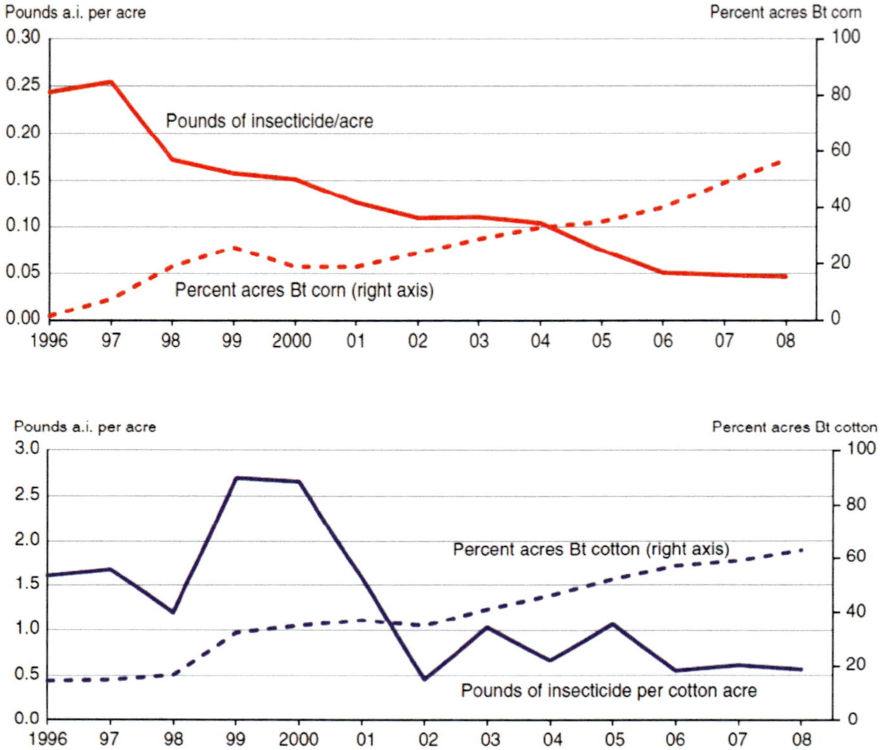

Figure 6: Insecticide use in pounds per planted acre and percent acres of tolerant (a) corn and (b) cotton since their introduction (reproduced from USDA/NASS)

Extrapolating the universe of the five major crops to 21 selected crops (including minor crops), we can see a more detailed picture of the distribution in the pesticides used. Figure 12 presents the overall percentages pesticide usage on crops in the US, where corn corresponds to almost 40% of the total pesticide use in national production, followed by soybeans and potatoes. Other smaller sectors, such as fruits and vegetables, also have their share with their respective collection of active ingredients that are taken into account further on:

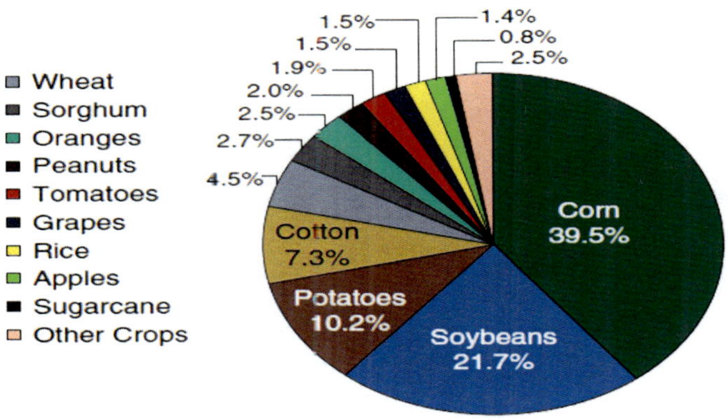

Figure 7: Percent use of pesticides in the 21 major crops in 2008 (reproduced from USDA/NASS).

On the rank of most used pesticides in the US, glyphosate, a systemic herbicide, has been the number one for quite some time. Herbicides are the most important class, followed by fumigants, fungicides and insecticides. Table 10 brings a brief rank of the most frequently used pesticides by type.

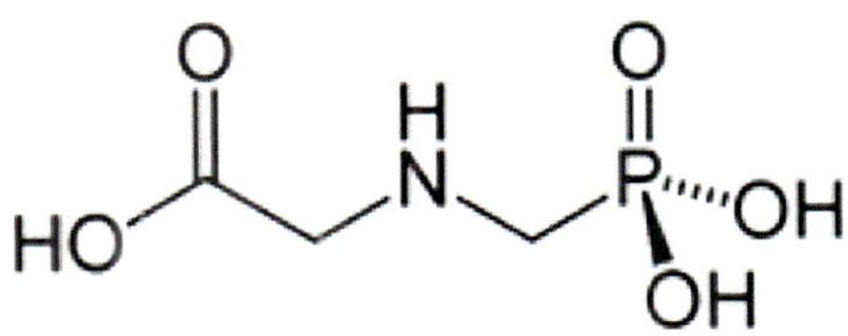

Figure 8: Glyphosate molecule.

Table 2: Most commonly used conventional pesticide active ingredient in agricultural sector in 2007 (adapted from USDA/NASS)

Rank	Range (millions of pounds)	Active ingredient	Type
1	180-185	Glyphosate	Herbicide
2	73-78	Atrazine	Herbicide
3	50-55	Metam Sodium	Fumigant
4	30-35	Metolachlor-S	Herbicide
5	28-33	Acetochlor	Herbicide

6	27-32	Dichloropropene	Fumigant
7	25-29	2,4-D	Herbicide
8	11-15	Methyl Bromide	Fumigant
9	9-11	Chloropicrin	Fumigant
10	7-9	Pendimenthalin	Herbicide
11	7-9	Ethephon	Plant Growth Regulator
12	7-9	Chlorothalonil	Fungicide
13	7-9	Metam Potassium	Fumigant
14	7-9	Chlorpyrifos	Insecticide
15	6-8	Copper Hydroxide	Fungicide
16	5-7	Simazine	Herbicide
17	5-7	Trifluralin	Herbicide
18	4-6	Propanil	Herbicide
19	4-6	Mancozeb	Fungicide
20	3-4	Aldicarb	Insecticide
21	3-4	Acephate	Insecticide
22	2-4	Diuron	Herbicide
23	2-4	MCPA	Herbicide
24	2-4	Paraquat	Herbicide
25	2-4	Dimethenamid	Herbicide

The graph on Figure 14 shows that Glyphosate contributed to 38% of the overall pesticide usage in 2008, becoming the most used herbicide by far, but also above insecticides and fungicides in the market.

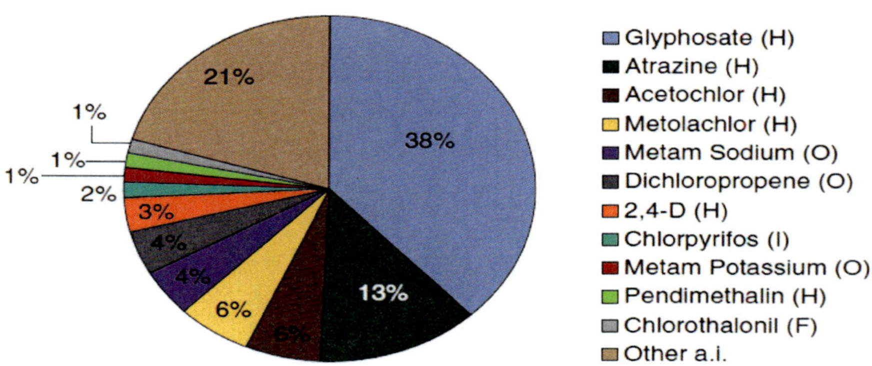

Figure 8: Percent use of pesticides in pounds in the major 21 crops in 2008 (reproduced from USDA/NASS)

Those data are broken down in the main categories of pesticides, i.e. the most used active ingredients within the herbicides, insecticides and fungicides types, respectively, for the same period, as shown in Figure 15 (a, b and c). Chlorpyrifos is the top insecticide in that period, and chlorothalonil the top fungicide.

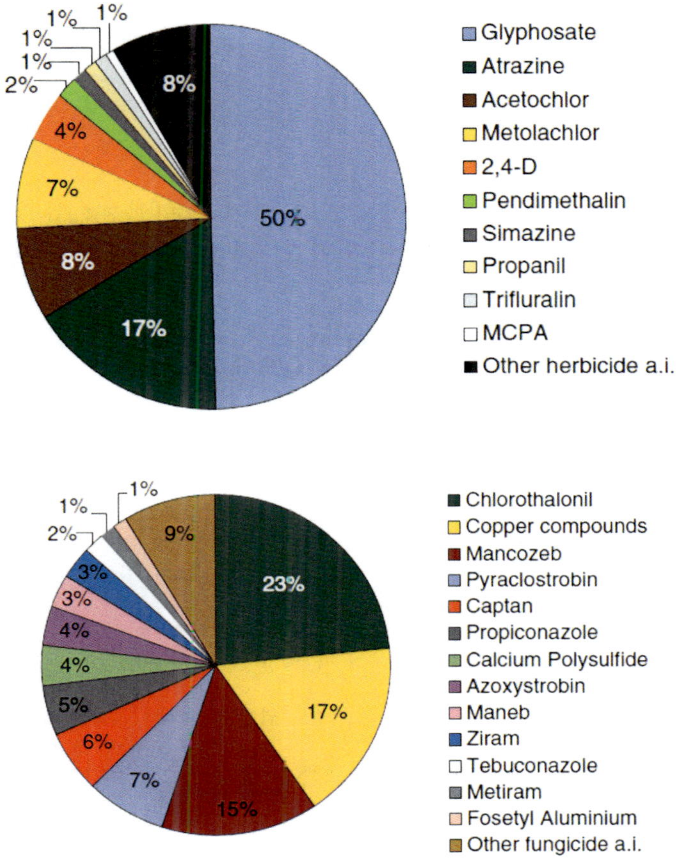

Figure 9: Percent use of active ingredients in the main categories of pesticides in 2008: a) herbicides, b) insecticides and c) fungicides (reproduced from USDA/NASS)

Overall, the historic data of pesticide use in the US reached a relatively stable level in the past decades, as can be seen in Figure 16. It should be expected that the agricultural sector adjusts itself with the regulatory enforcements, and evolves at increasing production per acre without necessarily drastically increasing the amounts of pesticide applied.

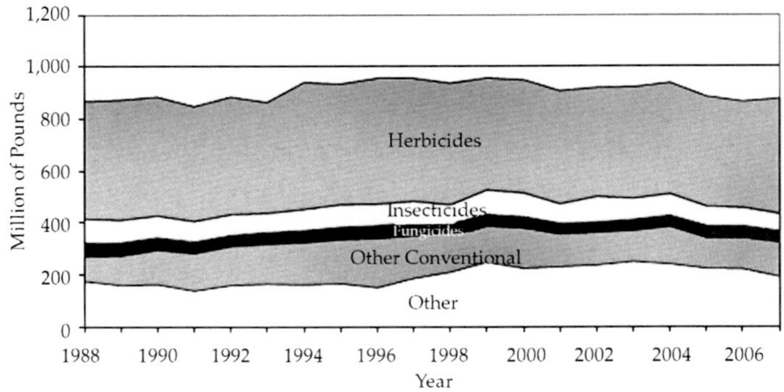

Figure 10: Annual amounts of pesticide active ingredient used in the US by pesticide type, estimates from the agricultural market sector in 1988-2007 (reproduced from NSAA/USDA).

The scenario explored so far considers the reality of agriculture in the US, due to its size, variety and importance, and also because it is where most of the chemical development in active ingredients took place, and there is a scientifically-driven regulatory body in place.

In Table 11 there is a comparative data set with the estimated pesticide use in the US versus the rest of the world, per type of active ingredient. The US consumes a relevant slice of the total amounts of pesticide used in food production.

Table 3: World and US amount of pesticide active ingredient used by pesticide type, estimates for 2006 and 2007 (reproduced from USDA/NASS)

Pesticide Type and Year	World Market		US Market		US percentage of World Market (%)
	Million lbs.	%	Million lbs.	%	
2006					
Herbicides	2,018	39	498	44	25
Insecticides	955	18	99	9	10
Fungicides	519	10	73	6	14
Other	1,705	33	457	41	27
Total	5,197	100	1,127	100	22
2007					
Herbicides	2,096	40	531	47	25

Insecticides	892	17	93	8	10
Fungicides	518	10	70	6	14
Other	1,708	33	439	39	26
Total	5,211	100	1,133	100	22

Note 1: Totals may not add due to rounding.

Note 2: Does not include wood preservatives, specialty biocides, and chlorine/hypochlorites.

Source: EPA estimates based on Cropnosis Limited (www.cropnosis.com), USDA/NASS (www.nass.usda.gov) and EPA proprietary data.

We can see that, as far as the data goes, the US alone consumes about f the total amount of pesticides by the rest of the world. The fact that the agricultural sector in the US is remarkably opulent, and dominates large sets of market share in commodities, can only corroborate with the importance of pesticides in food production, despite the many concerns they impose, and which will be discussed in the following chapters.

3

CHAPTER

THE IMPACTS OF PESTICIDES AND THEIR FATE IN THE ENVIRONMENT

Erick Soares Lins

INTERNATIONAL REGULATIONS AND CONCERNS ABOUT HEALTH AND THE ENVIRONMENT

WHO (The World Health Organization) ranked the 10 chemicals or group of chemicals that pose major public health concerns:

- Air pollution
- Arsenic
- Asbestos
- Benzene
- Cadmium
- Dioxin and dioxin-like substances
- Inadequate or excess fluoride
- Lead
- Mercury
- Highly hazardous pesticides

The evolution in agricultural practices is usually accompanied by regulation, and FAO is the body that serves as a reference to many countries.

FAO regulations address food safety (the commensurate supply to feed populations) and food security (the assurance that such food is not toxic or dangerous in short- and long-terms). FAO's 'International Code of Conduct on the Distribution and Use of Pesticides' states responsibilities for both governments and the private sector, and it clearly mentions in its Article 6 - Regulatory and Technical Requirements:

Governments should:

- Introduce the Necessary Legislation for the Regulation of Pesticides and make Provisions *for its effective enforcement, including the establishment of appropriate educational, advisory, extension and health-care services, using FAO guidelines as far as possible. In so doing, they should take full account of local needs, social and economic conditions, levels of literacy, climatic conditions and availability of appropriate pesticide application and personal protective equipment;*
- Strive to establish pesticide registration schemes and infrastructures under which products can be registered prior to domestic use and ensure that each pesticide product is registered before it can be made available for use;
- Conduct risk evaluations and make risk management decisions based on all available data or information, as part of the registration process;
- Use the principles described in the Manual on Development and Use of FAO and WHO Specifications for Pesticides for determining equivalence of pesticides;
- Promote the advantages of, and cooperate with other governments in, the establishment of harmonized (regionally or by groups of countries) pesticide registration requirements, procedures and evaluation criteria, taking into account appropriate, internationally agreed technical guidelines and standards, and where possible incorporate these standards into national or regional legislation;
- Establish a re-registration procedure to ensure the periodic review of pesticides, thus ensuring that prompt and effective measures can be taken if new information or data on the

performance or risks indicate that regulatory action is needed;

- Improve regulations in relation to collecting and recording data on import, export, manufacture, formulation, quality and quantity of pesticides;
- Collect and record data on the import, export, manufacture, formulation, quality, quantity and use of pesticides in order to assess the extent of any possible effects on human health or the environment, and to follow trends in pesticide use for economic and other purposes;
- Permit pesticide application and personal protective equipment to be marketed only if they comply with established standards;
- Detect and control illegal trade in pesticides;
- When importing food and agricultural commodities, recognize good agricultural practices in countries with which they trade and, in accordance with recommendations of the Codex Alimentarius Commission, establish a legal basis for the acceptance of pesticide residues resulting from such good agricultural practices in a manner that is consistent with the WTO requirements so as not to lead to technical barriers to trade.
- Pesticide industry should:
- Provide an objective pesticide data assessment together with the necessary supporting data on each product, including sufficient data to support risk assessment and to allow a risk management decision to be made;
- Provide national regulatory authorities with any new or updated information that could change the regulatory status of the pesticide, as soon as it becomes available;
- Ensure that the active ingredient and other ingredients of pesticide products being marketed correspond in identity, quality, purity and composition to the substances tested, evaluated and cleared for toxicological and environmental acceptability;
- Ensure that active ingredients, and formulated products for pesticides for which international specifications have been developed, conform with the relevant FAO specifications for

agricultural pesticides, and with WHO pesticide specifications for public health pesticides;

- Verify the quality and purity of pesticides offered for sale;
- When problems occur, voluntarily take corrective action and, when requested by governments, help find solutions to difficulties;
- Provide their national governments with clear and concise data on export, import, manufacture, formulation, sales, quality and quantity of pesticides.
- Technical assistance funding agencies, development banks and bilateral agencies should be encouraged to give high priority to requests for assistance from developing countries which do not yet have the facilities and expertise for pesticide management and control systems.

PROPERTIES AND CHARACTERISTICS OF THE MAIN PESTICIDE CATEGORIES

Figure 1: Pesticide application (image from visualhunt.com)

As shown in terms of the quantities used in crops every year, herbicides, insecticides and fungicides make the three major types of pest killing chemicals for agricultural use. More detail about the developments in each category will be given below:

Herbicides

Figure 2: Commercial glyphosate based herbicide

Control of weed is something that is documented to ancient times, and it evolved to become less empirical and more selective. A very important discovery was the one made by Bonnet in 1896, reporting that mustard (a broadleaf weed) could be controlled with copper sulphate solutions. Other inorganic compounds with known herbicide activity were developed following the plant nutrition theory by Liebig, and pushed by the increase in the use of chemical fertilizers. Some of the most relevant inorganic herbicides are:

- Ashes (believed to be used by the Roman Empire to control weeds on roadsides)
- Common salts (ex. NaCl, which is effective when applied at high rates. Use of such salts is mostly discontinued, except for copper sulphate for aquatic weed)
- Bittern (a solution with crystallized Bromides and Iodides form recovered from evaporated seawater)
- Copper sulphate
- Sodium arsenite (used until the 1960s)
- Ammonium sulfamate (low toxicity to mammals and highly water soluble)
- Ammonium thiocyanate
- Ammonium nitrate
- Ammonium sulphate
- Iron Sulphate

- Chlorates and Borates (nonselective, highly toxic to plants, and still widely used specially for deep-rooted perennial weeds)

Some organic herbicides were in place prior to 1945, petroleum oils especially, but their dosage was usually very high - up to 80 gal/acre in carrots. The costs, the damage to crops and the occurrence of PCBs (which posed an even greater environmental and health issue) led the use of oils to an end.

In 1941 the herbicidal properties of 2,4-D were reported, and the fact that it was more selective and required smaller quantities was of interest. Fuel oils were still used along railroads and roadsides to help the active ingredients of 2,4-D and 2,4,5-T, for instance, to better penetrate the soil.

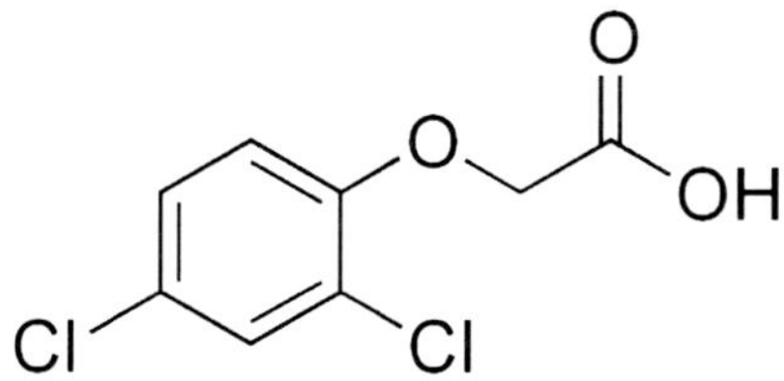

Figure 3: Structure of 2,4-D

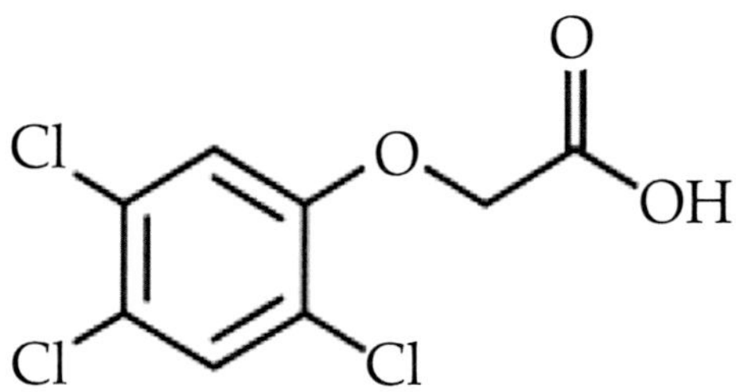

Figure 4: Structure of 2,4,5-T

Phenols, more specifically substituted phenols, are the pioneers in synthetic organic pesticides. In 1932, 3,5-dinitro-O-cresol, was introduced and in 1935 4,6-dinitro ortho-O-cresol, also known as DNOC, was patented for weed control. Different from other broad-spectrum phenols, DNOC presents high selectivity, along with effectiveness at low rates as well as being non-irritating and less corrosive. DNOC ended up rapidly replacing inorganic salts in North America and Europe. One of the problems presented by DNOC and similar compounds was its non-selectivity, not to mention it is corrosive to metals. The reported toxicity to mammals, with acute oral LD50 of about 50-100 mg/kg (WSSA, 1974) in rats and known adverse effects in humans decreased their use, as more effective alternatives came about.

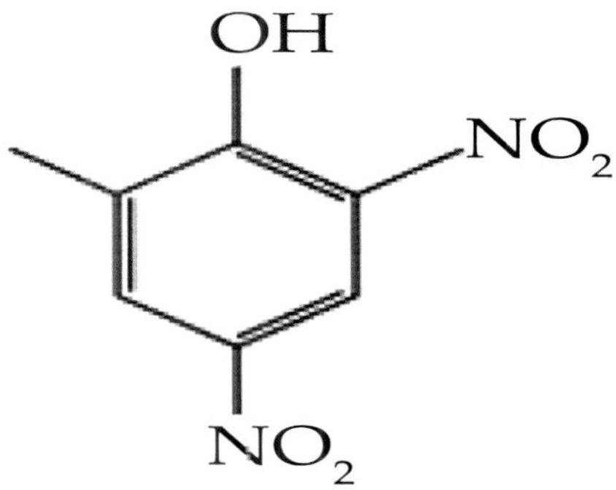

Figure 5: DNOC

After 1944

Some of the most important chemical groups of herbicides developed in the following years and their environmentally relevant properties are summarized below:

1. Phenoxy Alkalotic Acids (example structure shown in Figures 19 and 20)

- Ex.: 2,4-D, 2,4-DB, MCPA, MCPB, dichlorprop, mecoprop, silvex.
- 2,4-D is the great representative, considered the counterpart of DDT to weed control.
- These herbicides are more active in foliar surfaces than in soil application; they are toxic to herbaceous broadleaved plants and deciduous trees;
- This group is not very persistent in soil, being 2,4,5-T maybe the most persistent and can remain for a time between two months to one year. They are degraded by bacteria that use them as carbon source. A similar mechanism occurs in water, except in slower rates due to less microorganisms around. Microbial degradation attacks the side chain, following ring

hydroxylation and resulting in succinates and other organic acids. Photolytic degradation occurs with the removal of the side chain, hydroxylation of the ring, and subsequent polymerization of the polyhydroxy benzenes, resulting in humic acids. The rate of photolysis is lower with greater number of chlorine groups in the phenoxy ring.

- Toxicity in animals is considered low, ranging from 300 to 1000 mg/kg of oral acute toxicity in rats, and it has been reported that almost total elimination of the parent molecule in the urine of animals exposed to 2,4-D occurs within up to 48 hours. (Loos 1975)

2. Picloram

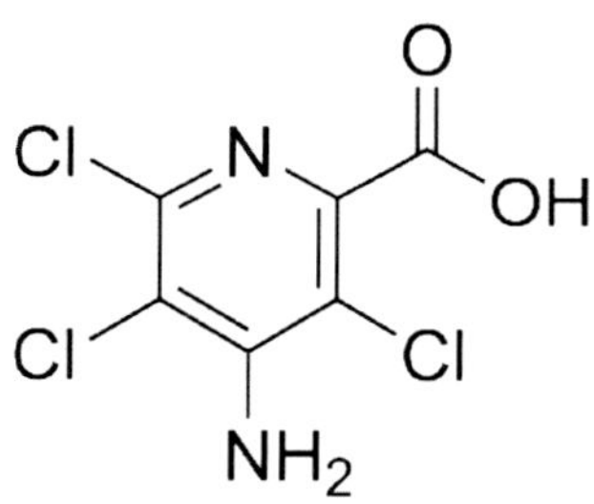

Figure 6: Structure of Picloram

- Usually applied in formulations as the dimethyl salt in mixtures with 2,4-D.
- It is a pyridine, similarly to the insecticide chlorpyrifos and soil nitrification inhibitors. It is more potent than phenoxy alkanoic herbicides, and is sufficiently selective to be used in cereal crops for weed control, especially broadleaf and brushes.
- Considered a "highly mobile herbicide", it can penetrate soil as deep as 45 cm as well as spread laterally in certain conditions. Its persistence in soil may be what makes it so effective for weed control, although it impacts the environment in longer terms as well as other non-targeted crops.
- Acute oral toxicity in rats ranges from 2000 to 8000 mg/kg, and is readily eliminated in urine or feces within 2 days from the ingestion. It is not highly lipophilic, so does not accumulate

in animal tissues. It is soluble in water, and may then contact plant tissues as non-targeted organisms, although suffering photolysis rapidly, being thus less persistent in water than it is in soil.

3. Benzoic Acids

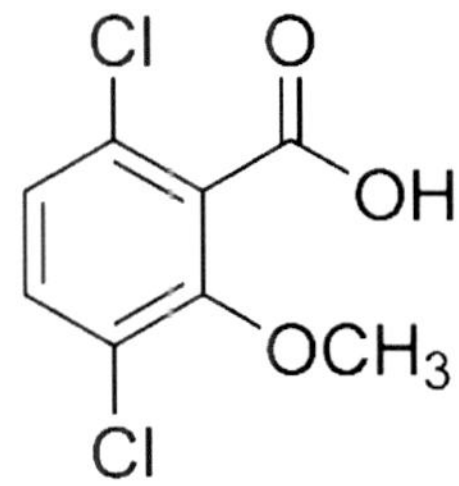

Figure 7: Structure of Dicamba - benzoic acid is the main moiety

- TBA (2,3,6-trichlorobenzoic acid) was first developed in 1948 and two important derivatives are dicamba (3,6-dichloro-o-anisic acid) and chloramben (3-amino-2,5-dichlorobenzoic acid). Also referred as synthetic auxins.
- Dicamba is effective against 2,4-D tolerant weeds, being useful in several crops. Chloramben is the most selective of the three, making it useful to control preemergence broadleaf weeds.
- Moderately water soluble, being considerably persistent and mobile. Reported half-life of dicamba and chloranbem is of less than one year (Frear, 19760. They suffer microbial degradation, apparently by decarboxylation. (MacRae and Alexander, 1965).
- Although highly toxic to broadleaved plants, they exhibit low toxicity to mammals, ranging from 500 to 1000 mg/kg for dicamba, 700 to 1000 mg/kg for TBA and 3500 mg/kg for chloramben (all data as acute oral toxicity in rats). Due to their properties, carryover to non-targeted vegetation and is of greater concern.

4. Triazines

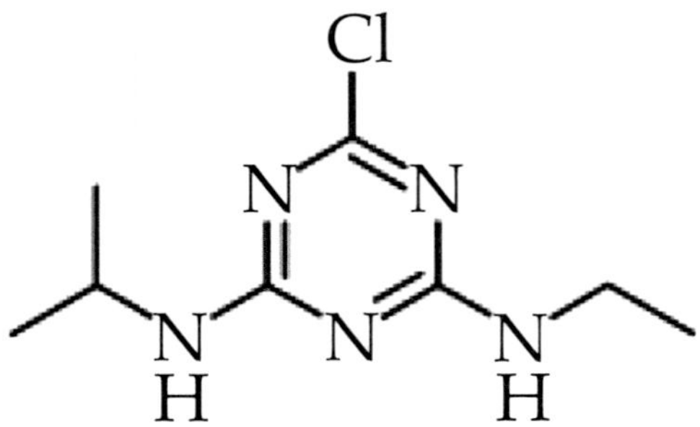

Figure 8: Structure of atrazine

- The group of Atrazine, Simazine, Chlorazine and Propazine. Basically, molecules with a heterocyclic aromatic ring, with different radical groups attached.
- Considered to have a major responsibility in the increase of corn yields in the US in the 1960s and 1970s.
- They are mostly moderately persistent in soil, depending on factors like pH, temperature, moisture and organic matter quantities. Triazines adsorb to soil clay and organic matter, especially in low moisture content and low temperatures, which partially deactivate them, thus requiring higher application rates to achieve desired weed control performance.
- Atrazine is known to photodegrade even when adsorbed to soil in certain conditions (Pape and Zabik, 1970). Triazines can still be chemically degraded to hydroxy-analogues, and microbial degradation occurs more slowly to cyanazine than to atrazine, for example. This has to do with adsorption rates of each molecule in soil components, which will make them more or less available for microorganisms to absorb and process.
- Triazines pose low toxicity to animals, with oral acute LD50 ranging from 300 to 50000 mg/kg. Toxicity to non-targeted plants is of greater concern, since they are usually quite persistent and mobile. Atrazine is itself a matter of long studies and discussion, being observed as the most common-herbicide present in water near corn-growing lands.

5. Ureas

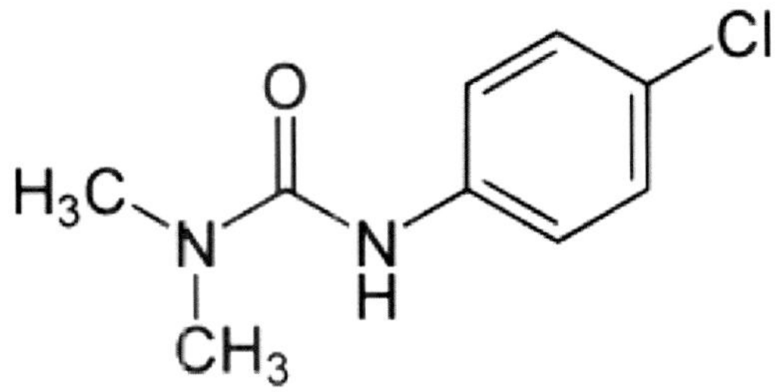

Figure 9: Structure of monuron

- Monuron was reported to have herbicidal activity in 1951, and later a series of compounds including diuron, fenuron and neburon, were developed.
- The group of ureas holds certain parallels to the triazines. Linuron, for instance, is used in similar manner as atrazine for weed control in corn and sorghum, but they possess different selectivities: high rates of linuron can damage corn, whereas soybeans and potatoes are more sensitive to atrazine. (WSSA 1974)
- Their persistence in soil is inversely proportional to their availability in soil solution. Fenuron is more water soluble than the others, so it is less persistent. Adsorption to clay or organic matter in soil is relative to chemical groups in the structure of the herbicide, so N-methoxy groups and ring bromines will give increased adsorptivity, when compared to N-methyl groups and ring chlorines.
- Ureas can degrade by photodecomposition with dealkylation and hydrolysis, or microbial activity.
- Their toxicity to animals is considered low, with oral LD50 in rat ranging from 1500 to 11,000 mg/kg. Elimination in feces is rapid.

6. Dinitroanilines

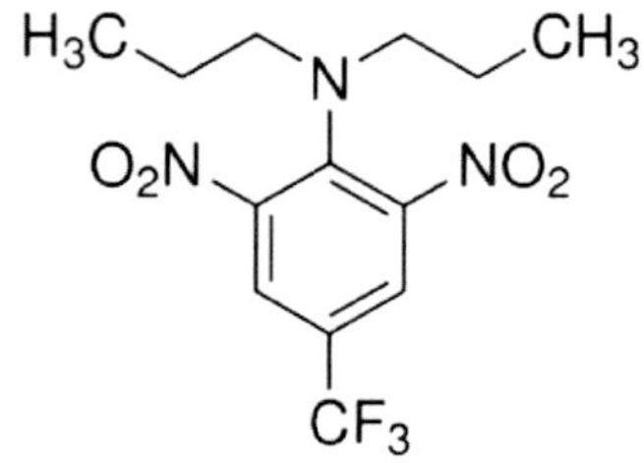

Figure 10: Structure of trifluralin

- Trifluralin is the most commonly used of the substituted 2,6-dinitroaniline herbicide compounds.
- The selectivity of Trifluralin made it useful for major crops such as cotton and soybeans.
- Dinitroanilines are very susceptible to photolysis, giving diverse products (Leitis and Crosby 1974). They can also be degraded by microorganisms via reduction of the amino groups and N-dealkylation reactions. Trifluralin is then considered non persistent and it is not mobile.
- They pose toxicity to fish when directly exposed, with LC50 of 0.058 ppm for bluegills in static fish pond range. Toxicity to other animals is considered low. (Probst 1975)

7. Thiocarbamates

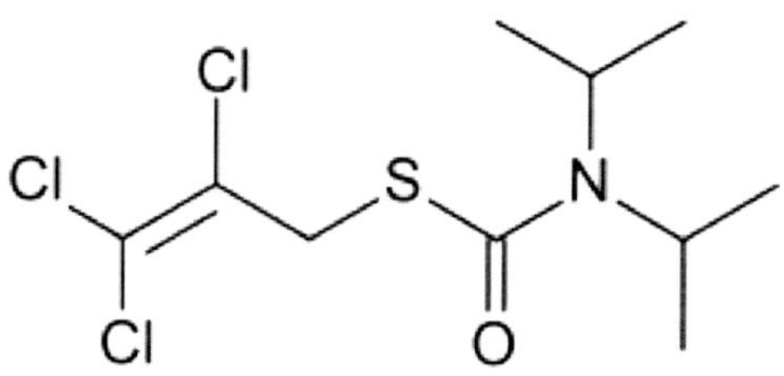

Figure 11: Structure of triallate

- A group of sulfur derivatives of carbamic acid. Some of the most popular compounds are EPTC, butylate, cycloate and diallate.
- They are applied to a variety of crops, usually beans, alfafa and peas, but also to beet, barley, corn and legumes, potatoes and root crops. Selectivity varies, with triallate being recommended

for the control of wild oats in wheat.

- Thiocarbamates are of low solubility in water and can be degraded in soil, so their effectiveness depends on the degree of adsorption. They usually persist for several weeks after application.
- Dithiocarbamates, e.g., CDEC and metham, are relevant products for the control of residues in foodstuffs. Like the parent thiocarbamate compounds, they are also non persistent and have low toxicity to animals, with acute oral LD50 of more than 800 mg/kg. EPTC, as mentioned, has an LD50 of 1600 mg/kg, and the others are in the range of 500 to 4000 mg/kg.

8. Bipyridiliums

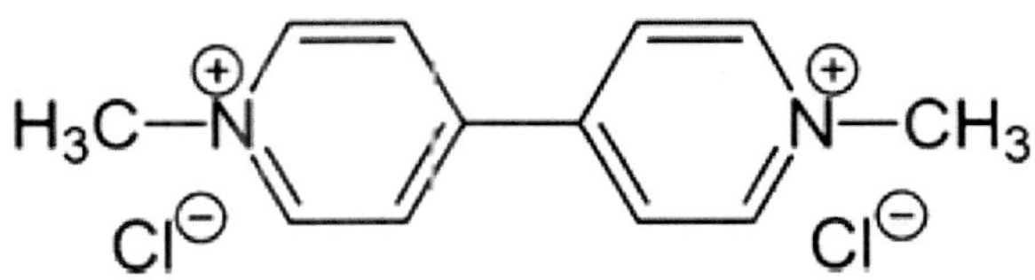

Figure 12: Structure of paraquat

- Also referred to as quaternary ammonium compounds, represented by diquat and paraquat. Herbicidal activity discovered in 1958.
- Both are contact herbicides, not active in soil. They are non selective, and mostly used as aquatic herbicide and for weed control in non crop lands. Paraquat is used as desiccant and defoliant too.
- In contact with clay and organic matter on soil, these compounds bind via cation exchange which readily inactivates them.
- Microbial and photolytic degradations occur to diquat and paraquat in exposed surfaces, but not in soil. The strong adsorption makes these compounds highly persistent, being found even after years since application (Calderbank and Slade 1976). Treatment with ammonium chloride can only partially desorb those compounds.
- Paraquat is considerably toxic to animals, with oral LD50 ranging between 100 and 150 mg/kg in rats. Human exposure

to paraquat has been shown with symptoms such as skin irritations and nose bleeds.

9. Organic Arsenicals

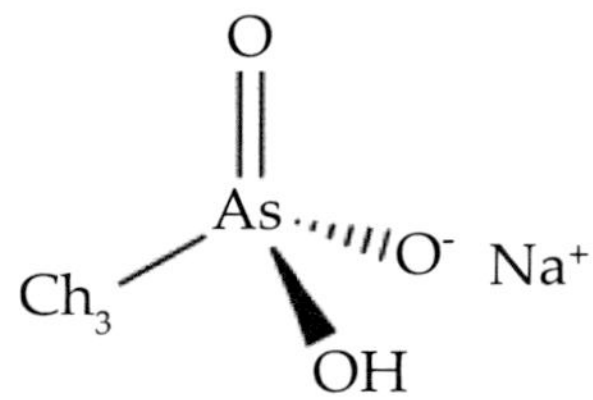

Figure 13: Structure of MSMA

- mostly derivatives of arsonic and arsinic acids. Some examples are: Disodium methanearsonate (DSMA) and Monosodium acid methanearsonate (MSMA), and related to Cacodylic acid.
- These derivatives are contact herbicides, more selective than cacodylic acids, although both generally require repeated applications for effectiveness. Once used in combination with 2,4-D/2,4,5-T mixtures as non selective herbicides. MSMA has been more recently used in cotton fields.
- MSMA is readily inactivated in contact with soil due to strong adsorption. Microbial degradation produces CO2 and arsenate, which will remain as an inorganic compound of the soil almost permanently.
- Toxicity to animals are unexpectedly high, given the fact that there is arsenic in these compounds. Oral acute LD50 in rats is over 800 mg/kg for MSMA. Accumulation of arsenate or inorganic arsenic residues and the potential environmental impact are a more concerning issue.

10. Glyphosate

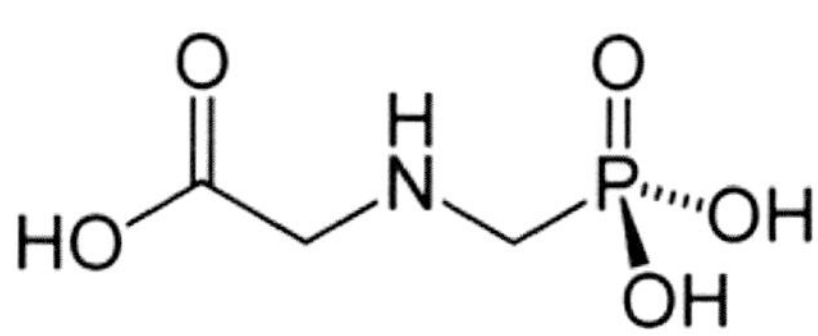

Figure 14: Structure of glyphosate

- N-phosphonomethyl glycine, a herbicide introduced in 1971.
- Non selective herbicide with high activity against a variety of weeds. It is used for preplant and post harvest control of weeds in several crops, in fruit or plantation crops, and aquatic weed control. Glyphosate is also present in diverse formulations for urban use, in gardening, and railroads.
- It is non persistent, being deactivated upon absorption and degraded by microorganisms in relatively short time. Its properties made it the most used herbicide in the US in recent years, and the advent of herbicide-tolerant genetically modified seeds bolstered the increase of its utilization.
- Oral LD50 in rats are over 4000 mg/kg. Chronic exposure studies with timespan of 2 years showed no adverse effects in rats or dogs, even with dietary dosage of 300 ppm. Glyphosate is classified by the Environmental Protection Agency (EPA) as a Group E chemical, i. e., a substance with no evidenced carcinogenicity in humans.

Insecticides

Figure 15: Paris Green mixture

Prior to 1945, insecticides at use were mostly of inorganic origin. One of the most famous was an arsenical mixture known as Paris Green, discovered in 1865. Other arsenates, especially Lead (PbHAsO4) and

Calcium ($Ca3(AsO4)$ and $CaHAsO4$) were developed in the years after. Arsenic insecticides are highly toxic to plants and animals, and are accumulative and persistent in the environment. Fluorides were also in use, namely sodium fluoride (NaF), and naturally occurring fluorspar (CaF2) and cryolite (Na3AlF6). This group stopped being used eventually, due to poor performance. Sulfur had its use, as well as certain substances extracted from plants, e.g., Nicotine and pyrethrum.

After the World War II, the era of synthetic insecticides has its start, and it leads to the enormous variety that can be found nowadays. Some important insecticide groups for environmental circumstances are:

1. Carbamates

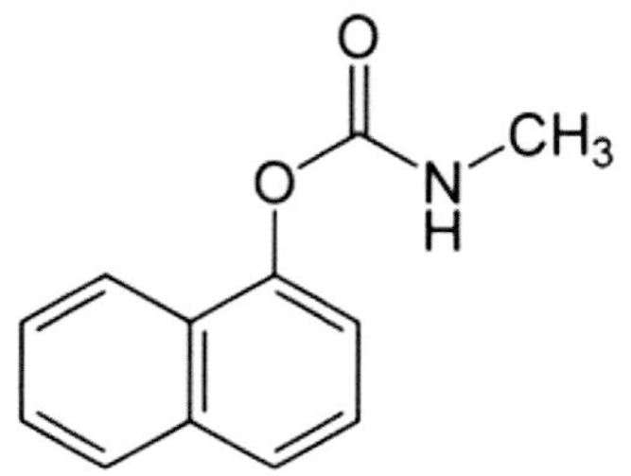

Figure 16: Structure of carbaryl

- Esters of carbamic acid.
- Carbaryl is one of the main representative of broad spectrum insecticides. It is used in the control of more than 50 insect species.
- Other carbamate insecticides include propoxur, carbofuran, and aldicarb.

2. Organophosphates

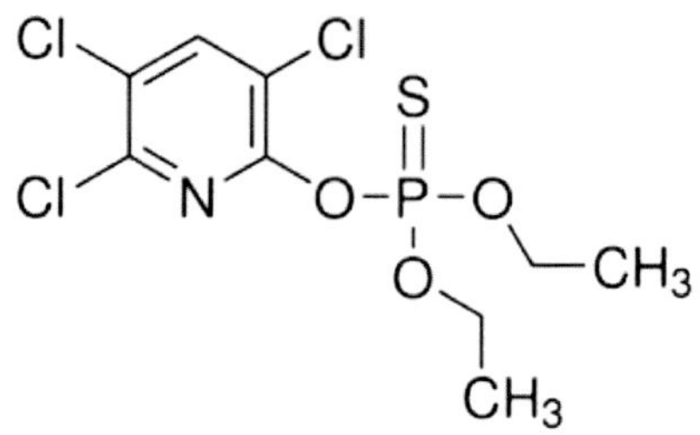

Figure 17: Structure of Chlorpyrifos

- Phosphoric acid derivatives whose development dates back to World War II.
- Can be divided in 6 subgroups, by their molecular structure: phosphates, phosphorothioates, phosphorodithioates, phosphorothiolate, phosphonates, and phosphoramidates.
- Chlorpyrifos, malathion and dimethoate are frequently used organophosphate insecticides.

3. Pyrethroids

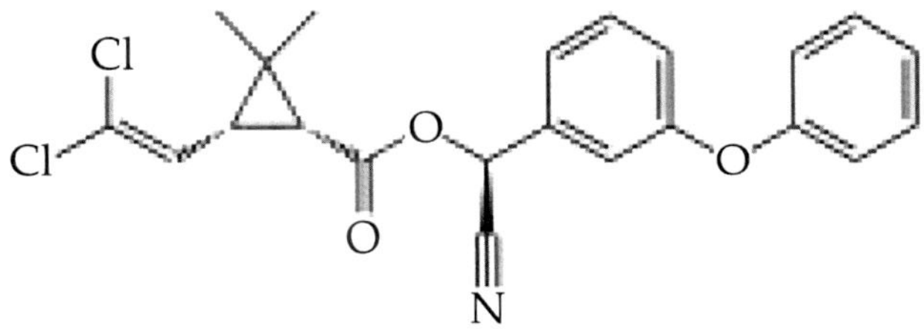

Figure 18: Structure of Cypermethrin

- It is a class of synthetic substances made from pyrethrum.
- They can be divided in two groups: Type I (pyrethrins, permethrin) and Type II (cypermethrin, fenvalerate). This categorization is based upon the mode of activity in sensorial neurons in insects.

4. Organochlorines

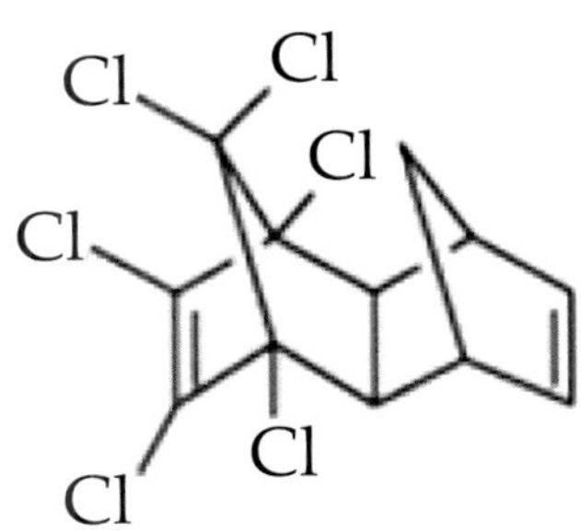

Figure 19: Structure of Aldrin

- DDT, aldrin, dieldrin and heptachlor.
- Banned since the 1970s, but with a known long time persistence. It is sparingly soluble in water, not susceptible to photolysis.

Highly lipophilic, these compounds tend to accumulate in animal tissue.

- DDT can undergo metabolic alterations, resulting in a myriad of products like DDE, TDE and dicofol. The latter possesses insecticidal activity.
- Methoxychlor, another member of this group, is known to have low cumulative power when compared to other organochlorine, although it is stable and persistent.

Fungicides

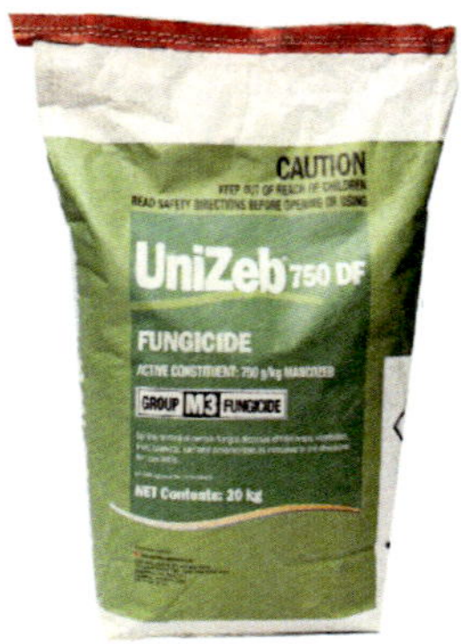

Figure 20: Mancozeb based commercial fungicide

1. Benzimidazoles

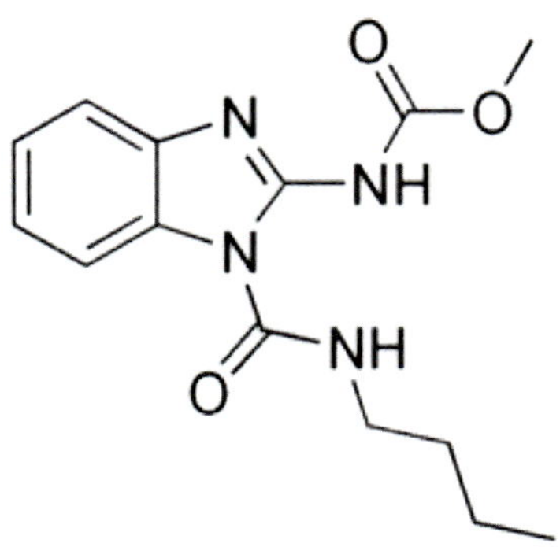

Figure 21: Benomyl molecule

- Also used as a veterinary medicine with anthelmintic action.
- Include Benomyl, carbendazim and Thiophanate methyl.

2. Dithiocarbamates

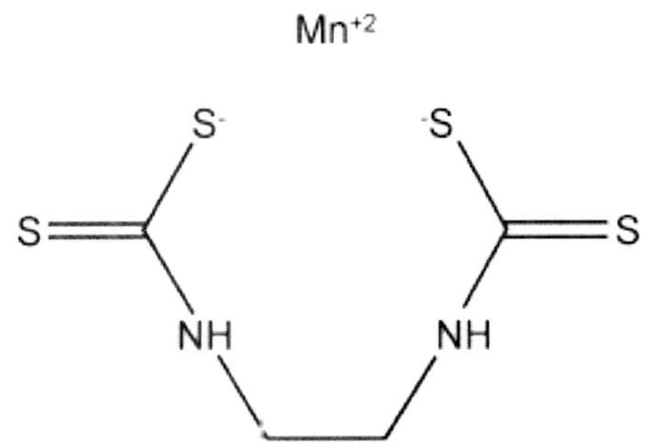

Figure 22: Maneb molecule

- May be divided in two main groups: dimethyldithiocarbamates (e.g., thiram, ziram) and ethylene-bis-dithiocarbamates (e.g., maneb, mancozeb)

3. Famoxadone

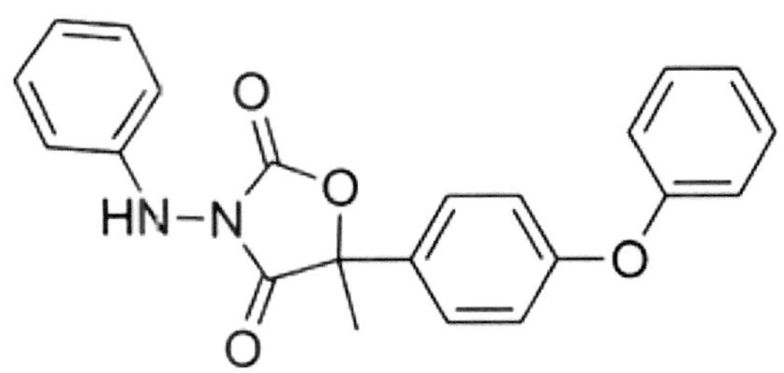

Figure 23: Famoxadone molecule

- A broad spectrum fungicide that belongs to the group of oxazolidinediones.
- It acts on the spores, inhibiting sporal differentiation and causes lysis of zoospores.

4. Fenamidone

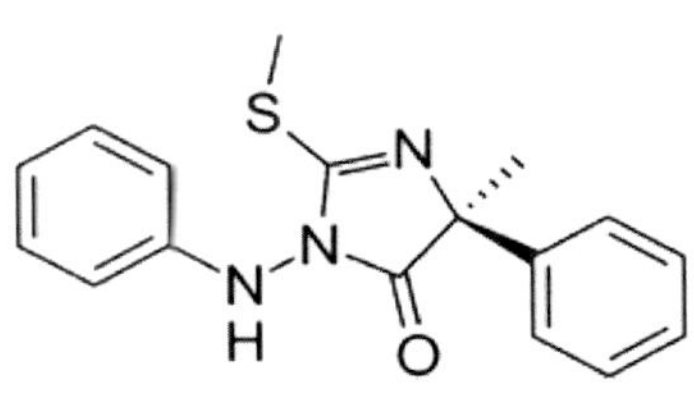

Figure 24: Fenamidone molecule

- Foliar fungicide, acts by inhibition of quinone receptors.

5. Chloronitriles

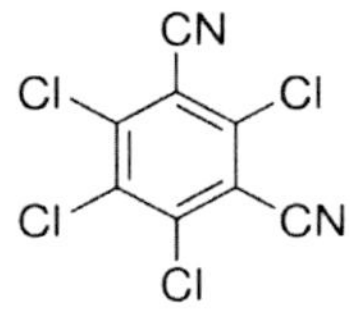

Figure 25: Chlorothalonil molecule

- Most widely used member of this group is chlorothalonil, which is used in peanut crops as well as for ornamental purposes.
- Chlorothalonil is considered nontoxic to mammals and birds, but it is highly toxic to fish, with LC50 of 0.25mg/L for rainbow trout and 0.3mg/L for bluegill.

6. Copper

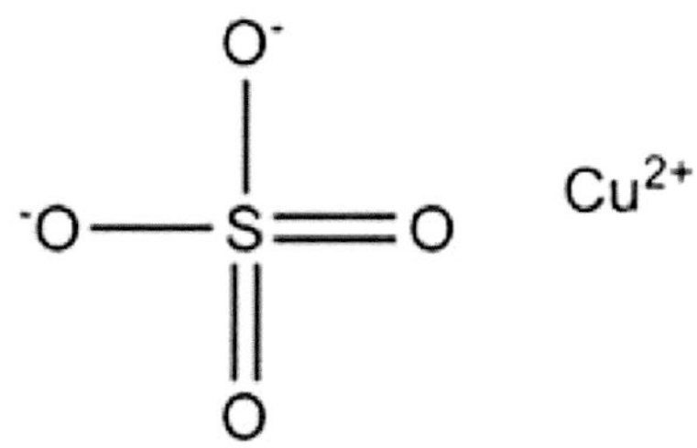

Figure 26: Copper sulfate salt

- Documented use as a fungicide dates from 1761, as a protection to wheat seed.
- Copper sulfate is a known herbicide, due to phytotoxic activity against weeds and algae.
- Used to control fungus in plants in the form of Bordeaux mixture. It is a broad spectrum protection product.
- It was observed a reduced crop growth, and decreased yield in potatoes.

7. Sulfur

- Yellow solid, not water soluble.

- It is believed to act as an enzymatic blocker in respiration. Besides antifungal activity, it is also phytotoxic. It was heavily used in North America until the 1970s.

8. Triazoles

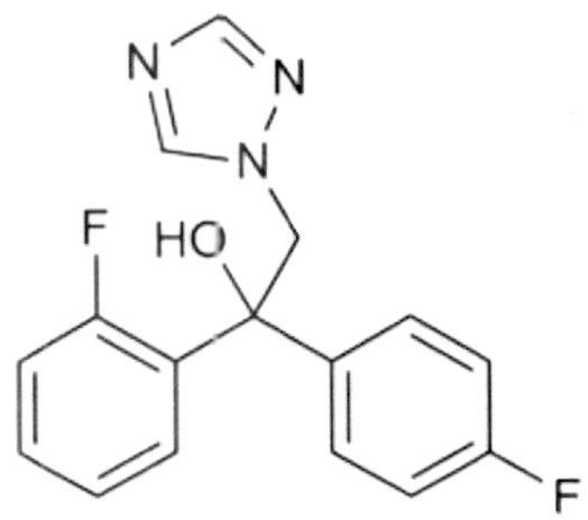

Figure 27: Flutriafol molecule

- Introduced in the 1970s, the group includes cyproconazole, flusilazole, flutriafol.
- Induce abnormal fungal growth by inhibiting C-14 demethylase enzyme, which plays a role in sterol production.

ENVIRONMENTAL FATE OF PESTICIDES

While DDT inaugurated a new age in how humanity fought pests, its wide use initiated a closer scientific look at the impacts the environment was getting from our agriculture, as well as possible adverse health effects to humans exposed to it. Considering that even the more fragile groups such as infants and the elderly had some sort of contact with chemical pesticides even in urban areas and their own households, investigation was welcome. The case of DDT concluded in its ban in the 1970s, having a relevant contribution in public awareness made by Rachel Carson's Silent Spring, released in 1962, a book that is always mentioned as a turning point in environmental matters. The book itself still suffers a heavy criticism from non-specialists, claiming that Carson's views and predictions were exaggerated, and the use of DDT had more benefits than harm. What most critics usually fail to notice, is that even before Silent Spring, lots of scientific studies were already being carried out and results showed that something was occurring: fish and birds mortality increases, reports of disturbances in lake trout reproduction, residues of DDT found in the wildlife, were observed in areas where DDT was sprayed. A few years later, articles reporting considerable levels of DDT residues found in seawater on the coast of California.

Some conclusions were taken from the whole body of research, as the methods evolved and the issue gained relevance: DDT was only environmentally persistent; it was not as selective as thought to be, and interfered in predator-prey relationship in diverse systems; it is bioaccumulative, meaning it remained quite stable while rising the steps in the food chain, until it reaches human consumption again; it is lipophilic, meaning it accumulates in fat and so residues can be found in milk, which poses a threat to infants and groups that depend on it as a major dietary source. Studies on the consequences of DDT application also showed that pest resistance was already becoming a problem, by simple evolutionary aspects: individuals who were able to survive DDT could pass their genetic characteristics to their offspring, which could be erroneously interpreted as DDT becoming "less potent" and so raising the application quantities or frequencies.

As said by Carson, "In nature, nothing exists alone".

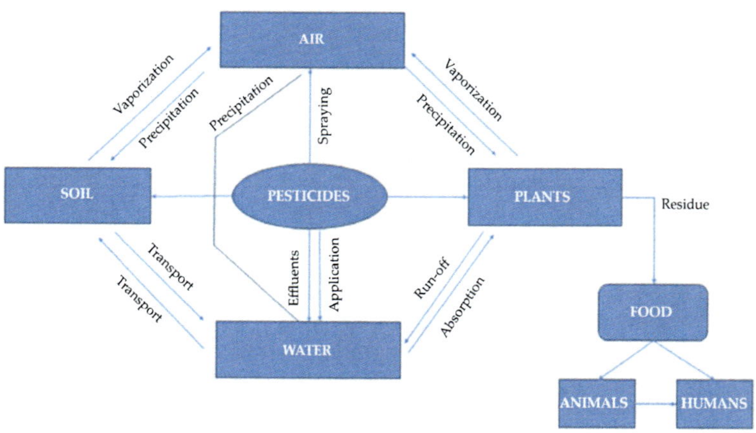

Figure 28: Environmental impact of pesticides (adapted from Murti CR and Nag D, 1991)

A broader classification of pesticides according to their characteristics is usually avoided, because of the risk of falling into oversimplification of this system. Molecules within one single group may have different behaviours depending on what is being looked at - their stability, lipophilicity etc. The impact caused by those products somewhat depends also on the aquatic systems *per se*, their nature and biology, as well as the actual application of such pesticides. Besides that, for instance, heavy usage on or close to water is one variable to be taken into account, as surface water monitoring has shown that water bodies located closer to agricultural areas have higher levels of pesticide contamination. Most of the time that contamination is unintentional, caused by run-off and spray drift, for example. Nonetheless, Table 12 below first elaborated by Edwards, provides a very didactic way to summarize what is known about certain relevant pesticide groups and their aquatic impacts.

Table 1: Characteristics of pesticides that cause hazards to aquatic systems (reproduced from Edwards CA, 1976)

Characteristic	Pesticides			
	Organochlorines	Organophosphates	Carbamates	Herbicides
High toxicity to aquatic fauna	+ + +	+ +	+ +	+
Solubility	-	+	+	+ + +
Potential for uptake and bioconcentration	+ + +	+	+	+ +
Persistence in aquatic systems	+ + +	+	+	+ +
Heavy usage on or close to water	+ + +	+ + +	+ +	+ + +

Note: "+" and "-" symbols are used as visual assignment of measurable potential of the characteristic in question. These labels were kept in order to maintain fidelity to the original work.

Each of these characteristics will contribute to an overall environmental impact that is not quantitatively described in such simple terms. Nonetheless, they are still useful for understanding the many variables in action.

- **High toxicity to aquatic fauna** - toxicity may be very specific to only a certain group of organisms or broad enough to be damaging for non-target species or even to most of the whole local ecosystem. Toxicity levels may be acute or chronic.

It is used to measure the pesticide risk to biota along with its persistence and lipophilicity, and it may affect diverse organisms physiological functions. Table 13 presents toxicity data for a number of common pesticides.

Table 2: Toxicity of Pesticides in Fish (Reproduced from Edwards, CA 1976)

	L.C.50 (ppm)	
Insecticides	**Bluegills**	**Rainbow trout**
Organochlorines (after 96 hours)		
Endrin	0.006	0.007
Toxaphene	0.004	0.008
Dieldrin	0.008	0.019
DDT	0.016	0.018
Aldrin	0.013	0.036
Chlordane	0.022	0.022
Methoxychlor	0.062	0.020
Lindane	0.062	0.060
BHC	0.79	Not available
Heptachlor	0.190	0.150
Organophosphates (after 24 hours)		
Azinphos methyl	0.022	0.014
Phosdrin	0.041	0.034
Fonofos	0.045	0.110
Malathion	0.120	0.100
Diazinon	0.052	0.380
Phosphamidon	Not available	5.0
Methyl parathion	5.7	2.7
Dimethoate	28.0	20.0
Carbamates (after 24 hours)		
Carbaryl	3.4	3.5
Zectran	11.2	10.2
Herbicides (after 48 hours)		

Trifluralin	0.019	0.011
2,4-D	0.9	1.1
Fenoprop	16.6	1.4
Diuron	7.4	4.3
Chlorfenac	19.0	7.5
Diquat	19.0	20.0
Dichlobenil	20.0	22.0
Sodium arsenite	44.0	36.5
Paraquat	Not available	62.0
Simazine	130.0	85.0
MCPA	100.0	Not available
Aminotriazole	100.0	Not available
Dalapon	115.0	Not available

Figure 29: Bluegill, one of the most common fish species used in toxicity studies

- **Solubility** - pesticides have different water solubilities. Highly soluble pesticides may easily diffuse in a water system, even permeating the soil and reaching groundwaters. These do not adsorb easily to sediments and organic matter, so they tend to be less persistent compared to those highly lipophilic. These in turn will have greater potential to accumulate in plant and animal tissue when ingested. Table 14 shows differences in the solubility of various pesticides.

Table 3: Solubility of Pesticides in water (Adapted from Edwards, CA 1976)

Pesticide	Amount (in ppm)
Organochlorines	
DDT	0.0012
Aldrin	0.01
Heptachlor	0.056
Methoxychlor	0.10
Dieldrin	0.18
Endrin	0.23
Toxaphene	0.40
Lindane	7.0
Organophosphates	
Parathion	24
Disulfoton	25
Azinphosmethyl	30
Diazinon	40
Phorate	50
Chlorfenvinphos	145
Malathion	145
Demeton methyl	330
Thionazin	1140
Dimethoate	2500
Carbamates	
Carbaryl	40
Carbofuran	700

Aldicarb	6000
Herbicides	
Simazine	5.0
Propazine	8.0
Dichlobenil	20
Atrazine	33
Diuron	42
Chlorfenac	200
2,4,5-T	280
Monuron	280
2,4-D	890
Aminotriazole	28%
Diquat	70%
Dalapon	80%

- **Potential for uptake and bioconcentration** - this is the potential a pesticide has of accumulating in animal and plant tissues. It also effectively determines the amounts that will be present throughout the food chain, and that may finally be brought to human exposure once again. Organisms in higher trophic levels in an aquatic system (e.g., fish) may absorb the contaminants directly from water and not only from predation, as shown by some studies in the 1970s. The mechanisms for bioaccumulation involve the lipid/water partition coefficient, which is a basic physical-chemical property of every substance. Pesticides with higher coefficients will be less prone to remain dissolved in water, thus accumulating in lipid media and tissues.

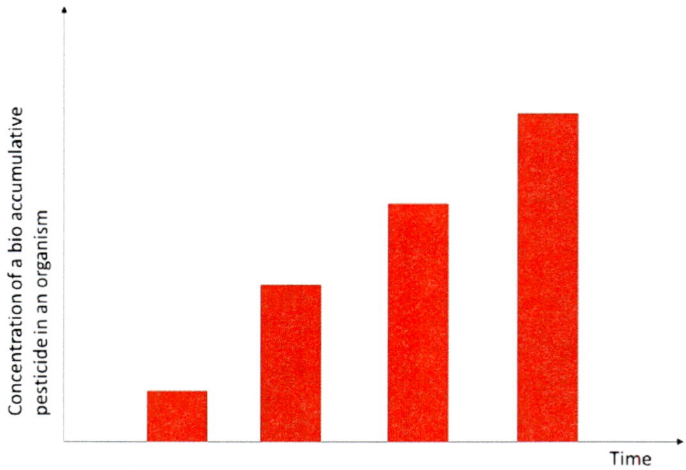

Figure 30: Schematic bioaccumulation, where the amount of pesticide residues within an organism increases in its lifetime

- **Persistence in aquatic systems** - it is a measure of how long the pesticide molecules can remain stable before undergoing some sort of chemical cleavage, metabolization by microorganisms or degradation. They can range between a few hours to years and even decades (e.g., organochlorines). More persistent pesticides may pose greater impact for the longer period they remain active in a system, since it gives them also a higher potential for uptake and accumulation. See Table 15 and 16 for examples of typical persistence of pesticides in water and in fish, respectively.

Table 4: Persistence of Pesticides in River Water (Adapted from Eichelberger and Lichtenberg, 1971)

Pesticide	**Percentage remaining**	
Organochlorines	**After 2 weeks**	**After 4 weeks**
BHC	100	100
Dieldrin	100	100
DDT	100	100
Endrin	100	100

Chlordane	100	85
Aldrin	80	40
Endosulfan	30	5
Isobenzan	25	10
Heptachlor	25	0
Heptachlor epoxide	100	100
Organophosphates		
Monocrotofos	100	100
Dimethoate	100	85
Ethion	90	75
Fenthion	50	10
Parathion	50	30
Malathion	25	10
Trithion	25	10
Carbamates		
Aminocarb	60	10
Propoxur	50	30
Zectran	15	0
Carbaryl	5	0
Fenuron	60	20
Monuron	40	30

Table 5: Persistence of pesticides in fish (adapted from Edwards CA, 1976, with data from Macek, 1970)

Pesticide	**Persistence time in fish**
Herbicides	
Diquat	< 3 weeks

Endothal	< 3 weeks
Simazine	< 3 days
2,4-D	< 1 week
Sodium arsenite	> 16 weeks
Dichlobenil	< 2 weeks
Organophosphate insecticides	
Diazinon	< 1 week
Dursban	< 1 week
Azinphos-methyl	< 1 week
Malathion	< 1 day
Parathion	< 1 week
Organochlorine insecticides	
Lindane	< 2 days
Methoxychlor	1 week
Heptachlor	1 month
Dieldrin	1 month
Camphechlor	> 6 months
DDD	> 6 months
DDT	> 5 months

4

CHAPTER

AQUATIC ENVIRONMENT AND EXPOSURE

Erick Soares Lins

HUMAN DEPENDENCE ON WATER

From the sites at the north of Scandinavia ca. 5300-4500 BC, to the Jomon ca. 12000-10000 BC in Japan, to the Natufian ca. 12500-9500 in the Levant region at the Mediterranean coast, all these settlements were subject to archaeological studies that found evidences of their existence. What they all have in common, is that the settlement was near a water body, either a river, a lagoon, and even seas. Portions of water meant resources, and these resources became starting material for fundamental developments, such as agriculture. In fact, one of the greatest revolutions in how we lived, i.e., the transition from a nomadic society to a sedentary one is intimately related to that observation.

Figure 1: Tigris River, Mesopotamia

The world's water supply is a matter of concern. Although the Earth's surface is 70% covered with water, an incredibly small part of that good for our direct consumption and utilization. In numbers:

The Oceans account for 97% of all Earth's water, and only 3% are fresh water. More than ⅔ of that fresh water is in the glaciers, and a little less than the remaining ⅓ is in underground aquifers, so hardly attainable. 1% of the total fresh water in our planet is actually at our immediate reach, in lakes, rivers, and streams.

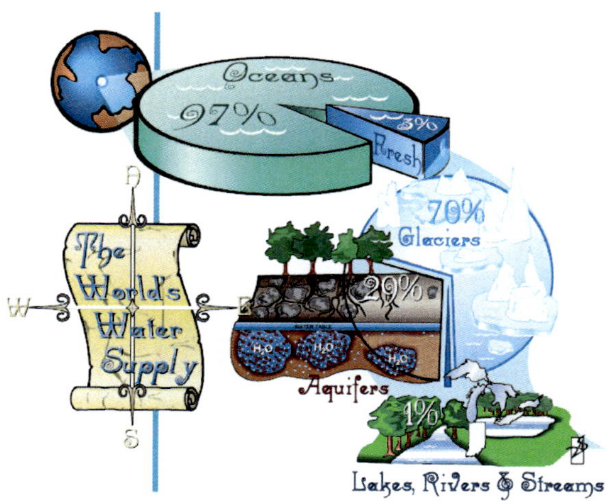

Figure 2: The World's Water Supply, illustration of the Earth's relative estimated amount of water (reproduced from Whitford F)

WATER BODIES

All that water keeps changing its state and availability in a continuous scheme of connected processes, known as Hydrological Cycle, or simply Water Cycle. The cycle links surface, ground and atmospheric water, and we shall see how environmentally important can be an interference in that cycle.

Basically, surface water is converted to water vapor by absorbing energy from the sunlight, thus being moved to the atmosphere - this is the **evaporation** step. When at higher altitudes and cooler temperatures, water vapor condenses forming clouds, which will eventually release water back down in the form of rain, hail, or even snow - that is the **precipitation**. Reaching the ground, water undergoes **infiltration**, entering the soil and percolating, either to reach ground water that will discharge in streams and lakes thus returning to the surface, or it will be absorbed by plants before that, which will use and release it to the atmosphere again by **transpiration**. Water is exposed to solar energy again, water vapor will form clouds, and the cycle goes on in its uninterrupted repetition.

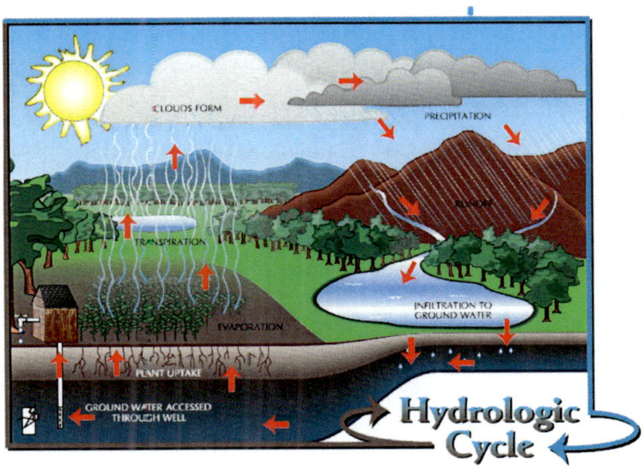

Figure 3: Hydrologic cycle (reproduced from Whitford F)

The collection of rivers and tributaries, streams, lakes and other water bodies in a specific region is known as watershed or drainage basin. Wetlands, forests, and other habitats are included in a watershed. Towns and cities become part of it simply by being there, as well as artificial water bodies like canals and water reservoirs. Some good examples around the world are the Mississippi river basin in the US (ca. 1,150,000 sq miles), the Nile River basin in northeastern Africa (ca. 1,300,000 sq miles) and the Amazon basin in South America (ca. 2,900,000 sq miles).

Figure 4: Amazon rainforest

In watersheds, surface and groundwater are connected. Rainfalls and melting snow will infiltrate the soil at a certain rate, as much as the porosity of the soil and the initial amount of water can hold. Any excess water in the top layers of soil, when saturated, will slow down the water percolation, and that is when surface runoff occurs. Runoff flows towards downslopes until it reaches lower sites, accumulating and turning the spot into a storage area (that could be a stream or a pond). Beneath the surface in that system, the soil is able to retain quantities of water within the minuscule voids of its solid parts. These are called aquifers, and they may consist of rock, clay, sand, fluvial sandstones and other components depending on the geologic formation. Aquifer can be wide as basins, take for instance the Guarani Aquifer in Brazil, which is estimated to occupy more than 450,000 sq miles and goes about 2,600 feet deep. It is estimated to contain 9,600 cubic miles of fresh water.

Aquifers are water saturated soil constituents, that is why most of the volume in an aquifer is not immediately attainable, but that does not mean the water is completely protected from contaminations. Unconfined aquifers, i.e., those that do not have a layer of less permeable rock separating them from the surface, can experience a higher rate of interchange of surface waters and the top layers of soil by vertical movements of water. That is where improper pesticide application may become a concern that extrapolates the visible superficial water bodies.

As presented by Funari et al., the characteristics of a waterbody make it more or less susceptible to pollution. The studies on aquifers are summarized on Table .

Table 1: Classification of aquifers with respect to vulnerability to herbicide pollution (adapted from Funari et al.)

Type of aquifer	Vulnerability degree (from Very Low to Very High)
Aquiferous network in fissured intrusive rocks	Very Low
Aquifer networks in sandstones and cobble-stones	Low
Multiaquifer system in Flysch** or sequenced formations	Low
Aquifer in morainic formations	Low
Aquiferous network in metamorphic rocks	Low
Multiaquifer in loose pyroclastic rocks	Medium
Aquiferous network in fissured dolomitic rocks with low Karst phenomena	Medium
Aquifer in fine-grained sands	Medium
Confined, semi-confined, or free surface aquifer with low permeability protection layer	High
Aquifer in volcanic or highly stressed intrusive rocks	High
Aquiferous network in fissured calcareous rocks	High
Aquiferous network in calcareous rocks with well developed Karst* phenomena	Very High
Free surface aquifer in alluvial sediments (from coarse to medium grained) with no protection layer	Very high

*Karst phenomena are processes related to the development of solution channeling and preferential paths inside calcareous masses;

**Flysch is the name of typical geological formations made up by lithological units of sandstones, marly limestones, silty and clay types with chaotic or irregular structures.

In terms of origins of pesticide contamination, different waterbodies are more subject to some exposure routes than to others, either from deliberate or accidental introduction. As summarized by Edwards:

Table 2: Aquatic systems that may be polluted by pesticides (reproduced from Edwards CA, 1976)

System	Main sources of pollution			
	Spraying	Run-off	Atmosphere (Precipitation)	Industrial Effluents
Ponds	++	++	+	-
Rivers	+	++	+	+++
Estuaries	+	+	+	++
Oceans	-	+	+++	+

Note: "+" and "-" symbols are used as visual assignment of measurable potential of the characteristic in question. These labels were kept in order to maintain fidelity to the original work.

A few important considerations about water bodies vulnerability rely on the characteristics of these systems and how they differ among each other.

- **Size** - the impact of a single pesticide contamination is certainly different in the oceans than it is in smaller system such as a pond, because of how diluted that pollutant will be in the end. On the other hand, atmospheric route (caused by rainfalls) can be more damaging to water bodies with larger surface areas.
- **Form** - flowing waters (e.g., rivers, streams) favour distribution of the pesticide along larger volumes of water when compared to stagnant bodies (e.g., ponds, lagoons), and that may dilute it as well. However, in the first case the contamination is not contained, so it can effectively reach other waterbodies at longer distances from the point source, and eve groundwaters. The depth and amount of sediment and organic matter will also determine how long and how bioavailable the pesticides will remain.
- **Location** - Not only those water bodies that are nearer agricultural areas are found to be more susceptible to pesticide contamination, there is also a weather influence. Regions with more rain incidence shall present larger run-off and even erosion, which will occasionally transport pesticides applied in land to water bodies in the proximities.

There is recent research showing evidences of anthropogenic pollutants have been found in the deep regions of the oceans, proving that highly stable molecules can be transported along the water system and finally reaching places very distant from their point source. Recently, PCBs and PBDEs were found in amphipods dwelling in the bottom of the oceans, in the Kermadec and the Marianas trenches. Bear in mind that production of PCBs and PBDEs has been prohibited around 40 years ago.

FACTORS AFFECTING POTENTIAL CONTAMINATION IN WATER BODIES

The largest portion of pesticide contamination in waters is attributed as nonpoint source, with the sole exception of direct dumping. There are key processes in the hydrological cycle that lead to runoff of sediments, nutrients and also pollutants, so understanding their principal concepts and controlling factors is important for a better modelling and estimates of environmental impacts:

- **Rainfall:** Rainfall impacts soil detachment and therefore the sediment availability to runoff. The rate, or intensity, of rainfall is the controlling factor.
- **Soil Permeability:** A measure of the amount of water the soil is able to pass under saturated conditions. It is a physical characteristic of soils, and mainly a function of texture, mineral and organic composition, associated with porosity, but also with the natural interconnections of such empty spaces.
- **Infiltration rate:** It is the rate at which surface water penetrates the soil, usually given in centimeters per unit time. Factors controlling infiltration are the capillarity in the soil, moisture, degree of surface compaction. Infiltration also varies according to rainfall intensity, initial moisture content of the soil and vegetation.
- **Interflow, or throughflow:** It is the movement of water through soil layers that is not downwards. Different layers (or soil horizons) have different permeability levels, so portions of water may run parallel to the ground. Eventually, interflow

emerges, usually at the bottom of valleys, and are themselves a potential gateway for contaminations in hydrological instances.

- **Surface runoff** : It is the water that remains at the surface even with portions infiltrated into the soil, as well lost by plants evapotranspiration. If precipitation rate exceeds infiltration rates, larger surface runoffs are produced, and these are the two main controlling factor in this case. Vegetation depletion along with soil compaction make a zone extremely sensitive to erosion caused by surface runoff.

- **Snow melt:** It contributes to surface runoff by a combination with spring rain, leading to erosion. Snowmelt can be a carrier of animal waste, sludges and pesticides that were left on agricultural soil during the winter time, that will at some point reach streams,

- **Groundwater:** Water that moves along the bedrock underlying the soil. It is fed by infiltrated surface water, and tends to flow towards river channels, therefore being a matter of concern in terms of chemical contaminations.

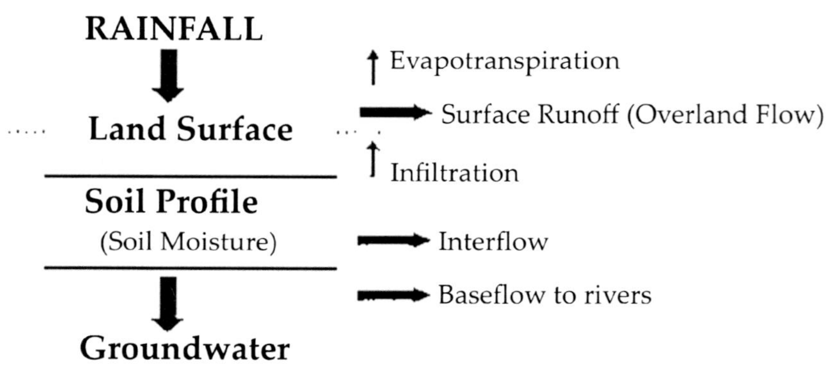

Figure 5: Schematic diagram showing the major processes that link rainfall and runoff - The role of sediment as a chemical pollutant is a function of the chemical load that is carried by sediments.

SOIL PROPERTIES IMPACTING WATER VULNERABILITY

Likewise, in the case of groundwater, its vulnerability arises from the combination of soil characteristics as well as the physical and chemical properties of the pesticides. The penetration path of a substance through soil depends on:

- Top soil properties: textural, mineralogical, microbiological and physico-chemical;
- Permeability and other less important hydraulic properties of the soil layers below the surface. The different levels of permeability of different layers will slow down or accelerate a pollutant downward penetration.
- The depth of the unsaturated zone (the piezometric surface);
- The amount of water at the surface, since water is the main carrier of any chemical pollutants. Rainfall, irrigation and run-off will play a major role in the percolation process.

Figure 6: Soil horizons and their different compositions

Unlike the detection of industrial pollutants, whose point-source usually can be traced to one (or more) specific discharge site(s), agricultural runoff is part of nonpoint-source contamination. Pesticides may come from crop lands, diverse house developments, or other uses near water bodies, and that is known as nonpoint-source contamination.

Water quality is directly related to sediments. As clearly defined by FAO, pollution by sediment has two major dimensions:

PHYSICAL DIMENSION - *top soil loss and land degradation by gullying and sheet erosion and which leads both to excessive levels of turbidity in receiving waters, and to off-site ecological and physical impacts from deposition in river and lake beds.*

CHEMICAL DIMENSION - *the silt and clay fraction (<63m m fraction), is a primary carrier of adsorbed chemicals, especially phosphorus, chlorinated pesticides and most metals, which are transported by sediment into the aquatic system.*

Small particles have large surface areas, thus being more able to have substances adsorbed and transported. Phosphorus and metals are largely transported in sediment particles in river flux, and many organic molecules, especially chlorinated, are too. The environmental fate of a pesticide is then estimated by how well it may be adsorbed by particulate matter, and that affinity is measured by the octanol-water coefficient, Kow. The larger the Kow, the higher hydrophobic (more lipophilic) character of the molecule, and higher the tendency to be associated with particulates. That lipophilicity is also what allows pesticides to enter the food chain starting with microorganisms that feed on small sediment particles and then bioaccumulate.

Sediments can be carried by flowing water, and those suspensions may contain adsorbed pesticides. These can be a result of runoff after application or even soil erosion, and being transported by water flow, they will eventually reach a repository where it tends to sit at the bottom. The common exposure routes by which pesticides can enter the hydrological cycle and reach water bodies are summarized below.

EXPOSURE ROUTES

Runoff and erosion are the primary routes for pesticides to reach surface water, but they are not the only ones. This implies that non-adjacent areas are susceptible to the impact caused by pollutants, perhaps not at the same level as those points closer to the contamination source, but still detectable. Other types of pesticide contamination follow the routes below:

- **Spray drift:** Small droplets can be carried by wind to hundreds of meters away from the application site, depending on the conditions. Spray drift usually happens when liquid pesticides are applied with tractor-mounted, aircraft, and even handheld spraying.

Figure 7: Aircraft spray

- Subflow: Soil applied pesticides can reach the cropland drainages, called field tiles, which will ultimately reach surface waters. It is enhanced with more rainfall, and weakly adsorbed pesticides, i.e., to strongly attached to soil particles or organic matter, can be carried by it.
- **Direct introduction**: Accidental spills or waste dump directly into surface waters.

Figure 8: Farmland close to a river

- **Runoff and erosion**: Heavy rainfall and the soil conditions (vegetation, slope, soil type, soil compaction etc) can result in runoff and erosion. Soil applied pesticides can reach surface water as they are carried along with detached sediment.

Figure 9: Rice paddy in Asia

- **Leaching:** Pesticides may be carried by water in a downward flow, through the soil, until it reaches groundwater. Soil properties such as permeability, and other factors like rainfall may determine how much of pesticides will leach. Leach will also depend on the pesticide's degradation rate and its ability to sorb to soil particles and organic matter. Leach can occur in Preferential Flow, where water will run faster through larger structural voids (e.g., wormholes, cracks) or Matrix Flow, where water will penetrate the smaller holes in soil, allowing pesticides to interact more time with soil particles.

A comprehensive report titled *The Quality of Our Nation's Waters*, released by the U.S. Geological Survey, released in 1999 showed important findings concerning nutrients and pesticides in water. Some of the findings taken from that document include:

Geographic distributions of pesticides follow patterns in land use and pesticide use:

- Herbicides accounted for highest frequency in streams in agricultural lands. Concentrations of Atrazine exceeded aquatic life guidelines in all agricultural streams and major rivers sampled.
- Other streams showed low frequency of herbicide detection, commensurate with the low use of herbicides in the respective lands.
- Insecticides were the major pesticides found in urban area streams. Almost all of the urban streams and major rivers sampled showed insecticide levels exceeding aquatic life guidelines. The most frequently found substances were diazinon, chlorpyrifos, and malathion.
- Most agricultural streams showed low to moderate levels of insecticides. Insecticide concentrations in most major rivers usually were lower than those measured in urban streams and exceeded aquatic-life guidelines in relatively few samples. Exceedances occurred, and in addition to the insecticides already mentioned, there were also detections of methyl azinphos.
- Organochlorine insecticides were highest in urban streams and where historical agricultural use was greatest. Although banned for decades, DDT, chlordane and dieldrin were found in bed sediment and fish. Almost all urban area streams showed contamination, and agricultural areas in which there has been heavy use of those active ingredients in the past.

Pesticide occurrence in streams:

- More than 90 percent of water and fish samples from all streams contained one or, more often, several pesticides. Those include products used nowadays as well as long persistent ones that have not been used in the past 30 years or so, e.g., DDT and organochlorines;
- About 50 percent of the wells sampled contained one or more pesticides. There were more detections in shallow ground water below both agricultural and urban areas, and less in deeper aquifers.

The most frequently found pesticides:

- Herbicides are the most common type of pesticide found in streams and ground water within agricultural areas.
- The top 15 pesticide compounds found in water are among those with the highest current use.
- Both agricultural and urban use contributed to the findings in major rivers and aquifers. Agricultural use herbicides (ex. Atrazine, Metolachlor) accounted for the most contaminations, followed by urban use herbicides (ex. Simazin). Insecticides also appeared in the monitoring, but at lower rates (ex. Diazinon).
- In fish and bed sediment, metabolites and derivatives of organochlorine pesticides were found, especially DDT and Chlordane.
- Findings are good indicators of pesticide use trends. There has been a decrease in the detection of DDT and organochlorines comparing to historical data. Other pesticides, such as Acetachlor, were found in increasing levels in recent period.

In a similar philosophy, there is also in place the Agri-environmental indicator, maintained by the European Commission. The European Union has a whole regulatory framework dedicated to the presence of pesticides in water, and it is composed by a number of directives: Water Framework Directive (WFD)(2000/60/EC), Pesticides Framework Directive (Directive 2009/128/EC), Environmental Quality Standards Directive (Directive 2008/105/EC), Drinking Water Directive (Directive 98/83/EC) and Groundwater Directive (Directive 2006/118/EC). These directives deal with actions to mitigate pesticide contamination, such as training and education for applicators, alternative pest management, it contains an harmonized list of hazardous pesticides not to be used, and establish maximum levels for total pesticides in water (0.5ug/L) and for single pesticides in drinking water (0.1ug/L), with some specified exceptions (e.g., Aldrin and Dieldrin must not exceed 0.03 ug/L in drinking water). The Commission report notes that agriculture is the main contributor to pesticide contamination in water bodies, and the products applied to crops may reach them by any of the several ways discussed, mainly spray drift, run-off and leaching. The report also reaffirms the severe environmental impacts caused by a high level of pesticides in water, which includes several damages to the fauna and flora, and ultimately the loss of biodiversity.

Along with that legislation, there are indicators set and monitored, being the following two the main indicators on that issue:

- Groundwater with pesticide concentrations above Environmental Quality Standards (EQS).
- Rivers with annual average pesticide concentrations above Environmental Quality Standards (EQS).

Some of the important findings of that European monitoring program, and conclusions thereof, are summarized below:

- About 7% of groundwater stations across Europe reported excessive levels of at least one pesticide. Atrazine and its metabolite Desethylatrazine were the most frequently found in levels exceeding the Environmental Quality Standards. Most of the groundwater at risk is located near or in agricultural land.
- Groundwaters presented a larger variety of pesticides detected, as it is shown in Figures 52 and 53, including 2,4-D, Alachlor, Aldrin, Alpha-Endosulfan, Alpha-HCH, Atrazine, Bentazone, Beta-HCH, Chlorfenvinphos, Chlorpyrifos, DDD p,p', DDE p,p', DDT o,p', DDT p,p', Desethylatrazine, Desisopropylatrazine, Dieldrin, Diuron, Endrin, Gamma-HCH, Isodrin, Isoproturon, Linuron, MCPA, Mecoprop, Prometryn, Propazine, Simazine, Terbuthylazine, Terbutryn and Trifluralin.
- Certain pesticide groups in rivers were detected and quantified

in more concerning rates than others. For instance, 35% of the rivers monitoring stations assessed presented exceeding levels of endosulfan throughout the year monitoring, as opposed to only 5% for the other groups.

- Fewer pesticides were reported in river waters, including Chlorfenvinphos, Chlorpyrifos, Cyclodiene, Diuron, Isoproturon and Trifluralin in some countries within the scope of the annual monitoring plan in 2009.

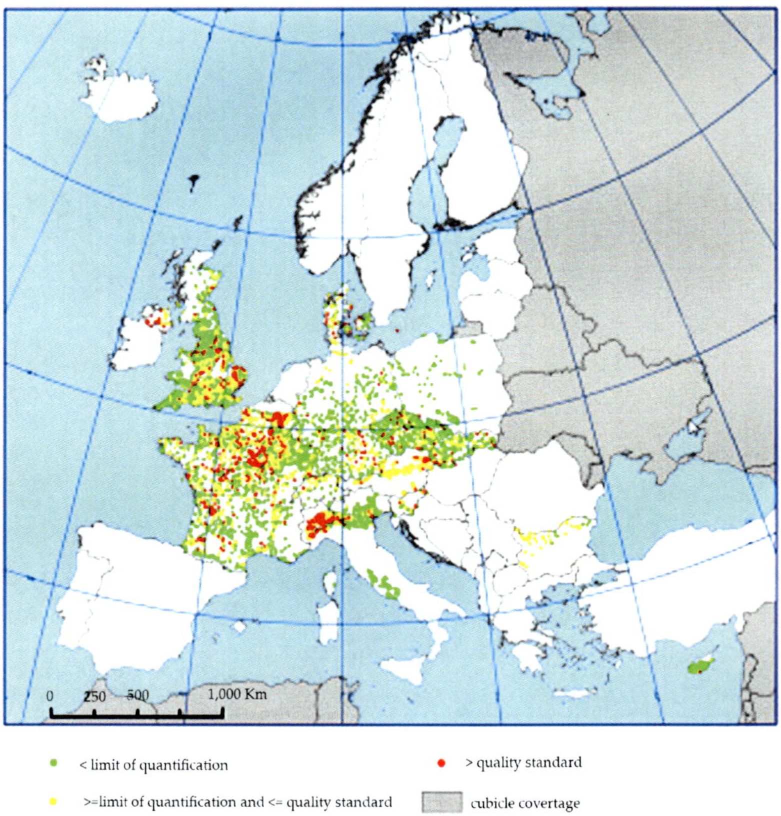

Figure 10: Occurrence and exceedance of selected pesticides (listed in Figure 53) in groundwater monitoring stations, 2010-2011 Source: European Environment Agency, WISE-SoE Groundwater

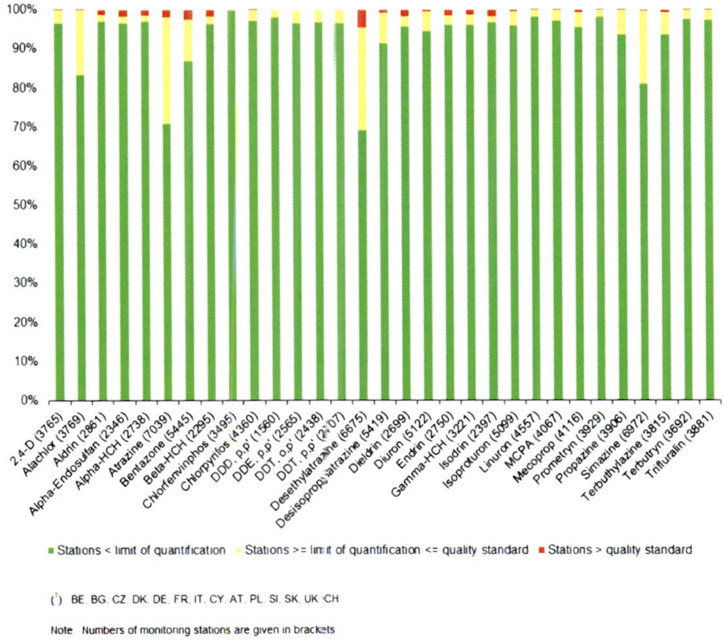

Figure 11: Occurrence and exceedance of Environmental Quality Standards (EQS) of individual pesticides in groundwater monitoring stations, (%), 2010-2011, (1) Source: European Environment Agency, WISE-SoE Groundwater

PESTICIDE MANAGEMENT IN DEVELOPED VS. DEVELOPING COUNTRIES

Figure 12: Pesticide applicator not wearing protective equipment

If the monitoring structure of the US and Europe were not considered ideal, although able to provide essential information, in 1985 FAO came to the conclusion that there were some important shortcomings in the way most developing countries handled their pesticide-related matters. This is what reads the International Code of Conduct on the Distribution and Use of Pesticides, adopted in that year:

> *"In the absence of an effective pesticide registration process and of a governmental infrastructure for controlling the availability of pesticides, some countries importing pesticides must heavily rely on the pesticide industry to promote the safe and proper distribution and use of pesticides. In these circumstances foreign manufacturers, exporters and importers, as well as local formulators, distributors, repackers, advisers and users, must accept a share of the responsibility for safety and efficiency in distribution and use."*

In other words, the inability of certain governmental instances in enforcing safety regulations in food production and consumption in their countries could not be an excuse to systemic control over consumers' and environment exposure to pesticides. The concerns that led to a recommendation that in practical terms affected all stakeholders in the production chain, even those abroad, were mostly related to known increased usage, but lack of surveillance of the residues. The points summarized by FAO as a description of the situation in developing countries are:

- Inadequate legislation and enforcement of pesticide regulations, including importation, use and disposal.
- Gifts of pesticides from donors that encourage inefficient use and abandonment of older quantities of the same pesticide.
- Stockpiling of pesticides, especially in countries with unstable governments, leading to abandonment of stockpiles in situations of insurrection and civil war. Examples exist where such a situation led to severe groundwater contamination and public health crises due to dumping of pesticides by untrained civilians.
- Storage and handling is a major problem, including leakage from old barrels and deliberate dumping of surplus pesticide

mixtures into water courses following application.

- Destruction of old stores of pesticides (due to deterioration of the active ingredient) is financially prohibitive (estimated at US$ 5000 per tonne) especially as stocks must be moved to a developed country for destruction. Consequently, old barrels deteriorate with leakage into surface and groundwater and/or dumping of stocks.
- Lack of training of users in pesticide handling and application, leading to improper application with environmental and public health consequences.
- Use of pesticides for inappropriate purposes, such as killing of trash fish.
- Use of old pesticide drums for drinking water, cooking, etc.

Figure 13: Applicator preparing the product

It is important to note that the "group" of developing countries includes largely different circumstances. On one side of that spectrum you have a few African nations in which agricultural pesticide use is almost nil, and on the other you find South and Central American countries whose economy relies heavily on the export of agricultural commodities, and which experienced a substantial rise on the use of several kinds of plant protection products in the last two to three decades. Although FAO's Code dates back to 1985, there is no reason to believe that a broad and diverse group like developing countries would have completely eliminated all of the problems listed above at this point.

Taking the Brazilian example, one of the major producing and exporting countries in the world, and one whose agricultural, although historically important, recently became the economical main driving force, with all the implications that arise from it. According to Andreoli (1993), Brazil was the third largest pesticide user in 1970, exceeded only by United States and France. Although having a pesticide regulation in place, he states that only 15% of the active ingredients in use were analyzed (presumably in food and environmental matrices) due to a lack of installed scientific capability, i.e., lack of analytical methods. This is an obstacle to the assessment of impact and exposure, and in consequence, to any implementation of public policies addressing consumers' health and environmental damage.

Today, the situation is quite different. According to Rigotto et al. (2014), since 2008 Brazil became leader of that rank, due to an expansion of more than 190% on the pesticide market in the last decade, a rate way greater than the world average for the same period. Starting almost at the same period, Brazil has an ongoing pesticide residues surveillance program, which evaluates all major crops (e.g. fruits and vegetables, soybean), including those imported (e.g. wheat, maize) as well as some minor crops (i.e. crops that lack specifically registered active ingredients). In the last cycle, 33 different products were analyzed for circa 200 pesticides and their metabolites. The surveillance program is able to determine the level of contamination and the origin of the produce, identify the pesticides and their compliance to MRLs (Maximum Residue Limits), either approved for use or not, being able to reach out to the companies and the very introduction of the pesticide into the production chain, which allows for further investigation and administrative measures once a non-compliance is unequivocally confirmed. In conclusion, Brazil has now a monitoring system in place, in which regulation enforcement and analytical capabilities are not the bottlenecks they were in the 1970s. The system might not be perfect, but it has proved to be effective to a great extent. Brazil has become also the leader in the retrieval of used pesticide containers, which is one of the issues mentioned by FAO.

When it comes to water contamination, Albuquerque et al. (2016) provided a more recent evaluation, consolidated with studies that aimed at that subject and their findings. States have their water monitoring programs in place, although global data is not readily available. The researchers could trace a clearer picture of the analytical capability at hand, methods characteristics, most commonly found pesticides (from 1998 to 2013) and their quantification levels. In that profile, herbicides are still the most frequently found, exceeding methods' LOQ (Limit of Quantification) in 14% of the samples. Some of the recurring substances in this group include triazines (e.g. Atrazine, ametryn and simazine), which are used in important crops like coffee, corn, and sugarcane. Clomazone and quinclorac, herbicides associated with irrigated rice were also constantly determined. Other relevant findings were metolachlor, alachlor, triuralin and metribuzin, which most likely linked to corn and soybean production. Among insecticides, fipronil, which is also used as veterinary medicine and in household products, was the leader, followed by organophosphates (e.g. ethion, chlorpyrifos, dimethoate, malathion and profenofos). Lastly, fungicides had varied determination rates, being Carbendazim the most common of them. The work features the calculation of a risk quotient for the 43 pesticides subject to that study. It is a calculation based on the empirical data about frequency and concentrations of the pesticides analyzed compared to water quality criteria internationally laid down (from environmental agencies worldwide and scientific literature). A higher RQ (Risk Quotient) implies higher environmental risk, and in that case insecticides presented themselves more of a concern than herbicides.

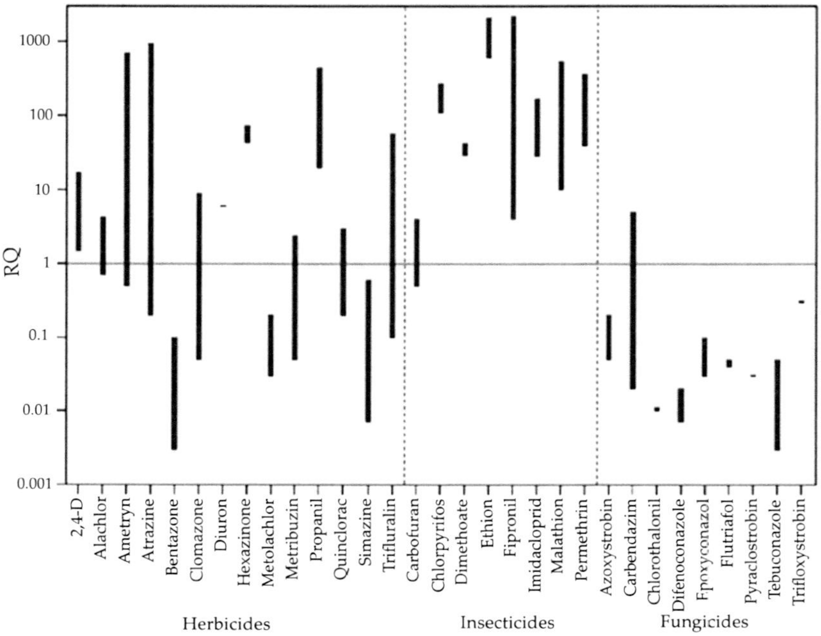

Figure 14: Risk quotients calculated for the concentration range for each pesticide under study (reproduced from Albuquerque et al, 2016))

ESTIMATES USED IN PESTICIDES RISK ASSESSMENT

Risk quotients are one part of a systematic overall ecological risk characterization for pesticides. Such characterization comprises exposure and toxicity data in a comprehensive way, intended for scientific reference and evaluation, and it is founded in four components:

- Risk Quotients

Risk quotient (RQ) is a general overview of the potential risk that a pesticide carries, and it is calculated as:

RQ =

Where EEC is stands for Estimated Environmental Concentration, which is an estimate of the quantity of pesticide that may reach non-target species and areas, and it is usually determined for ground and surface waters, soil, plants and wildlife. The 'Effects concentration' may be a toxicological end point such as LC50 for a particular organism.

RQ decision threshold is the unity. If RQ > 1, it means that exposure may produce adverse effects, observed in laboratory studies.

RQ has in itself an estimated value, so it cannot be considered alone to fully determine environmental impact of a pesticide. It is useful to lead to further and more detailed assessments.

- Level-of-Concern (LOC)

LOC values are compared to RQ values in exposure and risk assessment, and it is considered during the registration process of a pesticide. It has more of a regulatory purpose than RQ does. LOC are trigger ratios against which RQ values are compared in order to determine the risk level of a pesticide when non-target flora and fauna are exposed to it, acutely and chronically, and takes into account endangered and non-endangered species, as exemplified on Table 19.

Table 19 - Level of Concern values summarized to certain ecosystem scenarios (adapted from Whitford F)

End point and scenario		Application	Risk Quotient	Nonen-dangered	Endangered
Mammalian	**Acute**	Granular	LD50/FT2	0.5	0.1
		Spray	EEC/LC50	0.5	0.1
	Chronic	Spray	EEC/NOEC	1.0	1.0
Avian	**Acute**	Granular	LD50/FT2	0.5	0.1
	Dietary	Spray	EEC/LC50	0.5	0.1
	Chronic	Spray	EEC/NOEC	1.0	1.0
Aquatic	**Acute**	NA	EEC/LC50	0.5	0.05
	Chronic	NA	EEC/NOEC	1.0	1.0
Terrestrial plants	**NA**	NA	EEC/EC25	1.0	1.0
Aquatic plants	**NA**	NA	EEC/EC50	1.0	1.0

EEC = Estimated Environmental Concentration

EC25 and EC50 = 25 percentile and Half maximal effect concentration

FT2 =

LC50 = Concentration of the pesticide in the diet that will result in 50% mortality among test specimen

LD50 = Single dose of the pesticide in the diet that will result in 50% mortality among test specimen

NOEC = No Observed Effect Concentration

The RQ values, along with LOC, are then used as an objective measure to classify a scenario as low, medium or high concern. During the registration process, the conclusions of such assessment may change application rates and types (e.g. ground instead of aerial) as mitigation measures to protect exposed wildlife.

REGULATION

In addition to that, regulation are in place, although still subject to some criticism.

The concern around pesticides is being addressed and, obviously, it demands great investments in several fronts, which may be a challenge to many other developing countries and that is where cooperation with European institutions has its importance.

FAO considers the examples of European countries as counterparts of the scenario described above. Having a more complete data set, European countries managed to identify the pesticides contamination problems in the early 1990s, with a report issued by The Netherlands National Institute of Public Health and Environmental Protection (RIVM), which stated that *"groundwater is threatened by pesticides in all European states."* Measures to fight pesticide abuse and to lower environmental impacts set at that time included (FAO/ECE, 1991):

- Reduction in use of pesticides (by up to 50% in some countries).
- Bans on certain active ingredients.
- Revised pesticide registration criteria.
- Training and licensing of individuals that apply pesticides.
- Reduction of dose and improved scheduling of pesticide

application to more effectively meet crop needs and to reduce preventative spraying.

- Testing and approval of spraying apparatus.
- Limitations on aerial spraying.
- Environmental tax on pesticides.
- Promote the use of mechanical and biological alternatives to pesticides.

Measures applied by one of European Union's member states, Denmark, were very objective in terms of water bodies protection, such as prohibiting application within 10 meters of lakes, wetlands, water reservoirs, reassessing active ingredients (which meant restriction to certain pesticides), control and approval of spraying equipment, as well as taxing pesticides and promoting pesticide-free agriculture.

5
CHAPTER

PESTICIDES PROPERTIES AND ENVIRONMENTAL IMPACT

Erick Soares Lins

ELECT MATRICES FOR ENVIRONMENTAL INVESTIGATION

An important part of pesticides surveillance is done in water and sediments, even though fish tissue and microorganisms are often reliable contamination indicators. A knowledge of the behaviour and impact of persistent pesticides is fundamental, but there are challenges to it, including matrix and analyte properties. Some persistent pesticides tend to concentrate at different proportions in sediment, and the concentration of that suspended sediment can lead to inaccurate results. In other words, bad sampling procedures can lead to almost or completely invalid data. Table 20 depicts the proportions of old, persistent pesticides, and the load percentage in different suspended sediment ranges.

Table 1: Proportion of selected pesticides found in association with suspended sediment (After Ongley *et al.*, 1992)

Pesticide	log K_{ow}	% of chemical load at different concentrations (mg/l) of suspended sediment			
		mg/l = 10	mg/l = 100	mg/l = 1000	mg/l = 10000
Aldrin	5.5	15	55	90	100
Atrazine	2.6	0	0	2	20
Chlordane	6.0	30	75	95	100
DDT	5.8	20	67	93	100
Dieldrin	5.5	15	55	90	100
Endrin	5.6	18	57	90	100
Endosulfan	3.6	0	0	21	57
Heptachlor	5.4	13	48	88	100
Lindane	3.9	0	2	30	80
Mirex	6.9	75	95	100	100
Toxaphene[1]	3.3	0	0	12	47
Trifluralin	5.3	12	45	87	100
2,4-D	2.0[2]	0	0	0	4

[1] Toxaphene mixture; [2] Range is 1.5-2.5.

The potential for water contamination is one of the environmental fate studies that are assessed in a pesticide registration process. The recommended use, the ability to reach surface and groundwater and the pesticides overall behaviour under intense environmental scenarios, such as heavy rainfall, are considered for determining the risk of exposure to humans and non-target aquatic and terrestrial life.

Some of the most important methods used to estimate that impact are:

- Trigger values
- Computer modeling
- Laboratory and field dissipation experiments

TRIGGER VALUES

Mostly simpler calculations to estimate the impact of the pesticide associated with leaching and runoff, especially in terms of (i) Persistence, or the time in which the pesticide molecules remain stable and active in the environment, before undergoing transformations or ultimate degradation. The persistence of a pesticide is determined from studies like soil metabolism, hydrolysis and photolysis, and field dissipation; and (ii) Mobility, the ability of a pesticide molecule to move around and be translocated eventually reaching surface or ground water. Data are obtained from field dissipation and soil sorption studies. Mobility is intimately related to the molecule's sorptivity, since this property is what will determine how far and fast a pesticide contamination can travel or spread, by being more or less adsorbed to soil particles and organic matter.

Trigger values are separated in *Surface Water Triggers* and *Ground Water Triggers*. These values serve to place the pesticide in one of a number of categories that are based on the potential to contaminate water. For *Surface water triggers*, these categories are the result of a combination of three different ranges of half-life and three ranges of sorptive Koc values. Koc is the partition coefficient that measures the hydrophilicity versus the lipophilicity of a pesticide. It is known as 'organic carbon - water partition coefficient', and it is calculated as:

Koc =

Where Kd is the soil desorption coefficient, which is a measure of the mobility of a pesticide and is determined by:

Kd =

Table 2: Range values for pesticide persistence in soil based on Koc (adapted from Whitford F)

Sorption	Koc	Persistence	Half-life in soil
Low	< 500	Short	< 2 weeks
Intermediate	>500 and <5000	Intermediate	> 2 weeks but < 2 months
High	>5000	Long	> 2 months

These criteria make up a total of nine categories from combinations such as low sorption / intermediate persistence, or high sorption / short persistence etc. Low sorption pesticides are more likely to end up reaching water with runoff, while high sorption pesticides will stay bound to the soil and only reach off-target sites in the case of erosion.

Ground Water Triggers, which are also based on persistence and mobility criteria, relating the latter with the pesticide's leaching potential. The values are summarized on Table 22.

Table 3: Persistence and mobility criteria for estimating the potential of a pesticide to reach groundwater (adapted from Whitford F)

Persistence	Half-life	Mobility	
Aerobic soil metabolism	> 3 weeks	Koc	< 300
Field dissipation	> 3 weeks	Weak-to-moderate-acid	pesticide that is not attracted to soil particles
Photolysis	> 1 week	Water solubility	> 30 parts per million (ppm)
Hydrolysis	> 60 days (in sterile water)		

DYNAMICS OF PESTICIDES IN AQUATIC ENVIRONMENTS

Once introduced into the environment, a pesticide will may undergo a series of reaction and may have different fates. What ultimately happens to those chemicals will depend on soil characteristics and geographic profiles, but also on the molecules physical and chemical properties. Next we will explore into these properties, namely water solubility and adsorption, as well as their potential for bioaccumulation, and the processes of metabolism and elimination.

- Water solubility

Modern pesticides are mainly complex organic molecules, and their structures and functional groups play a role in determining its solubility in water. Although factors like the water pH, temperature and presence of salts and organic matter will affect that solubility, we can collect and verify how distinct pesticides within a group, or among groups, behave in terms of to which extent they will be able to stay dissolved in water.

A common method to determine the solubility of a pesticide is the one developed by Haque and Schmedding (1975). The pesticide under investigation is deposited inside the walls of a large container, in the form of a thin film. They add water to it and aliquots are analyzed at controlled time intervals, until equilibrium is reached.

Table 4: Water solubility of select pesticides (adapted from Khan)

Pesticide	Solubility
Atrazine	70 ppm
Monuron	230 ppm
2,4-D	700 ppm
Diquat	Very soluble
Amitrole	Very soluble
p,p'-DDT	0.0012 ppm
Lindane	10 ppm
PCP	20 ppm
Parathion	24 ppm
Leptoprophos	0.0047 ppm
PCB isomers	
2,4'- Dichloro	637 + 7 ppb
2,2,5 - Trichloro	248 + 4 ppb
2,2,5,5' - Tetrachloro	26.5 + 0.8 ppb
2,2',4,5,5' - Pentachloro	10.3 + 0.2 ppb
2,2,4,4',5,5' - Hexachloro	0.953 + 0.01 ppb

- Adsorption

Adsorption is the main process by which lipophilic pesticides are transported in the environment, because they bound to particulates and organic matter. The functional groups in the molecules may interact more strongly or more weakly to small pores in soil particulates, clay and organic matter.

A number of researchers have developed mathematical models to approach a pesticide's leaching power, and how fast or slow they may reach groundwater. These models vary in complexity, and take into account several physical and chemical parameters. For instance, Lakowski et al. proposed a calculation of the Leaching Index (LI):

LI =

Where *t1/2* is the half-life of a pesticide in soil, *S* is the solubility of a pesticide in water, given in parts per million (ppm), *Vp* is the vapour pressure in mmHg at 25°C, and *Koc* is the partition coefficient defined previously.

Another relevant work was the development of the equation known as the Groundwater Ubiquity Score (GUS), by Gustafson:

$$GUS = \log_{10} (t1/2\ soil) \times (4 - \log_{10} Koc)$$

This equation does not consider solubility in water neither the volatility of pesticides, since these are observed in concentrations that are considered negligible for the leaching process. There are summaries of pesticides LI and GUS values and their classification based on potential for contaminations.

With these several parameters considered, one can calculate the potential of contamination for several herbicides, as it is summarized on Table 24. It is important to notice that some of those parameters have ranges for values, taken from different experiment sets, making it important to decide whether using the extreme values or a mean for the calculation intended.

Table 5: Chemiodynamic properties of some herbicides, with minimum and maximum values reported for each parameter, whenever possible and applicable (adapted from Richardson, 1991)

Pesticide	Koc	t1/2 soil (days)	S (ppm)	Kh
Alachlor	150-342	4-38	242	1.3×10^{-6} - 3.2×10^{-8}
Ametryn	380-392	NA	190	1.2×10^{-9}
Atrazine	149-180	71-130	30	2.5×10^{-7}
Bromacil	34-60	150-350	815	3.7×10^{-8}
Chloramben	280	NA	700	NA

Chlorothalonil	2500	68	0.6	NA
Chlorpyrifos	6070-17000	54-63	2	NA
Chlorthal dimethyl	4000	NA	<0.5	NA
Cyanazine	168-200	10-108	171	$1.2x10^{-4}$
2,4-D	53-100	1-35	620	$6.2x10^{-3}$ - $5.6x10^{-9}$
Dicamba	1.5-2.2	14-433	6500	$2.3x10^{-6}$
Dinoseb	110-120	30	100	$3.0x10^{-7}$
Diuron	150-460	20-328	42	$5.4x10^{-8}$
EPTC	240-280	7-30	365	$5.9x10^{-4}$
Linuron	670-820	38-75	75	$2.5x10^{-6}$
Metolachlor	99-190	42-107	530	$3.7x10^{-7}$
Metribuzin	95-150	4-301	1200	NA
Monuron	80-100	166-170	230	7.6x10-9
Pendimen-thalin	16000	NA	0.3	NA
Picloram	13-48	30-330	430	$1.9x10^{-9}$
Prometon	79-350	NA	620	NA
Prometryn	610-810	60-94	33	$5.6x10^{-7}$
Propachlor	340-420	4-7	613	NA
Simazine	135-220	15-174	5	$3.4x10^{-8}$ - $3.2x10^{-10}$
2,4,5-T	8	33	150	NA
Terbacil	46	50	710	$8.2x10^{-9}$
Triallate	3.6	100	4	$6.7x10^{-3}$ - $7.9x10^{-4}$
Triclopyr	27	40	440	NA
Trifluralin	9990-13700	70-132	<1	$6.7x10^{-3}$

Kh is the partition coefficient between liquid and gas phases, or the partition in water and air.

The adsorption of a molecule is quantitatively measured by the Freundlich-type isotherm, an expression that takes into account some physical and chemical variables.

$= KC^n$

X: the amount of pesticide

m: the mass of adsorbent

K: the Freundlich constant, that relates to the number of chlorine atoms in the molecule and how they reflect in the adsorptive properties

C: the equilibrium constant of the pesticide

n: a constant related to the nature of adsorption

Dyson (2002) observed that the Koc value of a pesticide is affected by the soil pH, as results from studies on the herbicide mesotrione shown below.

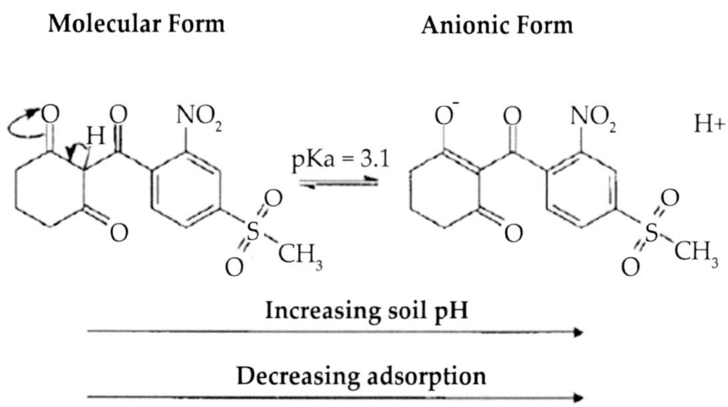

Figure 1: Soil behaviour of the systemic herbicide mesotrione

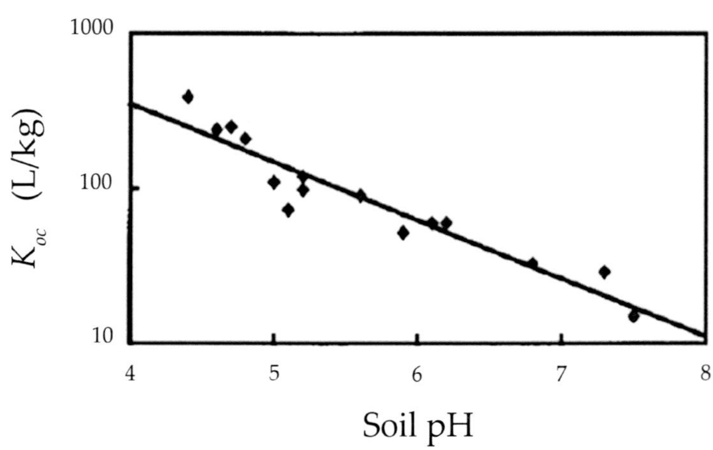

Figure 2: K_{oc} decreases with increasing soil pH for mesitrione

The amount of organic matter present in soil affects dramatically the availability and persistence of pesticides, due to the numerous adsorption sites that are generated during its chemical degradation processes. Organic matter is composed of plant materials such as cellulose, lignin, pigments, waxes and proteins, and their decomposition, both biotic and abiotic, in soil, will result in molecules with several different functional groups, or electrostatic interactions, as illustrated by Figures 61 and 62. Such groups act as binding sites for several molecules, including pesticides.

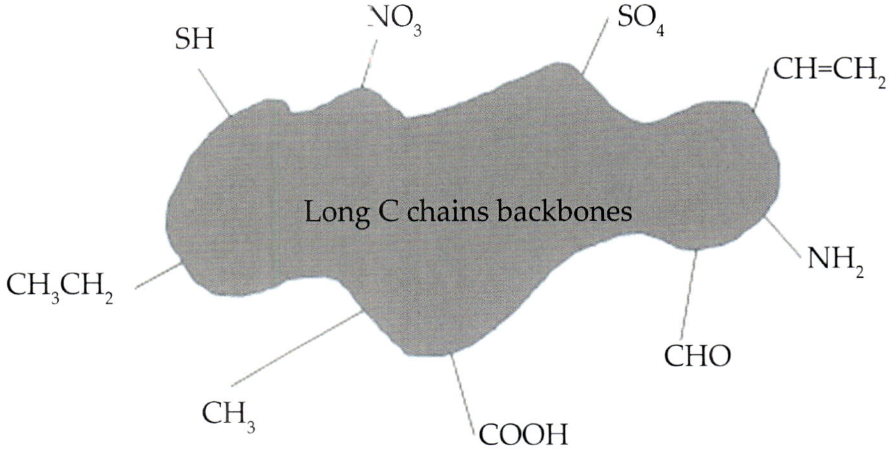

Figure 3: Examples of functional groups in organic matter to which pesticides can attach via chemical bonds or other interactions

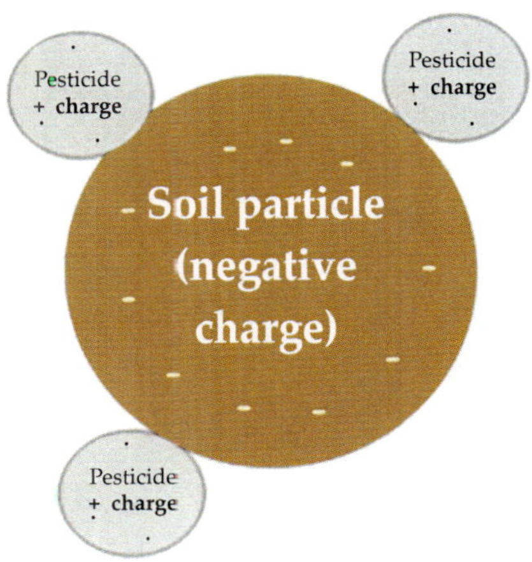

Figure 4: Positively charged pesticide particles adsorbed to negatively charged soil particle (reproduced from http://pest.ca.uky.edu)

Soil moisture can affect the overall toxicity as well, since the increase in water content may help displacing pesticide molecules that are adsorbed to soil particles and organic matter, thus making them available for organism intake. Table 24 shows that toxicity may be dramatically increased with higher moisture content for different pesticides.

Table 5: Toxicity of eight insecticides in dry and wet fine sandy loam to first-instar nymphs of the common field cricket, *Gryllus pennsylvaniccus* (adapted from Harris, CR, J. Econ. Entomol. 57, 946, 1964.

Insecticide	LD50 (ppm)		Difference (fold)
	0.0 % water in soil	12.1% water in soil	
Diazinon	53.85	0.40	134.6
Parathion	13.04	0.46	28.3
Trichlorfon	298.80	14.83	20.1
Dieldrin	4.44	0.27	16.4
Zectran	92.17	5.71	16.1
DDT	48.70	3.60	15.9
Heptachlor	1.14	0.09	12.7
Mevinphos	37.52	52.97	1.4

Microbial Degradation

Microbial degradation of pesticides is mostly carried out by fungi and bacteria. Common examples of fungi are *Penicillium, Aspergillus, Trichoderma* and *Fusarium*, and bacteria include *Pseudonomas, Flavobacterium, Arthrobacter,* and *Bacillus*. Basically, they are able to transform the residual pesticides into simpler carbon molecules and energy by a series of enzymatic reactions. The following scheme summarizes some of these known reactions and their products.

Table 6: Pesticides and respective degradation products from microbial activity (adapted from Matsamura 1982, Phillips et al, 2005, Singh and Walker 2006, Mandal et al, 2013).

Reaction	**Pesticide**	**Degradation products**
Hydrolysis	Carbaryl	1-napthol
	Chlorpyrifos	Diethylthiophosphoric acid
Oxidation	Aldrin	Dieldrin
	Parathion	Paraoxon
	Fipronil	Fipronil sulfoxide
Reduction	DDT	TDE (DDD)
	Parathion	Amino parathion
Dehydrochlorination	DDT	DDE
	y-HCH	y-PCCH (pentachlorocyclohexene)
Isomerization	Dieldrin	photodieldrin
Other reactions	y-HCH	a-HCH

- Bioaccumulation

It is dependent on the pesticide solubility character: those that are more lipophilic (more hydrophobic) tend to bioaccumulate more, since these are more likely to be dissolved into plant and animal tissues. The bioaccumulation determines how far a pesticide will go in the food web of an ecosystem.

Two methods provide quantitative idea for the bioaccumulation potential of a pesticide. The first one is the indirect method, and it is used to predict the extent of bioaccumulation by measuring the pesticide octanol/water partition coefficient (P or Kow). This coefficient is usually measured for many other types of organic molecules for different applications.

P=

C1 is the concentration of the pesticide in octanol

C2 is the concentration of the pesticide in water

Table 25 presents a series of partition coefficients determined for organophosphate pesticides based on this method.

Table 7: Partition coefficient of several organophosphates (adapted from Haque, Kearney and Freed)

Pesticide	**P**
Parathion	6430
Dicapthon	3790
Feniltrothion	2380
Dichlofenthion	137000
Ronnel	75300
Methyl Chlorpyrifos	20000
Chlorpyrifos	128000
Leptophos	2.02x106
Phosmet	677
Dialifor	49300
Phosalone	20000
Malathion	781
Dimethoate	508

The other method for determining the bioaccumulation factor is the direct method. Calculations of the bioaccumulation ration are pretty similar to the ones for partition coefficient, but the data comes from a model that actually takes into account the absorption and ingestion by some organisms, like algae, plankton, snails, mosquito larvae and fish.

Bioaccumulation ratio (BR) =

C1 is the concentration of the pesticide in the organism investigated

C2 is the concentration of the pesticide in water

Table 26 provides values obtained from a study conducted by Isensee et al. using an ecosystem model comprised of organisms such as daphnids (*Daphnia magna*), snails (*Physa sp.*), algae (*Oedogonium cardiacum*), mosquito fish (*Gambusia affinis*), using ^{14}C-labeled pesticides.

Table 8: Bioaccumulation ratio (BR) and water solubility of five compounds in various aquatic organisms. (Adapted from Isnee (1976), Isnee and Jones (1975) and Kanazawa et al. (1975))

Pesticide	Water solubility	Organisms	BR
Carbaryl	40 ppm (at 30oC)	Algae	4000
		Duckweed	3600
		Snails	300
		Catfish	140
		Crayfish	260
Atrazine	33 ppm (at 25oC)	Snails	2-15
		Algae	10-83
		Fish	3-10
HCB	Pratically insoluble	Algae	320-1570
		Snails	1360-3320
		Daphnids	770-1030
		Fish	1160-3740
Mirex	Insoluble	Algae	12200
		Snails	4900
		Daphnids	14650
		Fish	2580
TCDD	0.006 ppm (at 30oC)	Algae	2000-18600
		Duckweed	1200-5000
		Snails	1400-47100

Both methods are useful at predicting the degree of bioaccumulation of a pesticide. Is it clear that there is an interdependence between the water solubility and the bioaccumulation ratio, as it can be seen on Figure 62. Pesticides that are more lipophilic have higher bioccumulative potential, as opposed to those that are more soluble in water. There are models in place with which investigators attempt to correlate those two factors, resulting in plots such as the one illustrated here.

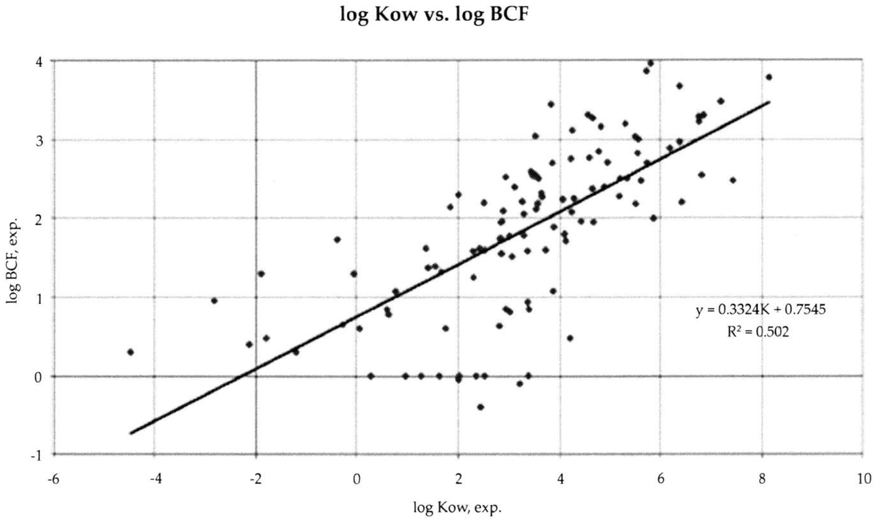

Figure 5: Correlation between logBCF and logKow for 112 pesticides (reproduced from The Ministry of Environment and Food of Denmark)

- Metabolism

The most important metabolic pathways that pesticides undergo are oxidation, reduction and hydrolysis. Following these reactions, metabolites may then conjugate with endogenous molecules, or be directly excreted. We shall see the chemical routes that pesticides may go when metabolized, focusing on the following groups:

- Organophosphates

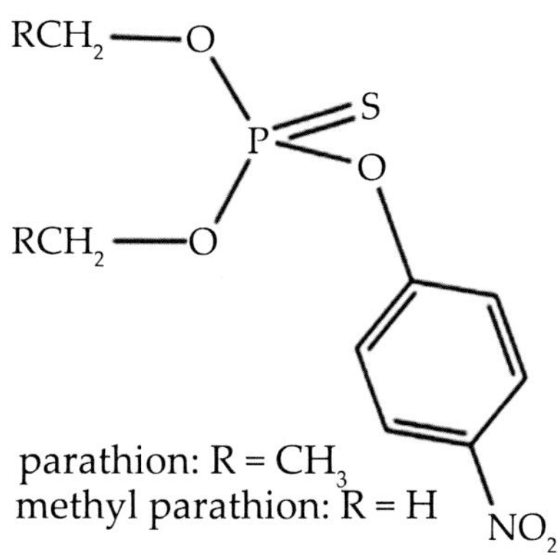

Figure 6: Organophosphate molecule and reaction sites - parathion and methyl parathion

Organophosphate pesticides such as the ones in the example may be subject to reactions in different sites, mainly those listed on Table 27.

Table 9: Reactions to which organophosphates undergo and respective products

Reactive site	**Reaction/products**
Sulfur atom	Forms sulfone and sulfoxide; Replaces sulfur with oxygen
Ester bonds	Attacked by specific esterases
Alkoxy bonds	Forms dealkylated products
Nitro group (in the case of parathion)	Reduced to amino groups

A visualization of alterations in malathion is depicted in Figure 64, in which carboxyesterase is responsible for the metabolic deactivation of the molecule, which is one of the resistance mechanisms that has been proposed for insects. Figure 65 shows a metabolic pathway proposed for diazinon in aquatic animals, where we can see the changes in functional groups.

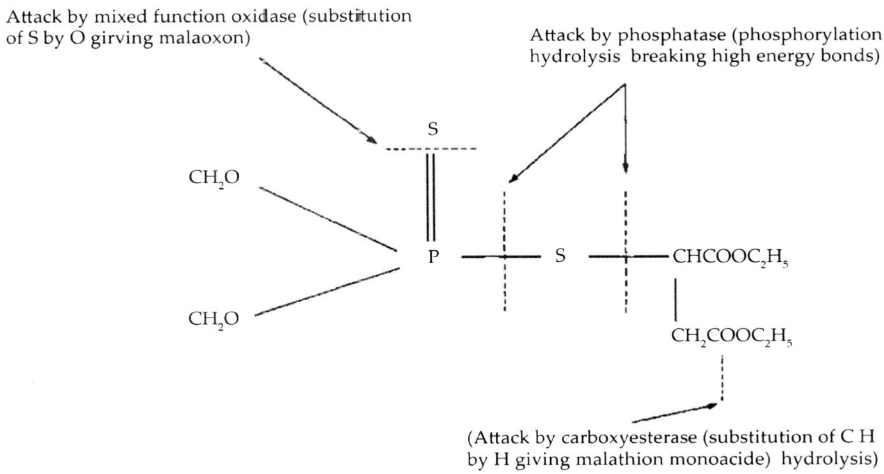

Figure 7: Metabolic reactions in malathion (reproduced from FAO)

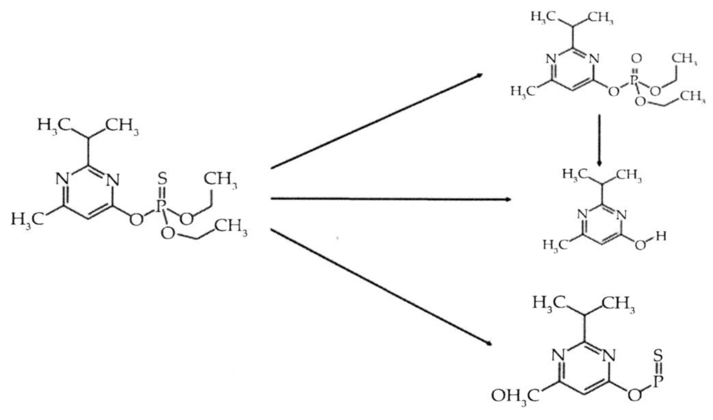

Figure 8: Metabolic pathways that Diazinon undergoes in aquatic animals (adapted from Hogan and Knowles 1968, 1972)

- Organochlorines

DDT has been found to be metabolized to DDD and DDE by several fish species, e.g., goldfish, dogfish, mosquitofish, cutthroat trout, young Atlantic salmon, and others. Table 28 presents the residual amount of DDT and its metabolites in some invertebrates studied.

Table 10: In vivo metabolsim of DDT as measured by the concentration of residues extracted from freshwater invertebrates following exposure to DDT (adapted from Johnson et al 1971)

Invertebrate	**% of total body residue**		
	DDT	**DDE**	**DDD**
Daphnia magna (Cladoceran)	73.4	19.7	6.6
Gammarus fasciatus(amphipod)	79.1	20.9	--
Palaemonetes kadiakennis (Decapod)	50.9	13.2	7.2
Hexagenia bilineate	14.9	85.0	--
Ischnura verticalis (odonata)	39.2	60.2	--
Libellula sp. (odonata)	56.3	28.4	--
Chironomous sp. (diptera)	80.0	19.1	--

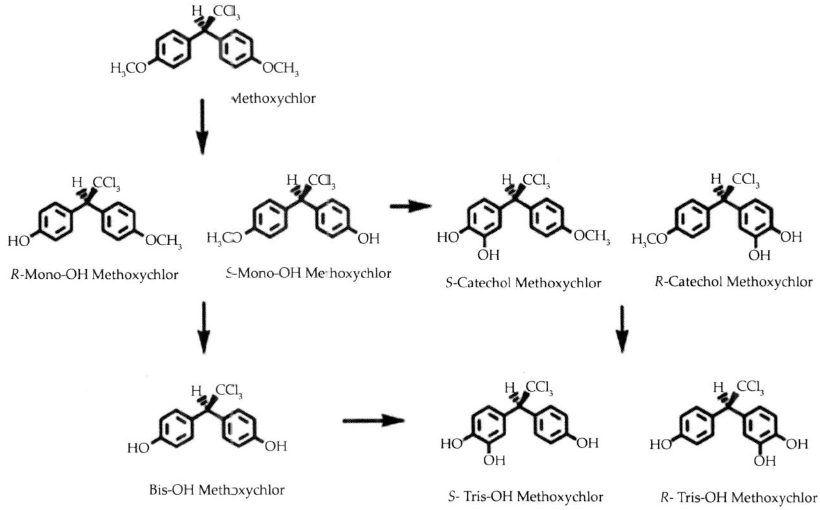

Figure 9: Proposed pathway of methoxychlor metabolism from in vivo studies; enantiomeric differentiation is shown (reproduced from Kupfer and Yu, 2012

- Cyclodienes

It has been found that cyclodienes undergo epoxidation of unsaturated bonds when metabolized. It is the case of Aldrin being converted to dieldrin, as presented in Figure 67.

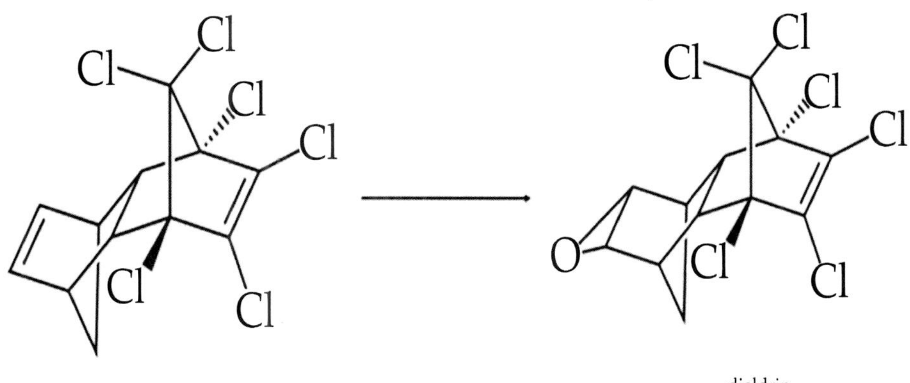

Figure 10: Metabolic conversion of aldrin into dieldrin; double bond in one of the rings in the molecules undergoes epoxidation

The metabolism of administered dieldrin in invertebrates is varied, and so the amounts of this substance found in their organism will be different, as detailed in Table 29.

Table 11: Residual levels of dieldrin in various invertebrates after exposure (data from Khan, Forte and Payne, 1976)

Invertebrate	**Total concentration of dieldrin (%) after 2 hours of exposition to aldrin**
Hydra littoralis (coelentrata, hydrozoa)	2.59
Dugensia (platyhelminthes, plannaria)	1.03
Asellus (arthropoda, crustacea)	2.30
Gammarus (arthropoda, crustacea)	7.07
Daphnia (arthropoda, crustacea)	5.69
Cyclops (arthropoda, crustacea)	4.13
Cambarus (arthropoda, crustacea)	8.48
Aeschna nymph (arthropoda, insecta)	24.9
Aedes aegypti larvae (arthropoda, insecta)	42.4
Anodonta (mollusca, pelecypoda)	19.0
Lymnaea (mollusca, gastropoda)	16.8

Endosulfan, another important cyclodiene, has been studied in terms of its metabolism, being found to be 98% eliminated within 14 days in goldfish.

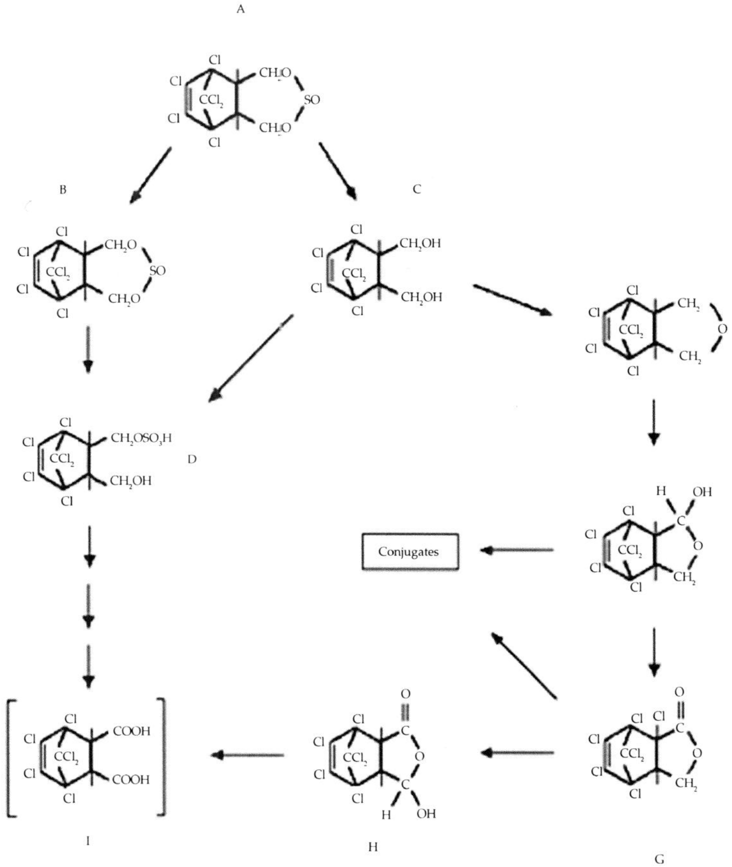

(Acapted from Stump and ABhauer, 1986)

A = endosulfan
B = endosulfan sulphate
C = endosulfan diol
D = endosulfan diol sulphuric acid ester
E = endosulfan ether
F = hydroxyendosulfan ether
G = endosulfan lactone
H = endosulfan hydroxylacetone
I = dicarbonic acid

Figure 11: Metabolic pathways proposed for endosulfan (adapted from Stumpf and Abhauer, 1986)

Pesticide elimination

The chemical properties that favor certain pesticide molecules to be absorbed by plant and animal tissues are well known and documented, as it was discussed previously. If we focus on the process of elimination we shall see that it has some complex mechanisms in itself.

Tooby and Durbin (1975) among other researchers concluded that, when organisms (e.g. fish) are exposed to insecticides, absorption and elimination occur simultaneously, and the balance in the rates of these processes will determine the fate of the substance.

The elimination of some organochlorines in fish follow the order (according to Macek 1970, Statham and Lech, 1973):

DDT < Dieldrin < Lindane

This increasing order of eliminated quantities after absorption in fish correlates well with the pesticides solubilities in water. While Lindane in goldfish was found to be 90% eliminated after 2 days, DDT which is more lipophilic, remains at ca. 50% after 32 days. Table 30 has a more complete picture of the quantities of each organochlorine found in tissues of fish at different time intervals.

Table 12: The concentration of insecticides in the tissues of fish at various times following their exposure to an initial concentration of 70 ppb of each insecticide (adapted from Gackstatter, 1966)

Organ	**Insecticide concentration (in ppm)**								
	DDT			**Dieldrin**			**Lindane**		
	0 days	**2 days**	**4 days**	**0 days**	**2 days**	**4 days**	**0 days**	**2 days**	**4 days**
Liver	11.00	3.15	3.25	14.75	4.15	4.3	9.60	2.15	0.80
Kidneys	7.45	1.50	1.60	4.10	1.60	1.6	3.05	0.90	0.40
Gills	5.80	4.50	5.55	5.00	6.70	7.2	6.40	2.75	1.05
Gall-bladder	0.30	17.55	14.10	1.65	NA	NA	2.40	1.50	2.50
Stomach	2.40	1.20	1.20	2.15	0.90	0.8	1.90	0.65	0.35
Intestine	2.80	1.10	1.60	4.75	3.80	4.1	5.50	3.40	1.50
Pyloric ceca	5.85	13.10	10.25	9.10	4.50	7.6	5.70	2.40	1.80
Skin	13.10	12.10	9.50	1.35	1.05	1.2	1.60	0.40	0.20
Brain	3.75	3.50	6.05	4.15	11.10	2.8	3.60	1.80	2.40

Muscle	0.95	0.55	1.15	0.95	0.45	0.4	1.20	0.25	0.15
Gonads	2.65	1.35	50.60	2.20	0.60	0.9	2.45	0.80	0.35
Fat	NA	58.45	NA	32.50	69.00	41.00	NA	NA	NA
Spleen	4.00	NA	3.00	NA	2.60	NA	NA	NA	NA
Carcass	2.20	2.45	2.60	2.45	2.50	2.9	3.10	1.60	0.70

Note: most of the values are averages of 2 fish. Bluegills (43 to 71 g wt) were used for DDT and Dieldrin, redear (33 to 39 g wt) for Lindane.

The persistence of pesticides was observed to follow that trend, as it is summarized in Table 31, presenting the time interval at which half of the amount of each of the pesticides administered to rainbow trout was eliminated.

Table 13: Persistence of pesticides in rainbow trout (adapted from Macek, 1970)

Half-life	Pesticide
< 1 day	Malathion
< 2 days	Lindane
< 3 days	Simazine
< 1 week	Diazinon, Dursban, Azinphos-methyl, Parathion, Methoxy-chlor, 2,4-D
< 2 weeks	Dichlorobenil
< 3 weeks	Diquat, Endothal
1 month	Heptachlor, Dieldrin
4 months	Sodium arsenate
> 5 months	DDT
> 6 months	DDD, Campheclor

Non-target organisms

Non-target organisms are, in a broad definition, any living being that is affected by the application of pesticides, not being the intended recipient. Plants, algae, fish, invertebrates, and ultimately humans, at some point fall into that definition when it comes to environmental consequences of pesticide application, even when it is unintentional.

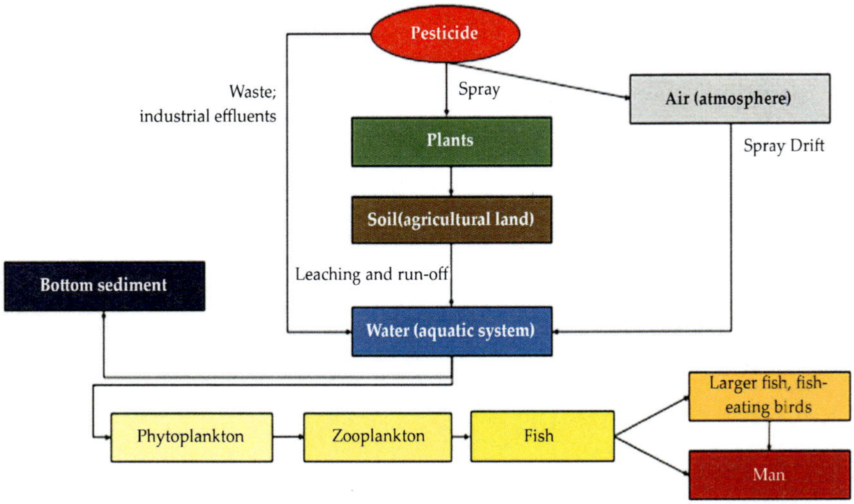

Figure 12: Possible pathways, sources and recipients of pesticide pollution. Adapted from Flint and Van den Bosch, 1981. Some routes were omitted in the figure for better visualization.

Several pieces of research have been carried out and published in an attempt to unveil the possible routes of pollution and the observed effects of certain herbicides in non-target plants, as well as their response to pesticide exposure. Tables 32 and 33 below summarizes some of those findings.

Table 14: Non-target effects of herbicides (This table is mostly from Obrigawitch et al. 1998)

Non-target Plant	**Exposure Mechanism**	**Effect(s)**
Field Crops/Grasses		
Alfalfa (*Medicago sativa*)	windblown soil	stunted vegetative growth, reduced seed yield and root weight, and reduced root nodulation
Barley (*Hordeum vulgare*)	crop injury	reduced seed yield
Buckwheat (*Fagopyrum esculentum Moench*)	crop injury	reduced seed size

Canola (*Brassica napus*)	Drift	delay in crop maturity and reduced seed yield
Corn (*Zea mays*)	drift, windblown soil	reduced seed yield
Cotton (*Gossypium hirsutum*)	Drift	flower number, fruit set, and boll size were unaffected
Red fescue (*Festuca rubra* L.)	Drift	seed yield, weight, and number decreased
Ryegrass (*Lolium perenne* L.)	Drift	seed yield, weight, and number decreased
Sorghum (*Sorghum vulgare*)	Drift	no apparent effect
Soybeans (*Glycine* max)	Drift	no apparent effect
Sunflowers (*Helianthus annuus*)	Drift	Reduced seed yield, visual injury, and reduced emergence
Wheat (*Triticum aestivum*)	crop injury	Visual injury and nutritional malfunction

Table 15: Response of non-target plants to pesticides (adapted from https://www.engg.ksu.edu/HSRC/karthipesticide.pdf)

Response	**References**
Complete death or stunted growth	(Norris and Freed, 1966a,b, and c; Dhillion et al., 1968)
Tolerance	(Hamner and Tukey, 1946; Pallas, 1963; Leonard et al., 1966)
Acclimation	(Dickinson et al., 1991)
Detoxification	(Lund-Hoie, 1969; Wichman and Byrnes, 1975; Burken and Schnoor, 1996; 1997)
Metabolism	(Burken and Schnoor, 1996; 1997; 1998; Schnabel and White, 2001)
Mutation and resistance	(Sandermann, 1992) Phenotype alteration (Sandermann, 1992)
Root or shoot uptake	(Wichman and Byrnes, 1975; Akinyemiju et al., 1983; Burken and Schnoor, 1996)

Translocation	(Lund-Hoie, 1969; Gao et al., 2000a; 2000b; Wilson et al., 2000)

Herbicides can cause a notable effect in water bodies that have dense vegetation. Experiments have shown that the destruction of vegetation results in larger mass decomposition and lower oxygen levels. These, in turn, may lead to fish mortality. In a form of restoring the balance, the lack of vegetation has shown to boost the concentration of phytoplankton, whose photosynthetic activity leads to restoration of the oxygen levels in water.

An advisory group commissioned by FAO assembled data on pesticides and their environmental side effects. The objective was to present alternatives to fight locusts after the ban on dieldrin, but impacts on non-target organisms were also cast, as shown in Tables 34, 35 and 36.

Table 16: Risk to non-target organisms at verified dose rates against the desert locust. Risk is classified as Low (L), Medium (M) or High (H). (Adapted from Graham)

Insecticide	**Environmental Risk**						
	Aquatic organisms		**Vertebrates**			**Non-target Arthropods**	
	Fish	**Arthropods**	**Mammals**	**Birds**	**Reptiles**	**Bees**	**Soil insects**
Bendiocarb	M	L	M	L	NA	H	M
Chlorpyrifos	M	H	L	M	M	H	NA
Deltamethrin	L	H	L	L	L	M	M
Diflubenzuron	L	H	L	L	NA	L	M
Fenitrothion	L	M	L	M	M	H	H
Fipronil	L	H	M	L	M	H	H
Lambda-cyhalothrin	L	H	L	L	NA	M	H
Malathion	L	M	L	L	NA	H	H
Teflubenzuron	L	H	L	L	NA	L	NA
Triflumuron	L	H	L	L	L	L	L

NA: insufficient data

The criteria used for the risk classification above is detailed below, divided in laboratory data and field data.

Table 17: Criteria applied for the environmental risk classification used in Table 34 - Laboratory toxicity data

Group	Parameters	Risk class			Reference
		Low	**Medium**	**High**	
Fish	Risk ratio (PEC/LC50)	< 1	1-10	> 10	FAO/Locustox
Aquatic arthropods	Risk ratio (PEC/LC50)	< 1	1-10	> 10	FAO/Locustox
Reptiles, birds, mammals	Risk ratio (PEC/LD50)	< 0.01	0.01-0.1	0.1	EPPO
Bees	Risk ratio (recommended dose rate/LC50)	< 50	50-500	> 500	PRG/ EPPO
Other terrestrial arthropods	Acute toxicity (%) at recommended dose rate	< 50%	50-99%	> 99%	IOBC

PEC: Predicted Environmental Concentration after treatment at the recommended dose rate

Table 18: Criteria applied for the environmental risk classification used in Table 34 - Data from field trials and control operations

Group	Parameters	Risk class			Reference
		Low	**Medium**	**High**	
Fish	Evidence of mortality	none	incidental	massive	PRG
Aquatic arthropods	Population reduction	< 50%	50-90%	> 90%	PRG
Reptiles, birds, mammals	Evidence of mortality	none	incidental	massive	PRG
Bees	Evidence of mortality	Not significant	incidental	massive	EPPO
Other terrestrial arthropods	Population reduction	< 25%	25-75%	> 75%	IOBC

Factors affecting bioavailability of pesticides

The amount of a compound that is present in an ecosystem surely affects it with proportionality at some level. However, even more critical than the total amount is its bioavailability. The bioavailable fraction of a chemical, and that includes polluting substances, is what will ultimately participate in processes such as uptake, metabolism and the responses that follow. Bioavailability is divided in two types: environmental and pharmacological.

Environmental Bioavailability

Figure 13: Tadpoles

As it has been presented earlier, pesticides may become bioavailable to organisms by three major means: from water, from bottom sediments and from other organisms. Factors such as hydrophilicity/lipophilicity of a pesticide, as well as its stability will influence its uptake from these environmental sources. The investigation of such impacts contains a series of uncertainties when extrapolating laboratory results to a greater ecological scheme, where interactions and behaviours must be accounted for.

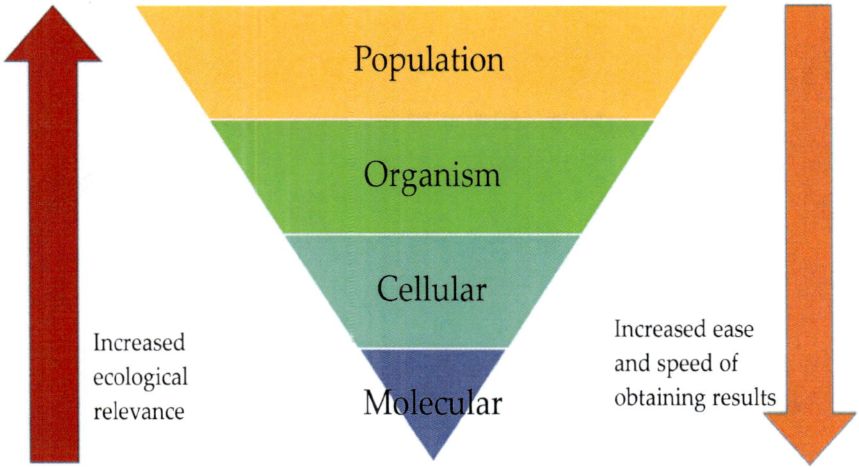

Figure 14: Ecological effects vs individual effects in investigating the environmental impacts of pollutants

Since compounds that bind better to sediments and organic matter will be more likely to enter the food chains, a way to estimate environmental bioavailability, or the environmental fate of a pesticide, is by finding the Total Organic Carbon Content (TOC) and the Dissolved Organic Carbon Content (DOC).

Other abiotic factors that affect bioavailability, and ultimately the general toxicity of pesticides that are present as environmental contaminants, are the water temperature, salinity, pH.

Temperature

It is well known that temperature variation can affect the vast majority of chemical reactions. Equilibrium states may shift, reaction rates may differ according to a lower or higher temperature in the system and surroundings. Endothermic reactions, i.e., reactions that are favoured by the increase in energy that is available, will be favoured when temperature is increased.

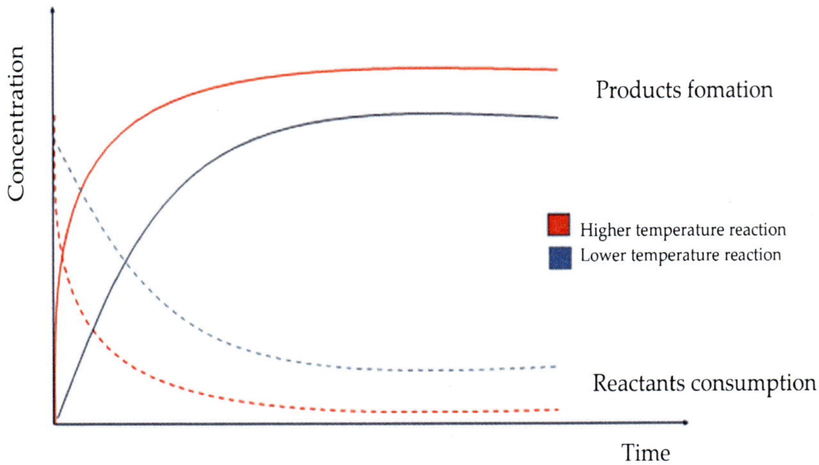

Figure 15: typical equilibrium curve for a chemical reaction (adapted from www.ausetute.com.au/equilibrium.html)

This still applies to complex biochemical reactions, such as those involved in the metabolism of organisms, including chemical conversion and detoxification pathways.

Likewise, transport processes may also become faster or slower with a change in temperature. Transport rates, mainly regarding influx and efflux through lipid membrane, may be speeded up for passive diffusion as well as with transporter proteins, although each process responds to it at a different degree. This, in consequence, will affect the uptake and excretion of chemicals of organisms.

It is important to note that, in actual environmental scenarios, even though temperature changes may seem small (usually no more than a few degrees centigrade), occur slowly and vary seasonally, macroscopic, long term consequences may be observed, such as species distribution and reproductive patterns.

There has been a number of studies correlating water temperature and acute toxicity of pesticides in fish. Apart from exceptions such as Methoxychlor and DDT, which showed decrease in acute toxicity in rainbow trout and bluegills in colder temperature, it has been observed that, in general, higher temperatures resulted in increased toxicity of pesticides in the species of fish studied. Table 37 presents some of the data acquired in these studies.

Table 19: 24 and 96 hour LC50 values of various pesticides to rainbow trout at different temperatures (adapted from Macek et al, 1969)

Pesticide	24 hour LC50 (ppb)			96 hour LC50 (ppb)		
	1.6 °C	7.2 °C	12.7 °C	1.6 °C	7.2 °C	12.7 °C
Aldrin	24	8.1	6.8	3.2	3.3	2.2
Azinohosmethyl	25	15	13	6.8	6.2	5.5
Chlorpyrifos	550	110	53	51	15	7.1
Dieldrin	13	3.1	3.1	2.4	1.1	1.4
Endosulfan	13	6.1	3.2	2.6	1.7	1.5
Endrin	15	5.3	2.8	2.5	1.4	1.1
Heptachlor	17	12	13	7.7	7.0	7.3
Methoxychlor	55	45	74	30	42	62
Naled	1300	620	240	340	220	160
Trifluralin	328	239	98	210	152	42

Salinity

The salt concentration in water environments may affect transport processes in organisms, especially regarding absorption of chemical through membrane in the epithelia, as well as the gills. Ionic and organic compounds may have altered their passive influx and efflux rates due to water salinity, as it changes the thickness of the diffusion membranes. Thus, the actual toxicity of a pollutant to which an organism is exposed, may be affected.

Research confirmed that the toxicity of dichlorvos and methyl parathion in mummichogs was affected with salinity change between 0.12 and 0.36%, even though the same experiments detected no change for DDT, endrin and heptachlor.

pH

Toxicity of a chemical is pH dependent, since the functioning of important metabolic and transport elements, such as membrane proteins and enzymes, have an optimal working pH, which is the physiologically relevant pH range of 7-8. Important chemical groups like primary amine

($-NH_2$) and sulfhydryl groups, present changes in their charges in the pharmacologically relevant range.

$$R\text{-}NH_2 \rightarrow R\text{-}NH_3^+ + OH^-$$

$$R\text{-}SH \rightarrow R\text{-}S\text{-} + H_3O^+$$

Pharmacological Bioavailability

The pharmacological bioavailability of a chemical is strongly dependent on the administration routes considered, sometimes differing significantly between direct ingestion, ingestion through food intake, respiratory ways, or skin penetration, to name examples. The bioavailability can be calculated as a ratio of extravascular and intravenous exposure, as stated by the equation:

F=

Where F is the absorbed dose fraction, Aev and Biv are the amounts of the substance in extravascular and intravenous, respectively, determined from a concentration vs. time plot, and Div and Dev are the intravenous and extravascular doses administered.

Chemical Uptake

Absorbance directly from water and by ingestion of other smaller organisms are the major ways by which chemical substances enter and climb the food chain regarding pelagic organisms (i.e., aquatic organisms that dwell and move in the water columns). Absorbance from sediments are relevant for organisms with roots and that dwell on the bottom of water bodies. The hydrophilicity/lipophilicity of the substance will also influence that mechanism.

In animals, the uptake may occur throughout the whole general surface of the organism, except when it is covered in an exoskeleton, which will unfavour penetration, or when it has gills and the guts, as these have the thinnest epithelia and thus absorbance will be easier. Gills and guts are very vascularized, so the path of a chemical towards the circulatory system in the animal is short.

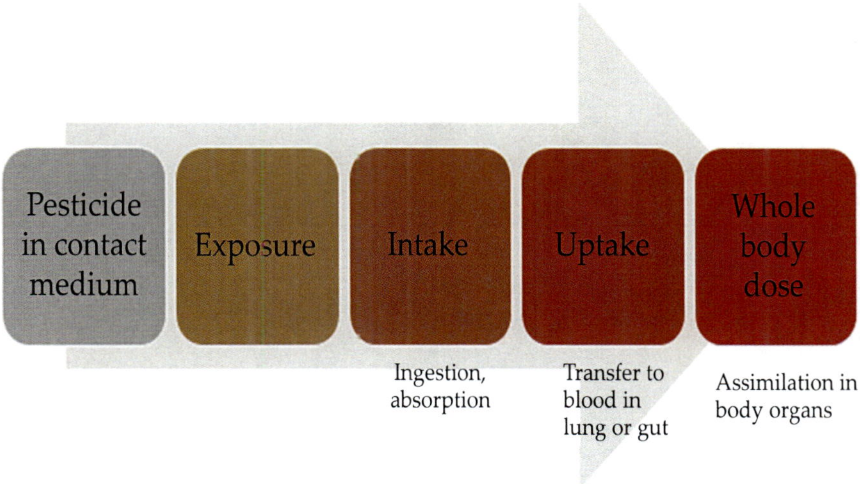

Figure 16: Accumulation of pesticide contamination in an organism. Adapted from www.integrated-assessment.eu/eu/guidebook/exposure_and_intake.html

The more hydrophilic a substance, more likely will it be to be absorbed via the gills and the gut, and absorbed by the general body surface at a smaller extent. This is the case of metal ions. Pesticides, however, are usually more lipophilic, so uptake will occur mainly by food intake and contact with sediments. The processes in the organism comprises the pesticide molecules being able to cross the cellular lipid membranes, especially in the intestine. Lipophilicity is directly and intimately related to the substance bioaccumulation and biomagnification. By definition:

- **Bioaccumulation** - the increase in accumulated level of a chemical in one organism, considering all sources (food, air, water etc);
- **Biomagnification** - the increase in the amount of a chemical across trophic levels.

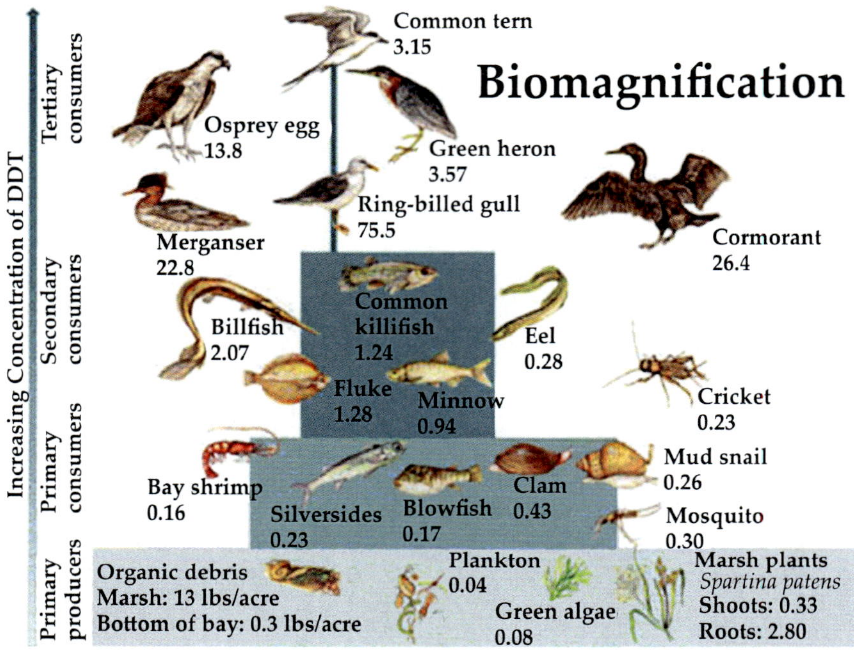

Figure 17: Biomagnification scheme for DDT in various exposed species. Originally published in Silent Spring)

Bioaccumulation (and consequently biomagnification) of a pesticide is directly related to its lipophilicity, which is quantitatively measured by its Kow, as discussed earlier. Thus, there is a proportionality between Kow and BCF – the log of each variable, to be more accurate.

BCF stands for Bioconcentration Factor, and it is calculated by

BCF =

Figure 75 shows a schematic scale-up of toxicant effects in environments. Individual organisms exposed to pesticides may suffer its toxicant effects in the genome, passing on the genetic characteristic to their spawn, so it is presented again in new individuals. That is why the toxicant effects point to both the "individual" and "genome", and the arrows feed each other in the scheme. Toxicant effects may then go up the hierarchy affecting entire populations of organisms, especially when it comes to potential reproduction issues and mortality rates. Such effects on populations end up affecting ecosystems, since species interactions will be unbalanced in some stance.

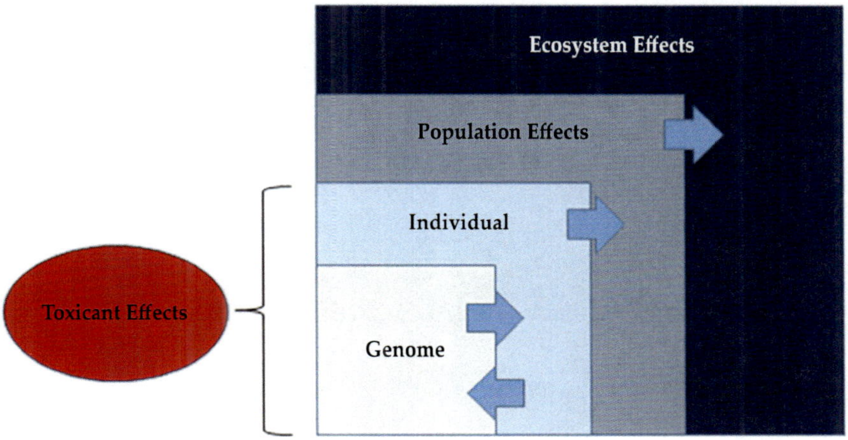

Figure 18: The hierarchy of toxicant actions (adapted from Nikinmaa)

Reproduction

Because reproduction in aquatic organisms is generally external, the development of the embryos is affected by the water surrounding it. Pesticides and other toxic agents may interfere with reproduction in a number of ways, mainly:

- Development of embryos (which may be under the scope of developmental toxicity)
- Production and accumulation of vitellogenin - the yolk protein
- Development of the gonads
- Fertilization.
- Reproductive hormone cycles
- Reproductive behaviour

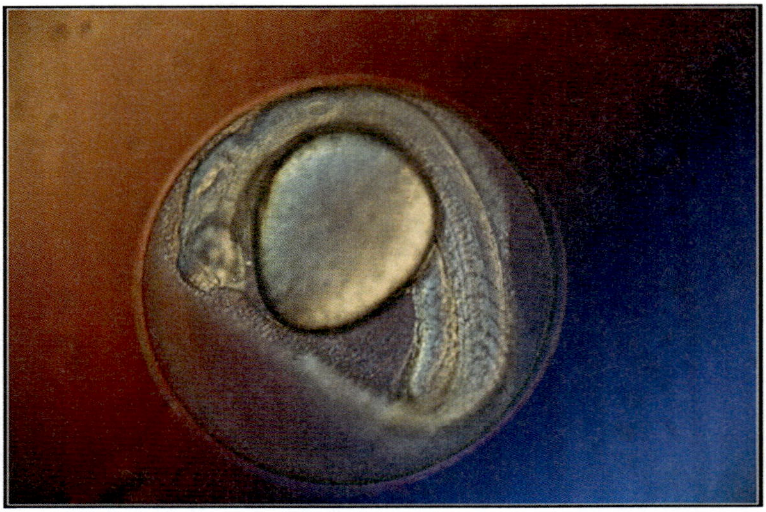

Figure 19: Embryo of zebrafish

The substances that affective the reproductive system are endocrine disruptors - molecules that bind to the hormone receptors, blocking its functions and inducing effects. For example, diethylstilbestrol is an estrogen agonist that mimics estradiol (a female hormone), and DDE can fit on the androgen hormone receptor.

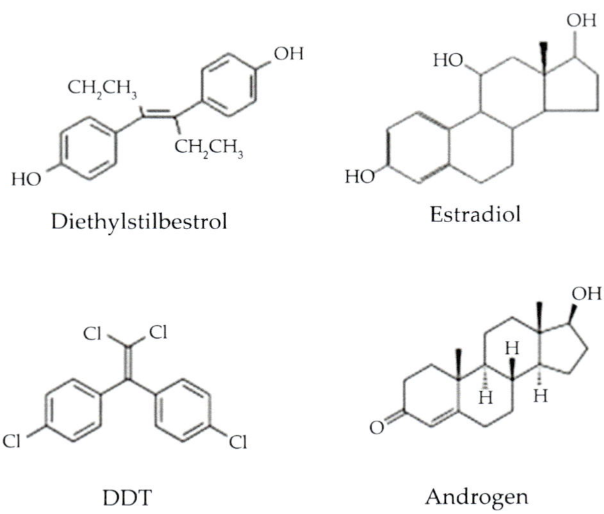

Figure 20: Diethylstilbestrol (DES) and Estradiol, DDT and Androgen

Organochlorines, phtalates, dioxins, and polychlorinated biphenyls (PCBs) are estrogenic substances. Table 38 presents the effects caused by classes of endocrine disruptors

Table 20: Effects of different groups of endocrine disruptorpesticides and their chemical structures (adapted from McKinlay et al., 2008)

Pesticides	**Chemical structure**	**Endocrine Disruptor Effects**
2,4-D (H)	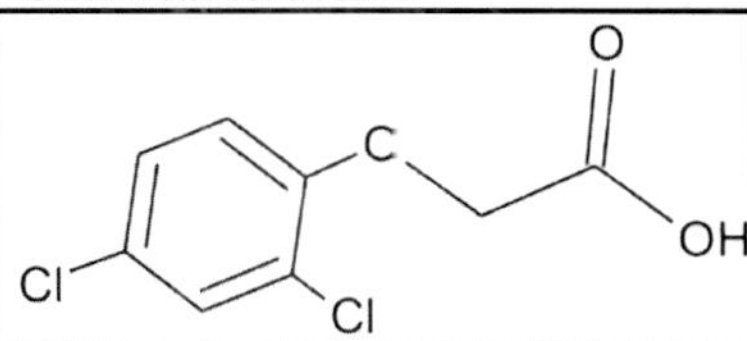	Synergistic androgenic effects when combined with testosterone
Acephate (I)	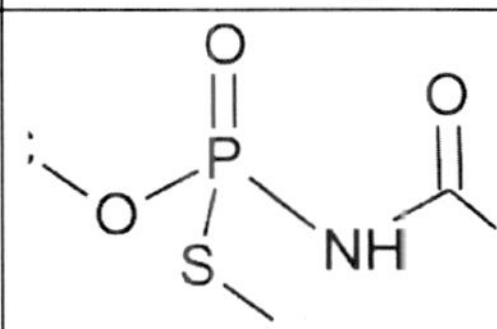	Disruption of hormone expression in the hypothalamus
Acetochlor (H)	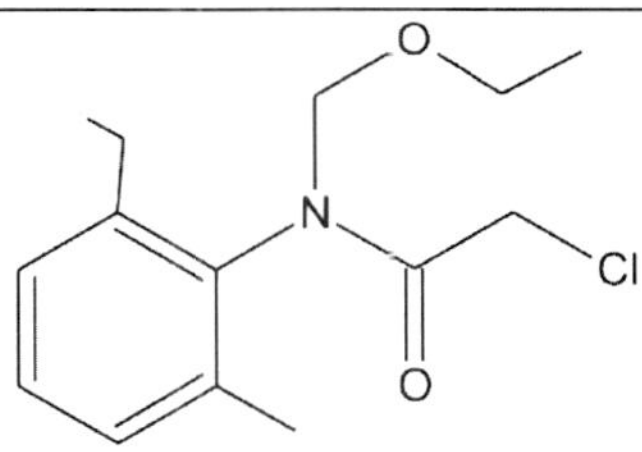	Interaction with uterine estrogen receptors, alteration of thyroid hormone dependant gene expression
Alachlor (H)	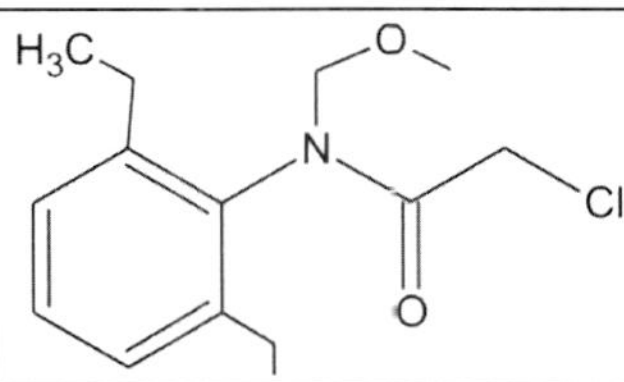	Competitive binding to estrogen and progesterone receptors. Interaction with the pregnane X cellular receptor, interfering with the production of enzymes responsible for steroid hormone metabolism
Aldicarb (I)	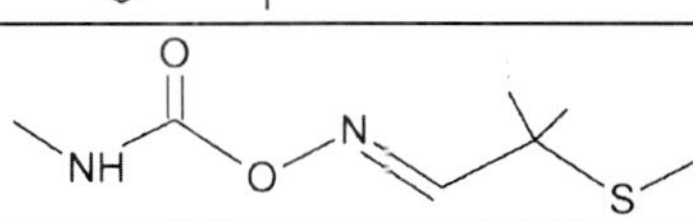	Inhibition of 17 beta-estradiol and progesterone activity
Aldrin (I)	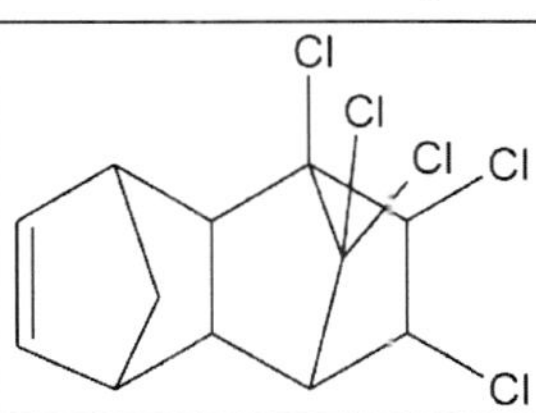	Competitive binding to androgen receptors

Pesticides	Chemical structure	Endocrine Disruptor Effects
Atrazine (H)	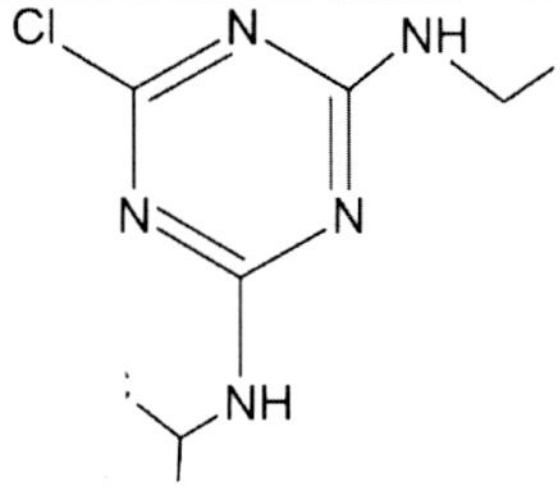	Androgen inhibition, weak estrogenic effect. Disruption of the hypothalamic control of lutenising hormone and prolactin levels. Induction of aromatase activity, increase estrogen production. Adrenal glands damages and reduction of steroid hormone metabolism
Bendiocarb (I)	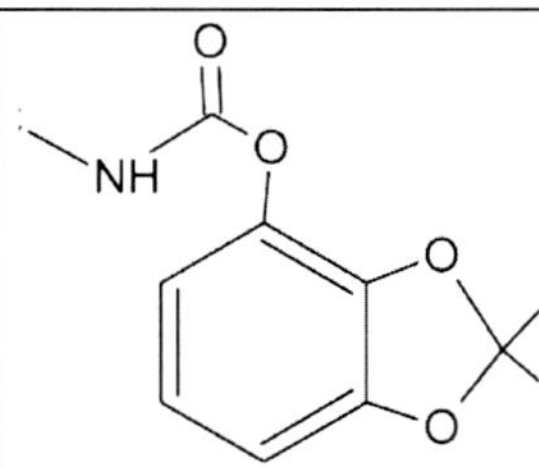	Weak estrogen effect
Benomyl (F)	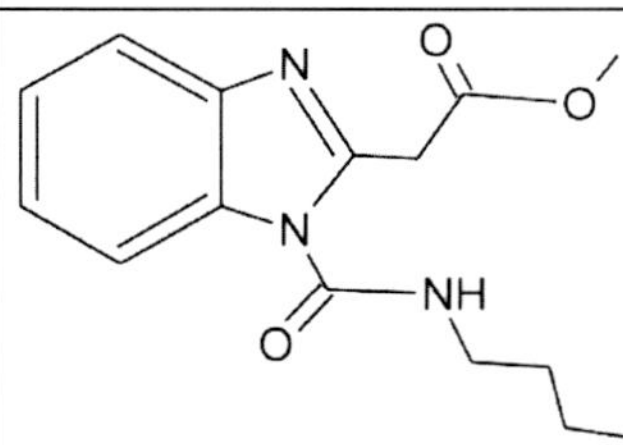	Increase of estrogen production and aromatase activity
Bioallethrin (I)	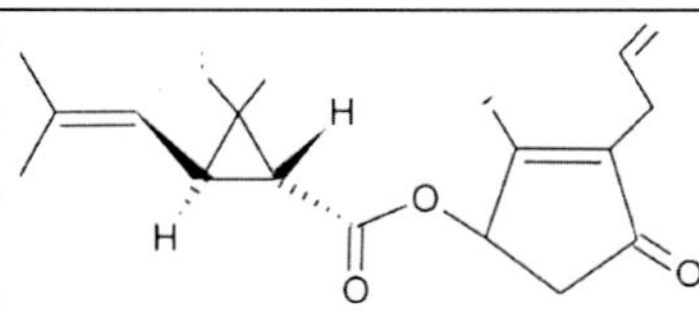	Inhibition of estrogen-sensitive cells proliferation
Captan (F)	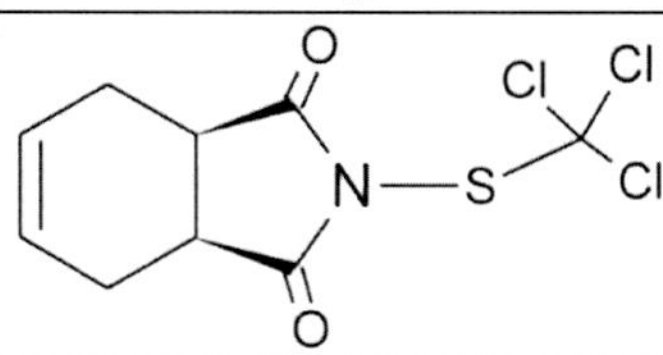	Inhibition of estrogen action
Carbaryl (I)	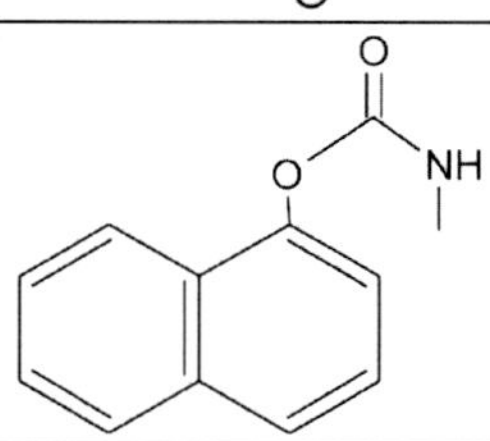	Weak estrogen effect

Pesticides	Chemical structure	Endocrine Disruptor Effects
Carbendazim (F)	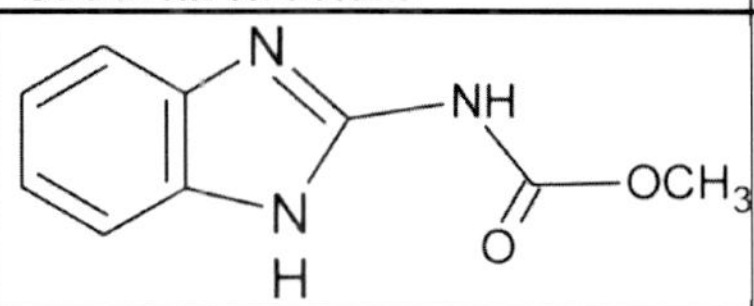	Increase of estrogen production and aromatase activity
Carbofuran (I)	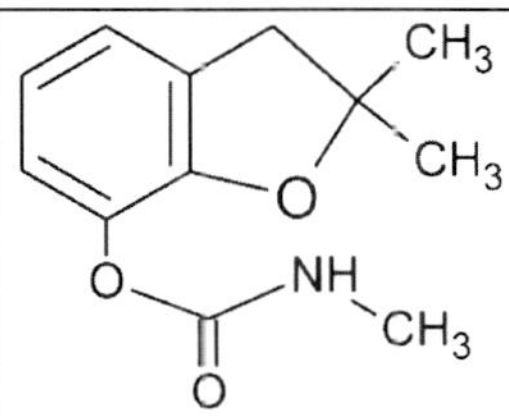	Increase of progesterone, cortisol and estradiol level and decrease of testosterone one
Chlorothalonil (F)	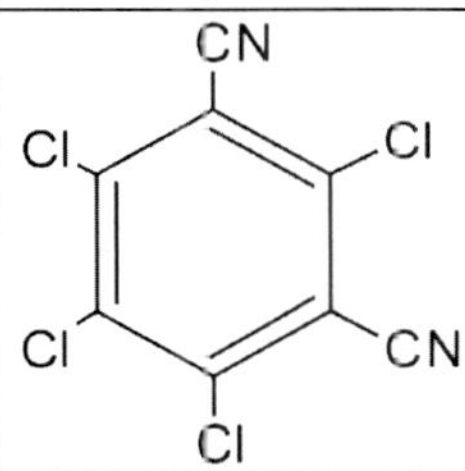	Activation of androgen-sensitive cells proliferation
Chlordane (I)	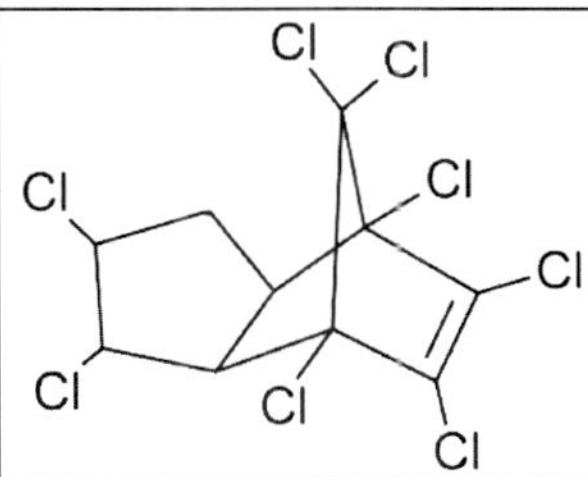	Competitive binding to androgen receptors Anti-estrogenic effect, inhibition of estradiol binding
Chlorfenvi-phos (I)	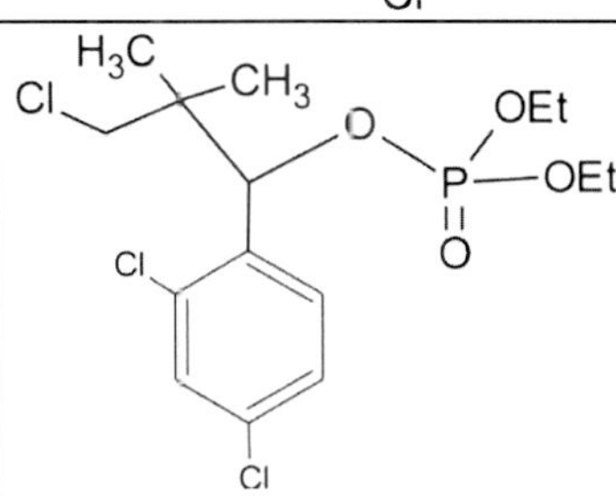	Weak estrogen effect
Cypermethrin (I)	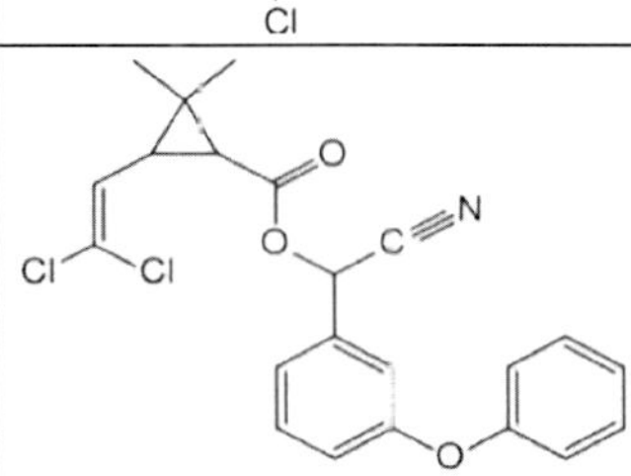	Estrogenic effect

Pesticides	Chemical structure	Endocrine Disruptor Effects
Cyproconazole (F)	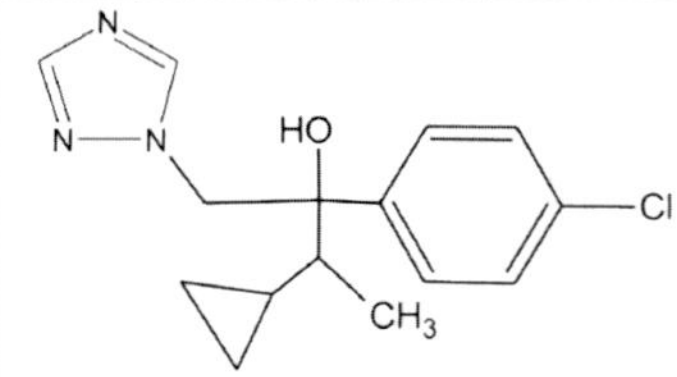	Inhibition of aromatase activity, decrease of estrogens production and increase of androgens availability
DDT and metabolites (I)	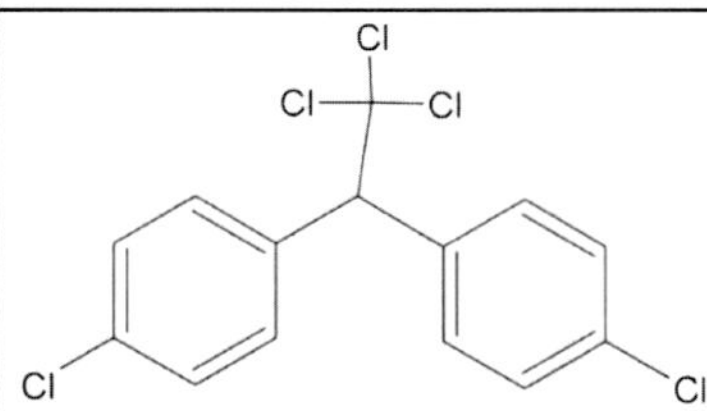	Competitive binding to androgen receptors, activation of androgen-sensitive cells proliferation. Stimulation of estrogen receptor production, estrogen receptor agonist and PR antagonist
Deltamethrin (I)	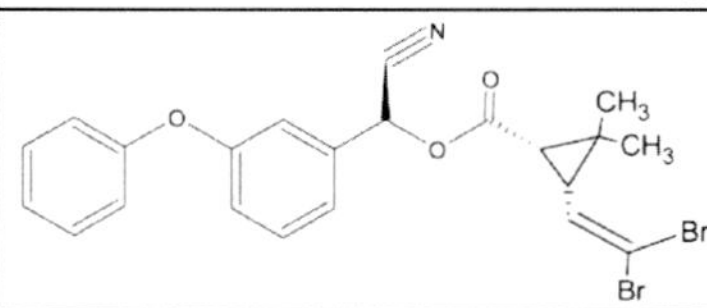	Weak estrogenic activity
Diazinon (I)	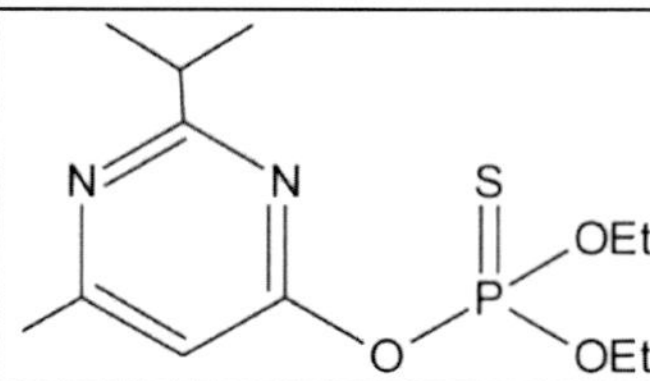	Estrogenic effect
Dichlorvos (I)	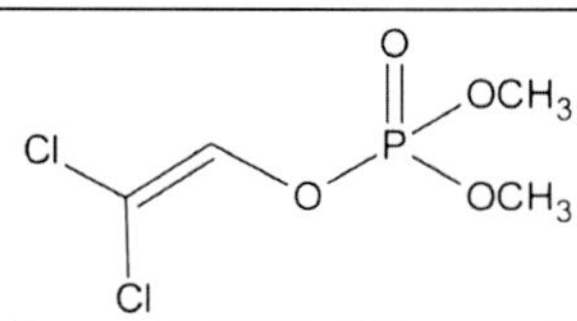	Weak androgen-receptor antagonist
Dicofol (I)	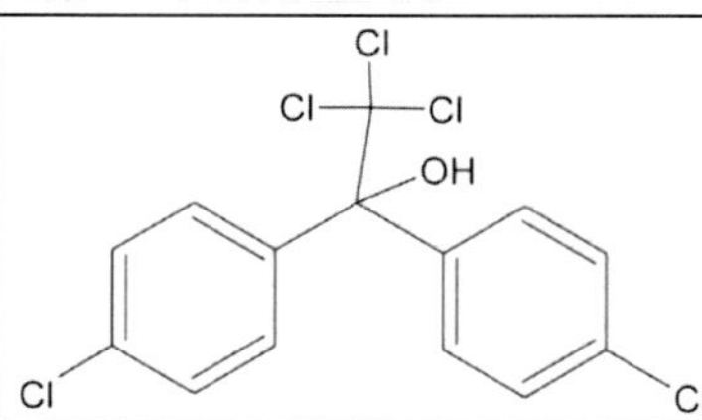	Inhibition of androgen synthesis, increase of estrogens synthesis, binding to estrogen receptor
Dieldrin (I)	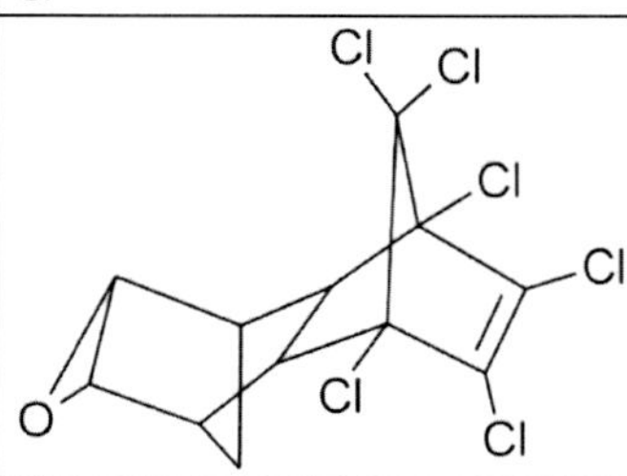	Competitive binding to androgen receptors, estrogenic effect, stimulation of estrogen receptor production

Pesticides	Chemical structure	Endocrine Disruptor Effects
Diuron (H)	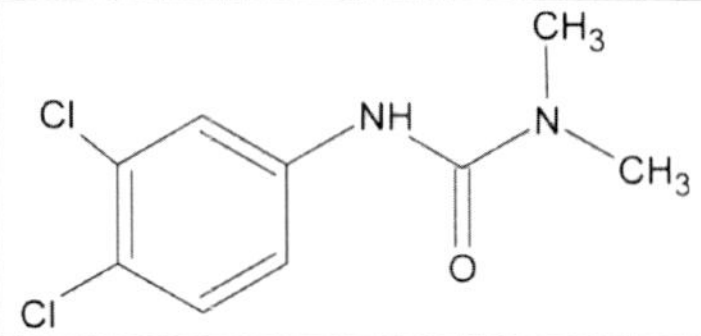	Inhibition of androgens action
Endosulfan (I)	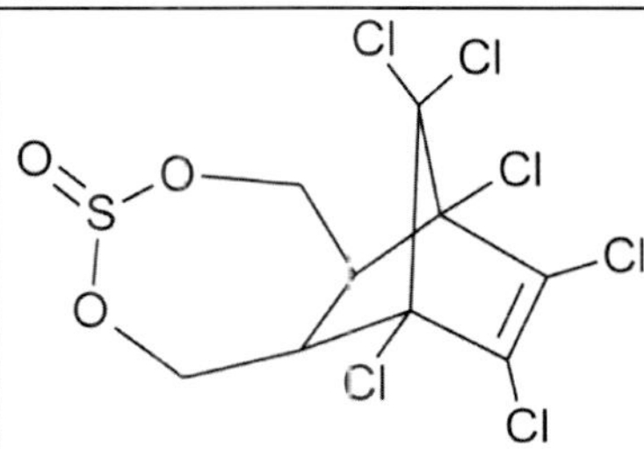	Competitive binding to androgen receptors, estrogenic effect, stimulation of estrogen receptor production, inhibition of aromatase activity
Endrin (I)	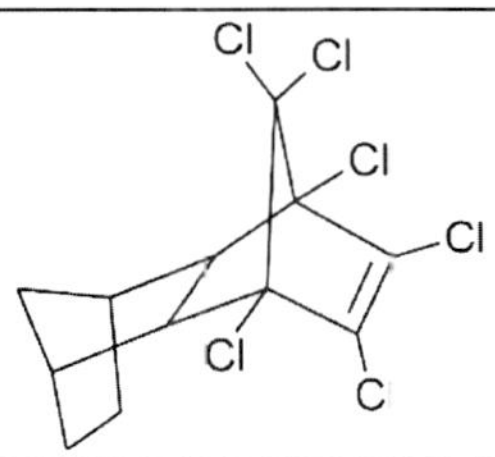	Competitive binding to androgen receptors
Fenoxycarb (I)	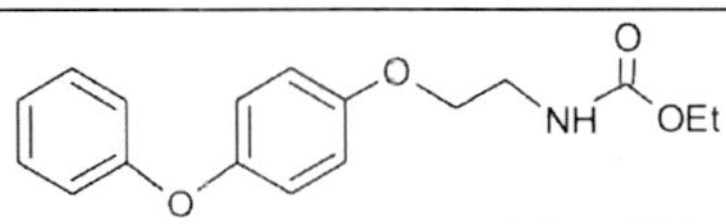	Interference with testosterone metabolism
Glyphosphate (H)	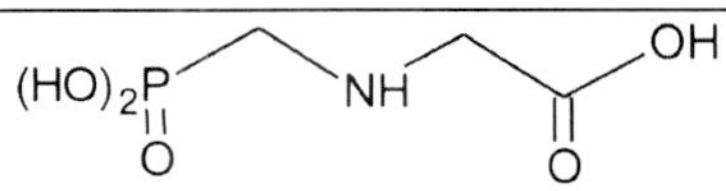	Disruption of aromatase activity, preventing the production of estrogens
HCB (F)	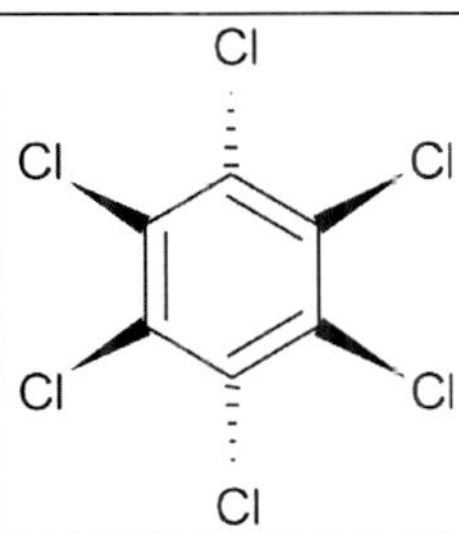	Severely disruption of thyroid hormone production. Enhancement of androgen action at low doses, but inhibition at high levels
Heptachlor (I)	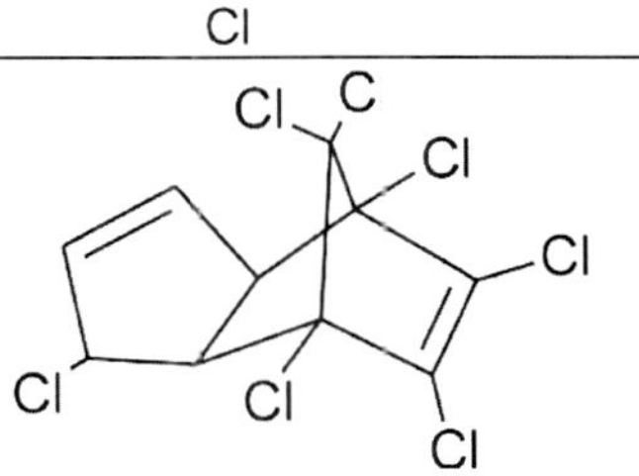	Binding to cellular estrogen and androgen receptors

Pesticides	Chemical structure	Endocrine Disruptor Effects
Methoxychlor (I)	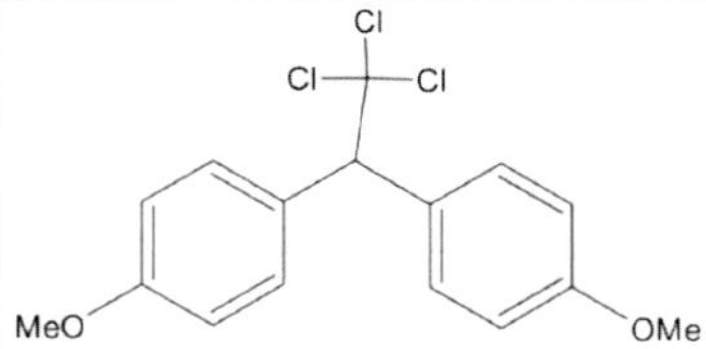	Strong estrogenic effect. Competitive binding to androgen receptor, interaction with the pregnane X cellular receptor
Mirex (I)	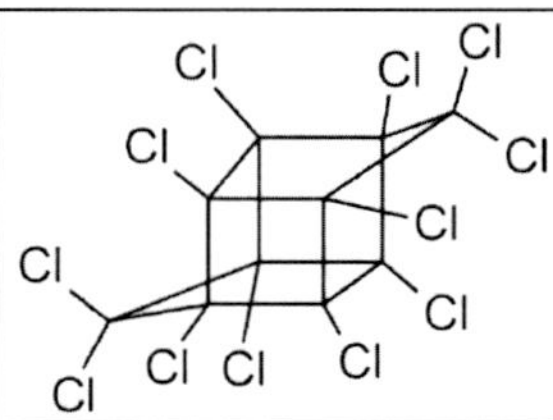	Weak estrogen effect
Molinate (H)	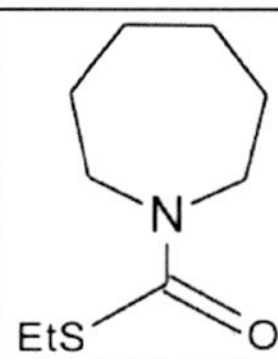	Reproductive tract damage, reduction of fertility
Parathion (I)	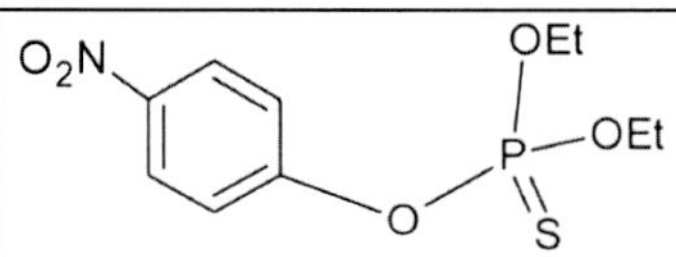	Inhibition of catecholamine secretion, increase of melatonin synthesis, inhibition of gonadotrophic hormone
Penconazole (F)	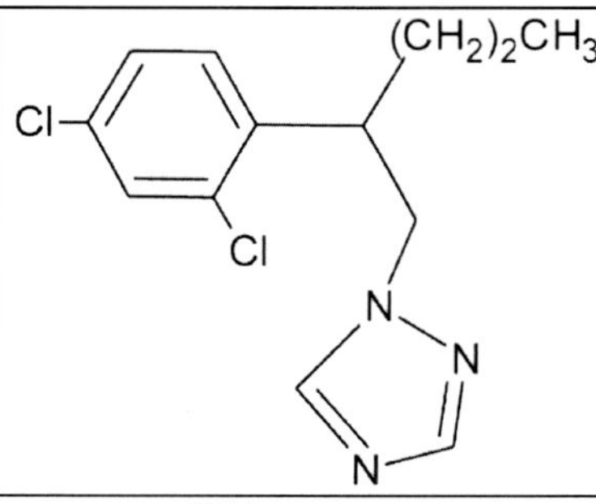	Weak estrogenic effect. Inhibition of aromatase activity, decrease of estrogens production and increase androgens availability
Pentachlorophenol (H, F, I)	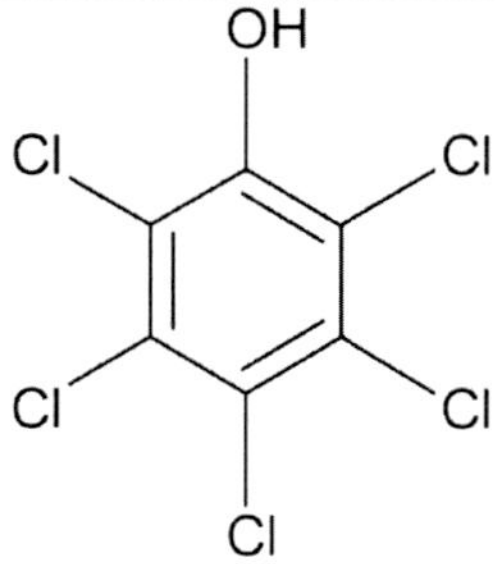	Weak estrogenic and anti-androgenic affect
Permethrin (I)	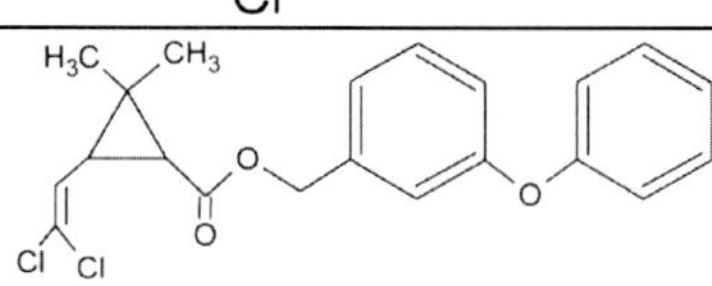	Inhibition of estrogen-sensitive cells proliferation

Pesticides	**Chemical structure**	**Endocrine Disruptor Effects**
Prothiophos (I)	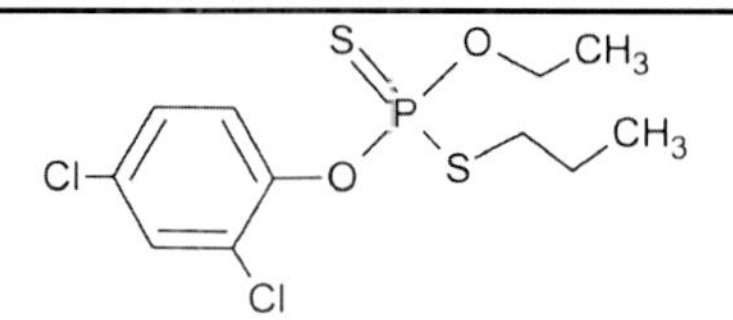	Estrogenic effect
Simazine (H)	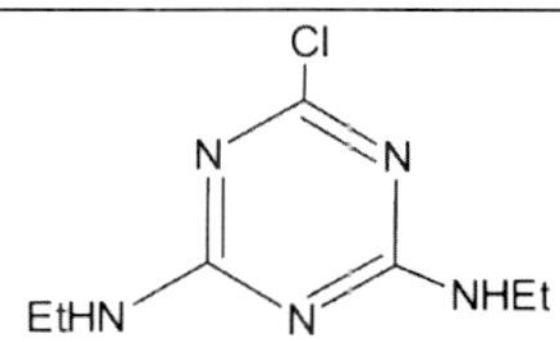	Induction of aromatase activity, increase of estrogen production
Tetramethrin (I)	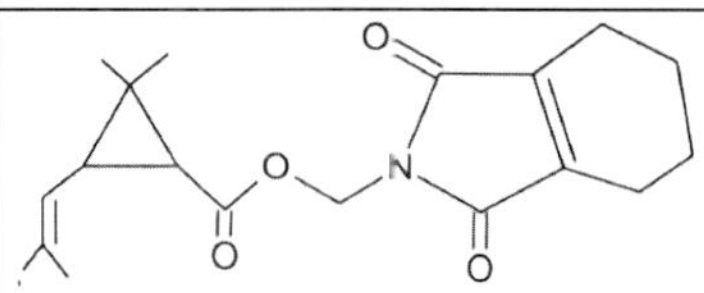	Estrogen-antagonistic effects in females only
Trifluralin (H)	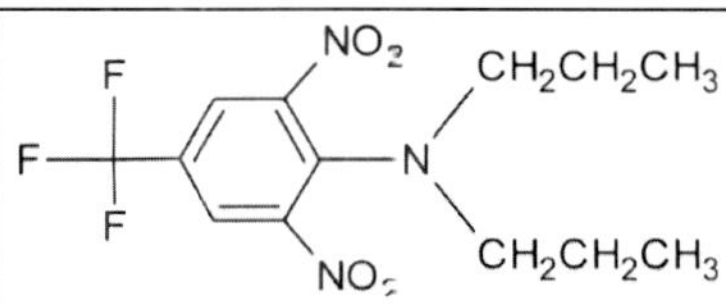	Interaction with pregnane X cellular receptor, interference steroid hormone metabolism
Vinclozolin (F)	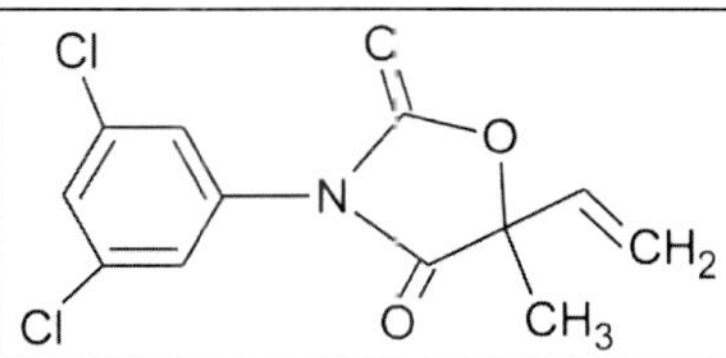	Competitive binding to androgen receptor Interactions with pregnane X cellular receptor, interference with steroid hormone metabolism.

(H): Herbicide, (F): Fungicide, (I): Insecticide

Neurotoxicity

The mode of action of most insecticides is by affecting functions in the nervous system, more specifically at neuromuscular junctions, targeting synaptic transmissions. Mechanisms affected when exposed to pesticides will differ according to the functions of the pesticide itself.

Organophosphate insecticides act by inhibiting acetylcholinesterase activity, thus blocking acetylcholine breakdown.

Pyrethroids and DDT act in a different way, targeting the sodium channels in the postsynaptic cells. These molecules prevent the ion channels to close after they were opened by the neurotransmitter (e.g. acetylcholine).

It is important to mention that these molecules affect the molecular mechanisms of any animals, and not only insects, even though these are the targets of such products.

Energy Metabolism

Pesticides, as well as other toxic substances, can affect the energy metabolism of organisms. It is known that energy can be produced by two pathways: anaerobically via glycolysis, and aerobically with glycolysis or beta-oxidation of fatty acids followed by the Krebs Cycle and oxidative phosphorylation, as summarized in Figure 78.

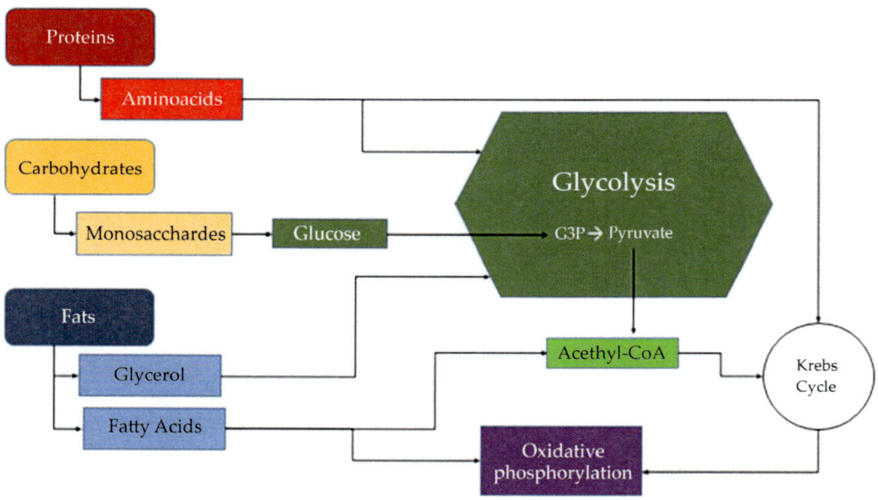

Figure 21: Energy production by organisms (adapted from Nikinmaa)

Compounds that affect energy metabolism may be divided in two main groups:

- Uncouplers of oxidative phosphorylation: hydrophobic weak acids that can move through the mitochondrial membranes and affect the pH gradient that exists, which in turn affects the phosphorylation of ADP into ATP.
- Disturbers of respiratory chain proteins: affect the electron flow (and opposite direction proton flow), unbalancing the electrochemical proton gradient that is needed for ATP synthase to function.

Membrane

Ion transport mechanism across membranes may be affected by pesticides. For instance, parathion has the ability to inhibit the sodium pump in cells, like some herbicides also do; it is believed that dephosphorylation of ATP caused by organophosphates, ends up affecting the sodium pump indirectly. The calcium pump is susceptive to the same molecules in similar fashion.

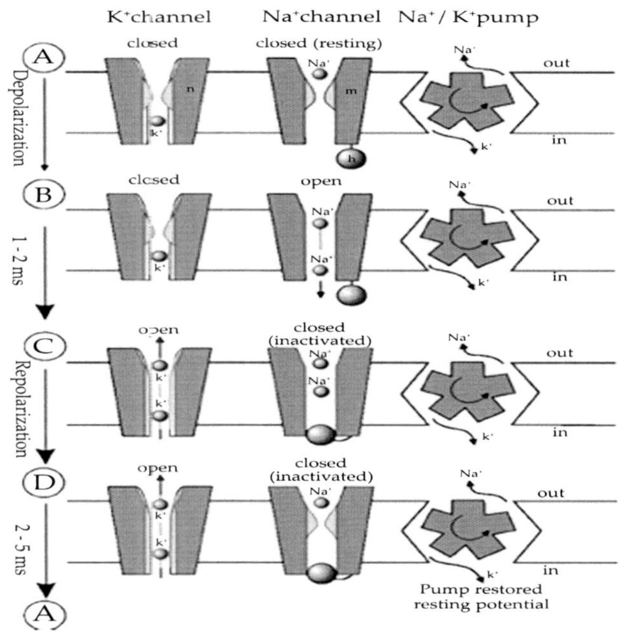

Figure 22: Schematic explanation of the functioning of sodium and potassium channels (reproduced from Davies 2007)

There is a gradient difference between the internal and external media: the extracellular fluid contains a high concentration of Na^+ and low concentration of K^+ ions, and it is the reverse inside the cell. Nerve stimulation causes the axonal membrane to become permeable to Na^+ due to the sodium channel opening (B). Sodium channel closure or inactivation (C) results in an efflux of K^+ due to the opening of potassium channels. An ATP driven sodium-potassium pump maintains the ion gradient across the axonal membrane (D). The continued transmission of the impulse across the *synapse* involves the release of a chemical transmitter, which becomes attached to receptor sites at the postsynaptic membrane where it depolarizes the membrane to generate another action potential.

Immunotoxicology

The immune system of organisms, either plants or animals, is programmed to prevent or diminish the impact external harmful agents. Different organisms have different immune functions and mechanisms, varying also in complexity. Environmental pollution leads to immunosuppression in vertebrates and invertebrates. Pesticides are known to affect immune systems, apparently in two ways: interfering with the innate phagocytosis and the acquired lymphocyte function.

EFFECTS OF HERBICIDES IN PLANTS

According to current knowledge, a particular herbicide may have more than one action mechanism at different physiological sites. A few of these toxic processes are summarized here.

Contact

Act destroying the integrity of the cell membrane, more specifically promoting the dissociation of bonds in the phospholipids and the membrane proteins. Macroscopically, leafs show signs of cell leakage by turning dark brown followed by necrosis.

Examples: Diquat, Paraquat, Sodium arsenite, Cacodylic acid, 2,4-D

Figure 23: Leafs with signs of necrosis

Mitotic Inhibition

Affect the plant's growth meristems, in root tips, young leaves, buds and the vascular cambium of stems and roots. Many of these herbicides are applied to soil, thus reaching routes and susceptive to leaching and run-off. For mechanistic insight, certain "symptoms" of poisoning have been documented: 1) high mitotic index due to arrested cell division at prophase or telophase, 2) inhibition of the formation of cell walls, resulting in multinucleated cells after division, and 3) individual chromosomes migrating to micronuclei, due to spindle misalignment.

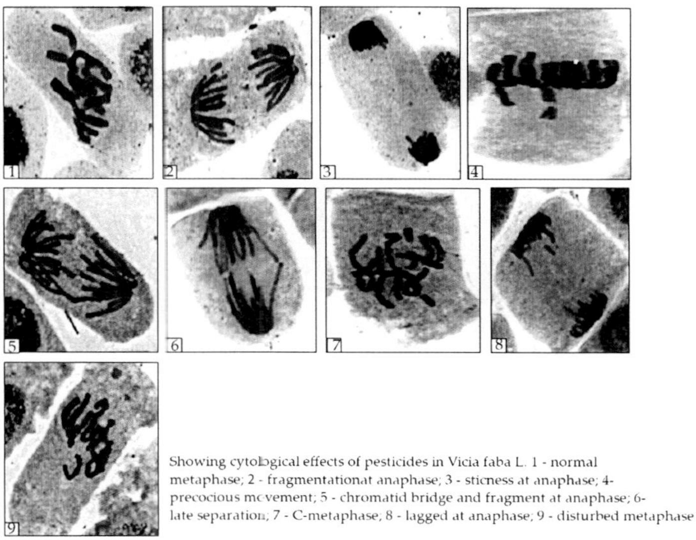

Figure 24: Cytological effects of pesticides in Vicia faba (adapted from Pandey RM, 2008). Details in the text

In Figure 81, we see Mitosis in *Vicia faba*, disrupted by various pesticides (including endosulfan, dieldrin and aldrin:

- normal metaphase,
- fragmentation at anaphase,
- stickiness at anaphase,
- precocious movement,
- chromatid bridge and fragments at anaphase,
- late separation at anaphase,

- C-metaphase
- Laggard at anaphase
- Disrupted metaphase

Photosynthetic Inhibition

Figure 25: Detail of a leaf

Since photosynthesis is a complex, multi-step process, it is natural to predict that different herbicides may affect it in different manners. The main ones are:

1) blocking the electron flow from Photosystem II,
2) competing with Ferredoxin-reducing substance receiving the electron flow from photosystem I, and
3) uncoupling cyclic photophosphorylation.

Photosynthesis consists of multiple mechanisms that ultimately convert solar energy into usable energy - which made researchers look more carefully into it in recent years considering the several attempts in developing an efficient and inexpensive energy production system that reduces the levels of CO_2, by designing an artificial version of it. Nature had a few million years to perfect it.

The photosynthetic process occurs in the cells' chloroplasts, more specifically in the grana, which contain chlorophyll molecules, and it can be divided in the groups of Light Reactions and Dark Reactions.

In the Light Reactions, photons are absorbed by chlorophyll (Photosystem II) and their energy is then transferred to the molecule, pushing it into an excited state by the energy absorbed by its electrons. One electron is ejected, and the gap left behind is filled by an electron coming from the reaction that generates O_2 from the cleavage of H_2O in the surroundings. The excited electron eventually releases that excitation energy in a chain of reactions named cytochrome system that yields ATP - these are called photophosphorylation reactions. This electron is taken by the Ferredoxin-reducing substance, which reduces $NADP^+$ to NADPH (Photosystem I). In the Dark Reactions, ATP and NADPH then have the roles, respectively, of: first, phosphorylating a 5-carbon sugar molecule; second, reducing CO_2 to combine it with the 5-carbon molecule, later producing 2 molecules of 3-carbon sugar.

Light reaction

$ADP + Pi + NADP + H_2O \quad ATP + NADPH + H^+ + O_2$

Dark reaction

$CO_2 + ATP + NADPH - H^+CH_2O + ADP + Pi + NADP$

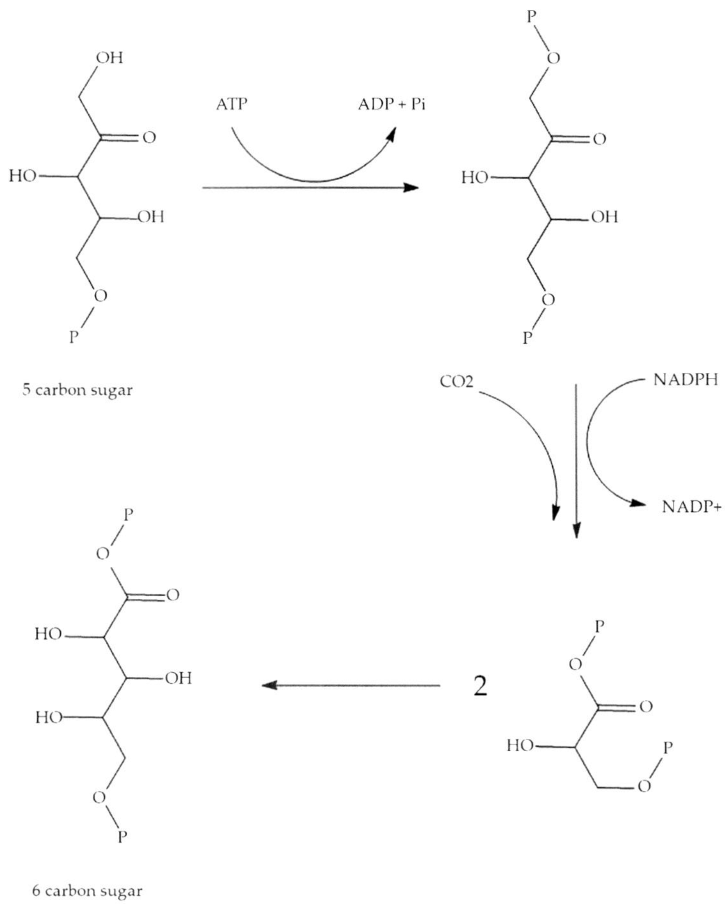

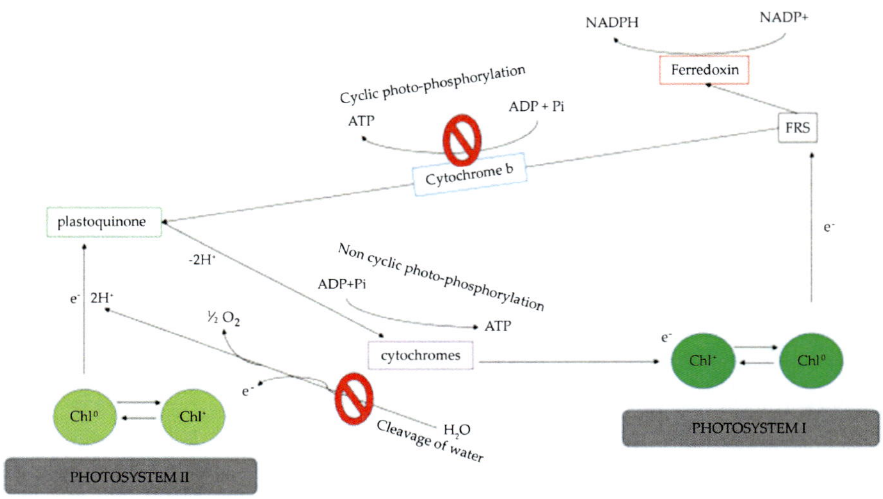

Figure 26: Photophosphorylation scheme, showing the two processes inhibited by pesticides

The two mechanisms that most herbicides are known to undergo in order to inhibit photosynthesis are: i) blocking the electron transfer in Photosystem II; and ii) taking the electrons that the Ferredoxin-reducing substance would be receiving from Photosystem I.

In the first scenario, some herbicides are believed to be able to deactivate the enzymatic process that cleaves the water molecule. The practical result of that is that the oxidized chlorophyll in Photosystem II can never be reduced back again, so the whole process gets limited, and eventually ceases, which in turn will stop also Photosystem I and the dark reactions. Diuron and atrazine are known to act in this manner.

The Hill reaction tested the role of an inhibitor to Photosystem II. A suspension of chloroplasts is exposed to light along with an electron accepting molecule to mimic NADP, in this example the DPIP - dichlorophenol indophenol dye). The change in color indicates the reduction provoked by the electrons that were ejected from the chloroplasts. In the presence of diuron, however, that process is suppressed proportionally to the concentrations.

In the second one, certain herbicide molecules are able to intercept the electrons coming from

Photosystem I that were supposed to be taken by the Ferredoxin-reducing substance. This interference prevents NADP to be reduced to NADPH, which would lead to a reduction in CO2 fixation. This is the mechanism of action of diquat and paraquat. Notwithstanding, these herbicides present a high toxicity level to the plants exposed, which arises from the reduction their molecules undergo from Photosystem I, turning them into free radicals. These free radicals produce hydrogen peroxide inside the cells, which is highly increases the membrane permeability, until it completely disrupts the cellular structure.

Examples: 1,2,4-triazinones, diquat, paraquat, uracils, alkyl anilides, chlorinated phenyl ureas.

Respiration

Respiration is one more process in plants that can be directly affected by herbicides, by, in practical terms, disturbing the processes of energy release and energy utilization. Chemical energy that is gathered from the conversion of carbohydrates and lipids to ATP, NADPH, CO_2, water and heat is made unavailable by the suppression of such processes.

The 3 processes of respiration that can be inhibited by herbicides are presented below.

i) Glycolysis

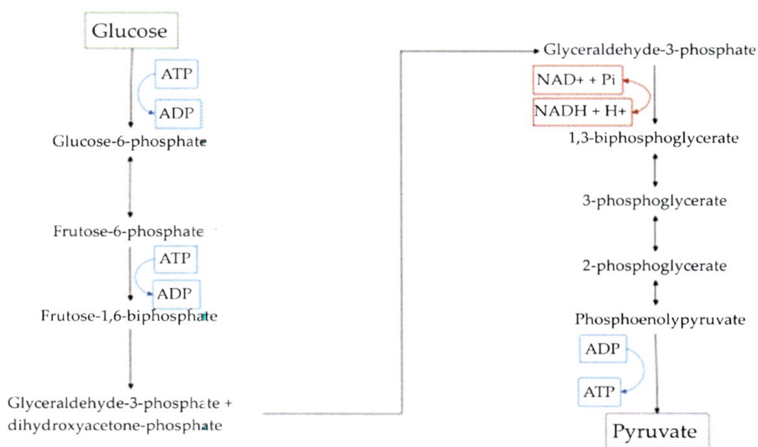

Figure 27: Glycolysis simplified scheme; enzymes were not mentioned for better visualization

Starch undergoes hydrolysis to yield sugars, which are then broken down to pyruvic acid. The dehydrogenation reaction feeds the production of NADPH with hydrogen atoms and electrons.

ii) Krebs cycle

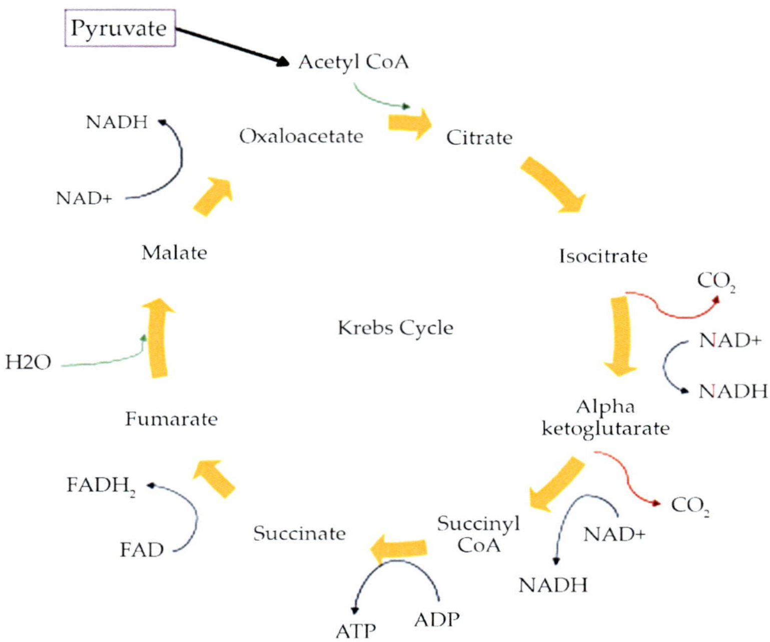

Figure 28: Krebs Cycle

Before the actual cycle takes place, a reaction involving coenzyme ASH dehydrogenates pyruvic acid (the end product of glycolysis), producing CO_2, acetyl CoA and NADH. Acetyl CoA combined with oxaloacetic acid is the species that actually plays a role in the Krebs Cycle, where sequential dehydrogenation reactions will happen producing NADH, $FADH_2$ and ATP at different ends, as well as CO_2 that results from the acetate that started the cycle (from acetyl CoA).

iii) oxidative phosphorylation

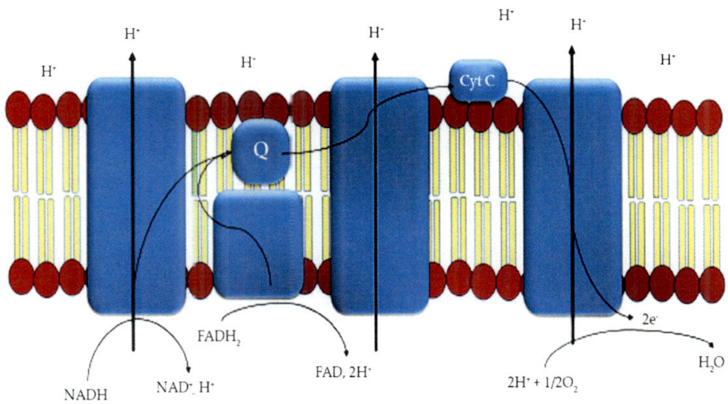

Figure 29: Electron transport chain in oxidative phosphorylation (adapted from Khan Academy)

Cyclohexanedione and aryloxyphenoxypropionic acid herbicides are two classes known to disturb plants metabolisms by interacting with acetyl-CoA carboxylase.

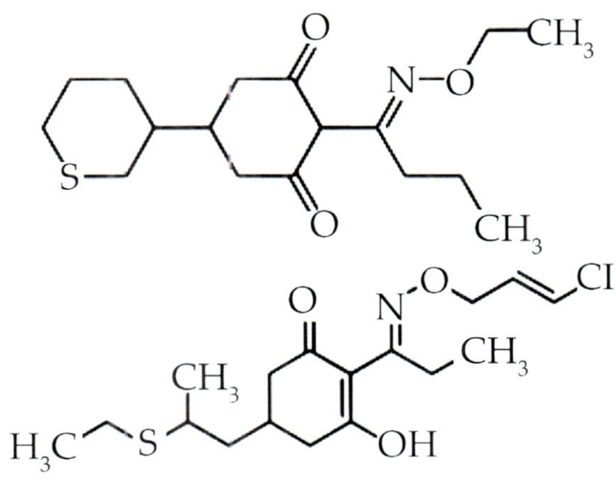

Figure 30: Chemical structures of cyclohexanedione herbicides - left: cycloxydim, right: clethodim

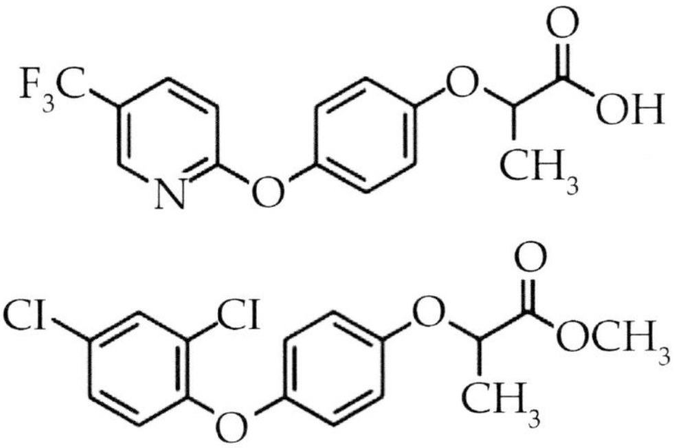

Figure 31: Chemical structures of aryloxyphenoxypropionic acid herbicides - left: fluazifop, right: diclofop

The Krebs cycle generates NADH and $FADH_2$, two pyridine nucleotides that can still be used to produce the high energy compound ATP. This occurs by means of a cytochrome or electron transport system that ends with oxygen as the final electron acceptor in forming water. Coupled phosphorylation side reactions occur in this process at 3 points, and they are the ones inhibited by acting herbicides, by uncoupling them. Normally, two ATP molecules are produced if the pair of electrons introduced come from FADH2, and three if the electrons come from NADH + H^+. The P/O ratio is thus between 2 and three for a coupled mitochondrial system. If the system is affected by an uncoupling agent, phosphorylation rate decreases, producing less than two ATP molecules in the process and so the P/O ratio is reduced.

Since respiration processes at molecular level are very similar among species, herbicides with that action are usually very toxic also to non-target organisms.

Examples:

- inhibitors of respiration: picloram, dicamba, propanil;
- uncouplers of oxidative phosphorylation: 2,4-D, 2,4,5-T, dinoseb.

Effects on Nucleic Acid Metabolism and Protein Synthesis

Herbicides are known to affect protein synthesis in cells in a number of ways. The herbicides that have been found to influence this process include Phenoxy acids (e.g., 2,4-D, 2,4,5-T), Benzoic acids (e.g., dicamba), and Picolinic acids (e.g., picloram). Researchers have been investigating them, and arrived to some important findings:

- 2,4-D mimics the auxin indole acetic acid, which occurs naturally, and thus plays a similar role in the protein synthesis mechanism;

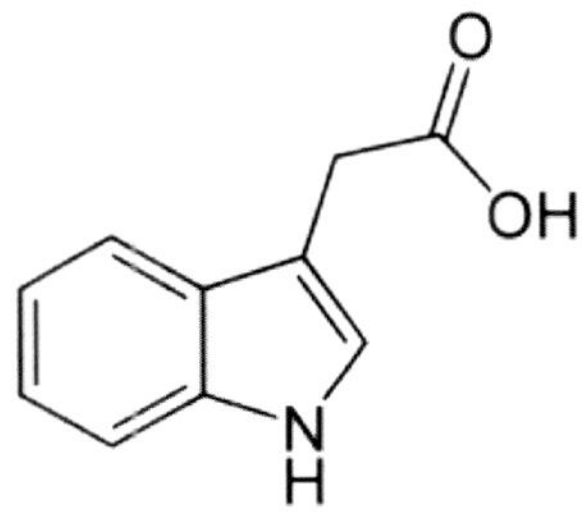

Figure 32: Auxin indole acetic acid structure

- 2,4-D, 2,4,5-T, dicamba and picloram have been found to increase the levels of DNA and RNA, as well as protein contents, in cells of wheat and in cucumber roots exposed to them;
- 2,4-D and dicamba were found to decrease the amounts of extractable nucleohistone proteins in cucumber roots (a susceptible matrix), but not so much in tolerant wheat.

This leads to the conclusion that the DNA-RNA transcription process is the one that is affected by auxin herbicides and natural hormones. These initially act on a cytoplasmic factor, in this case a hormone acceptor, later entering the nucleus. Once in the nucleus, those molecules interfere with the RNA polymerase or the histones. Histones are proteins that shield the DNA, and herbicides such as the ones investigated are capable of removing them, thus leaving exposed portions of the code for transcription. RNA polymerase is the enzyme responsible for RNA synthesis and plays a major role in the nucleus. The transcription processes result in three forms of RNA: i) Messenger RNA (m-RNA), which is a copy of the DNA code and acts in the cytoplasm; ii) Ribosomal RNA (r-RNA), which is responsible for shaping the template for protein synthesis in combination with m-RNA; and iii) Transfer RNA (t-RNA), which combines with amino acids in an enzyme complex (aa-AMP-enzyme), whose movement towards the active polysome is what constitutes the translation process. Proteins are the result of peptide chains that get longer as amino acids are linked in a particular sequence determined by the m-RNA template.

Figure - Protein synthesis scheme

NATURAL AUXINS AND AUXINIC HERBICIDES

Auxinic herbicides are described as "herbicides with growth regulator activity or growth regulators with herbicidal action." The first compounds of this class were MCPA and 2,4-D, produced during World War II. These led to the development of a wide variety of related substances throughout decades, as some examples shown in Figure X. These molecules are intended to act on the growth mechanisms of certain plants and weeds, as it is normally done by natural endogenous hormones such as indole-3-acetic acid (IAA), which was once known as "auxin" and other similar ones that were discovered later. The auxin effects include elongation, division and differentiation in cells, root initiation, leaf senescence (aging), and response to wounds. Auxinic herbicides are often used in cereal crops for broadleaf weed control.

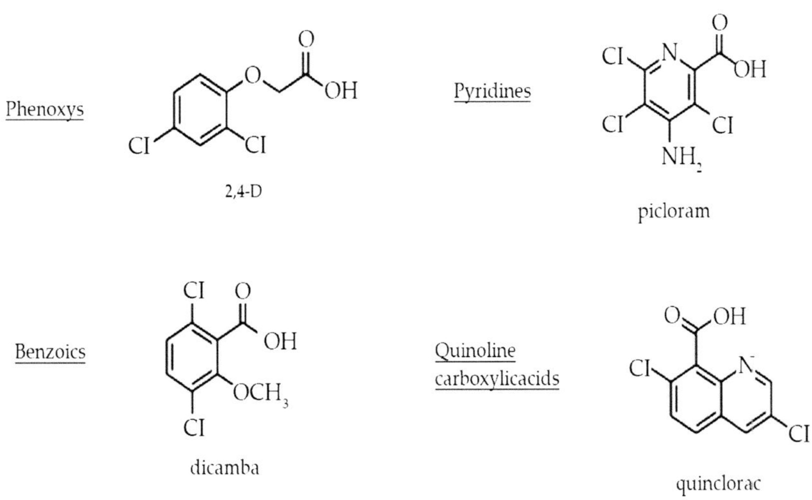

Figure 33: examples of molecules in each of the major grous of auxinic herbicides

Sterling and Hall summarized the characteristics that relate auxinic herbicides to natural auxins, or to be more specific, IAA:

- Dose-response profile: although showing clear toxic effects at high doses, auxinic herbicides can actually act as growth regulators and even stimulate plant cell growth at low doses;
- Mechanism: both IAA and auxinic herbicides promote growth by cell elongation instead of cell division;
- Replacement: it was found that, in plant cell culture medium, auxinic herbicides replaced IAA as the hormone supplement for proper cell plant growth. This behaviour has been noted for 2,4-D, picloram, chlorambem and dicamba;
- Differential sensitivity: different tissues have shown different degrees of sensitivity for auxinic herbicides, varying along roots, buds, callus and active meristems, as well as different physiological stages of growth;

HERBICIDES RESISTANCE

Resistance to pesticides is maybe one of the main reasons why usage may be increased in agricultural lands. Farmers eventually notice that their products do not have the same "power" as earlier. As stated by Devine and Eberlein, there are three main mechanisms that result in resistance:

1. Target-site mutations

   A typical target site is a protein named D1, which has a binding site for plastoquinone and so plays an active role in the electron transfer in photosystem II. Mutation occurs so that the binding site remains accessible. The mutation is referred as Ser264-to-Gly, and the resistance factor may differ among classes of herbicides (see Table 39 and 40).

2. Enhanced metabolic detoxification of herbicides

   Corn and other crop species have demonstrated to be able to improve metabolism speed in order to transform herbicides in non-active compounds. S-triazines, such as atrazine and simazine, which act by binding to plastoquinone inhibiting the electron transport in photosystem II, are inactivated by metabolic degradation in resistant plants. Substituted ureas, e.g. diuron, follow this same process.

3. Sequestration (or compartmentalization) of herbicides away from the target site.

Table 21: Resistance of chloroplasts from s-triazine-resistant weeds containing the Ser264-to-Gly mutation to classes of photosystem II electron transport inhibitors. (adapted from Roe)

Herbicide group	**Resistance factor (Resistant/susceptible in terms of I_{50} ratio)**
s-Triazines	100-1000
as-Triazines	10-100
Uracils	10-50
Biscarbamates	10-50
Heterocyclic ureas	10-100
Phenylureas	1-50
Hydroxybenzonitriles	1-3
Bentazon	0.5 (negative resistance)
Phenols	0.1-0.3 (negative resistance)

The mutated protein then does not bind to the herbicides, but still functions binding plastoquinone.

Table 22: Mutations identified in psbA code, which lead to resistance to photosystem II electron transport inhibition herbicides. (Adapted from Roe RM, 1997)

Species	**Amino acid substitution**	**Resistance factor**	
		Atrazine	**Diuron**
Most higher plants	Ser_{264} to Gly	1000	1
Portulaca oleracea	Ser_{264} to Thr	>400	>280 (Linuron)
Chlamydomonas	Ser_{264} to Ala	100	10
Anacystis	Ser_{264} to Ala	10	100
Synechococcus	Phe_{211} to Ser	7	2

Chlamydomonas	Val_{219} to Ile	2	15
Chlamydomonas	Ala_{251} to Val	25	5
Chlamydomonas	Phe_{255} to Tyr	15	0.5
Chlamydomonas	Leu_{275} to Phe	1	5

INSECTICIDES MODE OF ACTION

There is a large body of knowledge regarding the action mechanisms of insecticides, and more and more research is being performed in order to uncover what is still not clear. Like plants and herbicides, there is also great concern about the resistance that some bugs have shown to insecticides. The latest information about this subject is made publicly available at the Insecticide Resistance Action Committee (IRAC) website www.irac-online.org.

One of the main mechanisms for insecticidal substances to act is in the different parts or reactions of the nervous system, suppressing neurotransmitters, reacting with neuron axons, or binding to receptor sites. Organophosphates and carbamates, which are heavily used classes of insecticides in agricultural lands, are known to act as acetylcholinesterase inhibitors. Basically, these molecules react with the neurotransmitter acetylcholine transforming it in inactive molecules, choline and acetic acid. This process occur physiologically when the excitation is done, and any excess of the neurotransmitter must be taken care of.

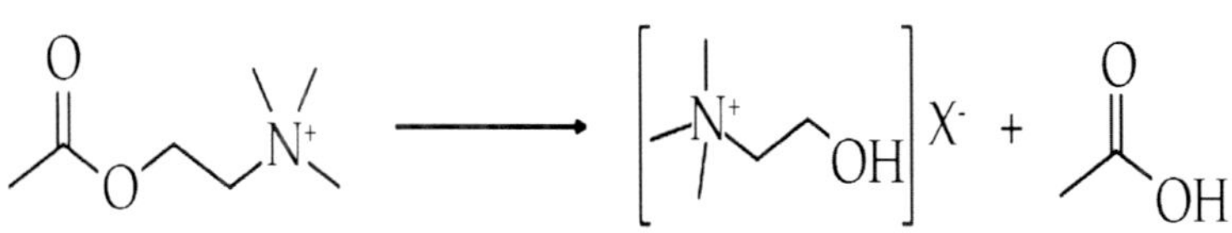

Acetylcholine Choline Acetic acid

Figure 34: Acetylcholine reaction

Likewise, organophosphates and carbamates act in an analogous fashion.

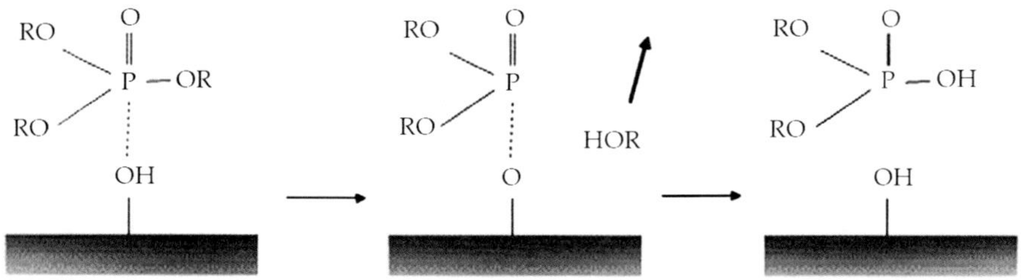

Figure 35: Reaction of AChE with organophosphate insecticides (reproduced from Yu SJ, 2015)

In terms of toxicologic responses, the inhibition of acetylcholinesterase by organophosphates is considered irreversible, due to the long period necessary for that reversal to take place, as exemplified in Table 41.

Table 23: Inhibition of AChE by organophosphorous insecticides in rats (adapted from DuBois, KP, 1963)

Compound	**Time to maximum inhibition (in hours)**	**Time for complete reversal (in hours)**
Trichlorfon (dimethyl)	0.25	6
Azinphosmethyl (dimethyl)	0.5	24
Disulfoton (diethyl)	3	120

Other action mechanisms of action of insecticides include, for example: binding to nicotinic acetylcholine receptors, interference with GABA-gated and glutamate-gated chloride channels, interference with octopamine receptors, respiration disruption, chitin synthesis disruption.

Table 42 presents the typical half lives of a number of insecticides exposed to the environment

Table 24: Approximate half lives of select insecticides in soil (Adapted from Yu SJ, 2015)

Insecticide	**Approximate half-life (in days)**
Chlorinated Hydrocarbons	
Aldrin	1-4 years
Chlordane	2-4 years
Chlorbenzilate	10-35 days
DDT	3-10 years
Dieldrin	1-7 years
a-Endosulfan	60 days
b-Endosulfan	800 days
HCH	2 years
Heptachlor	7-12 years
Isodrin/endrin	4-8 years
Lindane	20-50 days
Methoxychlor	120 days
Toxaphene	10 years
Organophosphates	
Acephate	3 days
Chlorfenvinphos	14-161 days
Chlorpyrifos	36-46 days
Chlorthion	36 days

Demeton S	54 days
Diazinon	11-21 days
Dichlorvos	17 days
Dimethoate	2-4 days
Disulfoton	290 days
Fenitrothion	4-28 days
Fenthion	34 days
Malathion	4-6 days
Methyl Demeton S	26 days
Methyl parathion	45 days
Monocrotophos	1-5 days
Parathion	180 days
Phorate	2 days
Trichlorfon	140 days
Carbamates	
Aldicarb	30 days
Bendiocarb	5 days
Carbaryl	7-28 days
Carbofuran	26-110 days
Fenobucarb	6-30 days
Methiocarb	15 days
Propoxur	44-59 days

Pyrethroids	
Bifenthrin	97-156 days
Cyfluthrin	5 days
Cyhalothrin	7-21 days
Cypermethrin	4-12 days
Deltamethrin	11-19 days
Esfenvalerate	75 days
Permethrin	18-23 days
Tefluthrin	< 31 days
Tralomethrin	44-82 days
Neocotinoids	
Acetamiprid	1-2 days
Imidacloprid	8-48 days
Nitenpyram	1-15 days
Thiacloprid	9-27 days
Thiamethoxam	11-26 days
Formadimines	
Amitraz	2 days
Chlordimeform	20-40 days
Juvenoids	
Methoprene	10 days
Hydroprene	A few days

Fenoxycarb	51-75 days
Benzoylphenylureas	
Diflubenzuron	< 7 days
Hexaflumuron	50-159 days
Diacylhydrazines	
Chromafenozide	22-136 days
Halofenozide	42-267 days
Methoxyfenozide	23-268 days
Tebufenozide	7-66 days
Thiadiazines	
Buprofezin	80-104 days
Others	
Bensultap	3-35 days
Cartap hydrochloride	3 days
Chlorfenapyr	1.4 years
Cyromazine	3-140 days
Fipronil	68-139 days
Flubendiamine	4-60 days
Hydramethylnon	7-28 days
Indoxacarb	3-23 days
Pymetrozine	2-69 days
Pyridaben	< 21 days

Thiocyclam	1 day
Spiromesifen	5 days
Spinosad	2-2.8 days
Cy1Ac	0.8-3.2 days
Acaricides	
Acequinocyl	< 3 days
Bifenazate	< 1 day
Clofentezine	28-56 days
Etoxazole	19 days
Fenazaquin	45 days
Fenpyroximate	26-50 days
Propargite	40-67 days
Spirodiclofen	0.5-5.5 days
Tebufenpyrad	20-30 days

PHOTODEGRADATION

Ultraviolet radiation, the region of the spectrum between 150 and 400nm, is the main contributor to photolysis, which is a largely relevant degradation process for pesticides in the environment. Since the ozone layer filters wavelengths below 295nm, sunlight that effectively reaches the ground and surface waters is restricted.

A general scheme for photolysis is:

$A + B \rightarrow \text{(light)}\ A+B^* \rightarrow (AB) \rightarrow A' + B$

Where

A is a pesticide molecule

B is the photosensitizer, a molecule that is excited when absorbs light (becoming B*)

A' is the photochemical product

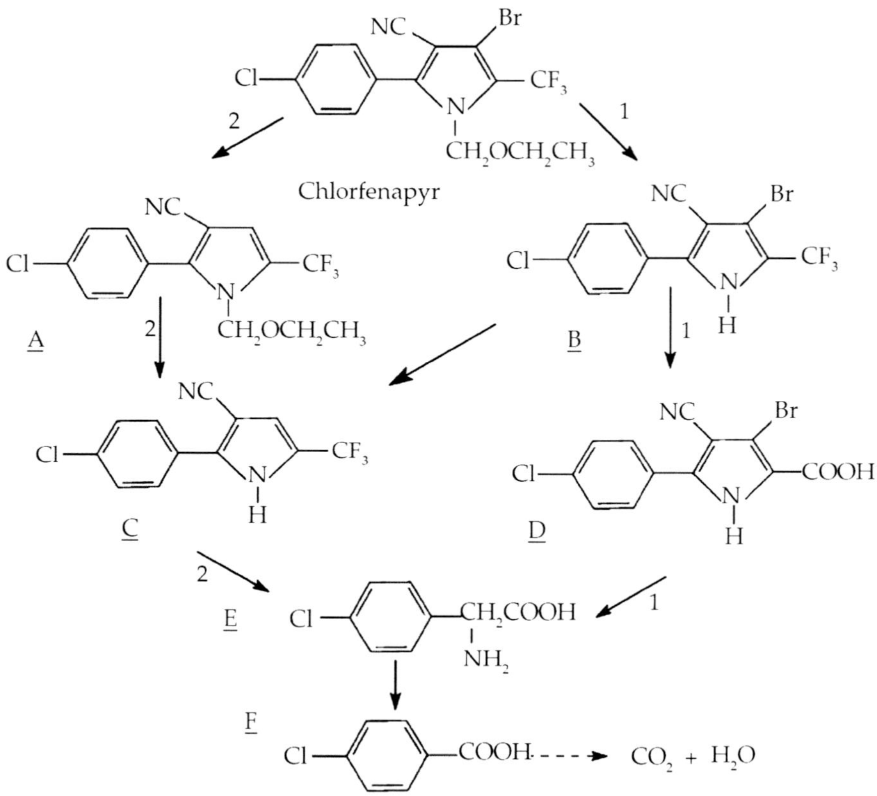

Figure 36: Proposed photocatalytic degradation pathway of chlorfenapyr in suspension of TiO2 irradiated by UV. (reproduced from Cao Y, 2006)

6
CHAPTER

BIOMARKERS AND RISK ASSESSMENT

Erick Soares Lins

BIOMARKERS DEFINITION

An important component of any risk assessment scenario. Such approach is widely used to clinical trials and physiological risk investigations, and it is also applicable to the study of the consequences of environmental pollution, including exposure to pesticides.

The World Health Organization (WHO) describes the term biomarker in such a way that it considers *"almost any measurement reflecting an interaction between a biological system and a potential hazard, which may be chemical, physical or biological. The measured response may be functional and physiological, biochemical at the cellular level, or a molecular interaction."*

Biomarkers are used then in the attempt to properly identify hazards and determine diseases and toxicological effects presented by organisms exposed to it. The objective of a well conducted risk assessment is to objectively qualify and quantify the severity of those dangers, instead of mere speculation. A good example is the monitoring of acetylcholinesterase as a function of exposure to organophosphate pesticides, since we saw in Chapter X that this enzyme suffers suppression from certain groups of insecticides.

Monitoring biomarkers is essential for a more complete investigation of the effects of a pollutant in the environment, from immediate responses to long term effects.

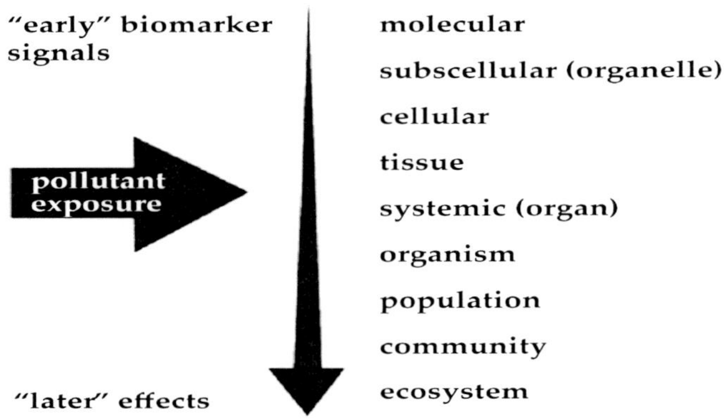

Figure 1: Response hierarchy of a biological system to the exposure to a pollutant (reproduced from Ghisi)

Biomarkers are divided in three main categories regarding their purpose in an assessment:

- *biomarker of exposure:* an exogenous substance or its metabolite or the product of an interaction between a xenobiotic agent and some target molecule or cell that is measured in a compartment within an organism. They are useful to identify any links and relationships between exposure and internal quantification or dosimetry;
- *biomarker of effect:* a measurable biochemical, physiological, behavioural or other alteration within an organism that, depending upon the magnitude, can be recognized as associated with an established or possible health impairment or disease. They provide objective information of alterations and symptoms as a consequence of absorption of a chemical, allowing one to investigate dose-response relationships;
- *biomarker of susceptibility* - an indicator of an inherent or acquired ability of an organism to respond to the challenge of

exposure to a specific xenobiotic substance. They are useful in identifying the degree of the measured response for individuals exposed to a chemical.

Biomarkers must then be properly selected, taking into account many of their properties and the context they are in, and validated against scientific criteria. The criteria to be considered when selecting a biomarker include its Specificity and Sensitivity. The procedure to monitor the biomarkers response must also be subject to Accuracy and Precision evaluations.

The validation criteria defined by the WHO for biomarkers follow:

- The marker is common to individuals in a population or subgroup;
- The marker has a defined variability across the non-exposed population or subgroup;
- The marker is common among a group of species;
- The marker reflects the host-chemical interaction, qualitatively or quantitatively;
- The marker has appropriate specificity and sensitivity to such interaction, thus providing unequivocal analytical information;
- The measurement of the marker is reproducible at short and long terms.

The activities of GST [Glutathione S-transferase], CAT [Catalase], lipoperoxidation [LPO], and Acetylcholinesterase [AchE] are probably the most commonly studied systems to play the role of effect biomarkers regarding pesticide exposure at biomolecular level. Histopathological determination is also a powerful tool in order to link physiological alterations to toxic activity of pesticides. The advantage of enzymatic biomarkers is that effects may identified even before physiological symptoms are visible.

Acetylcholinesterase activity is largely studied for the effects of insecticides in aquatic wildlife. Research with clam (*Meretrix casta*) exposed to organophosphorus insecticides showed different levels of inhibition: chlorpyrifos was the most toxic, followed by dichlorvos and methyl parathion (Devi et al. 2005). Other results are summarized below:

Table 1: AChE inhibition responses to different insecticides (data from Yu p342 - get refs)

Species	Insecticide	Inhibition of AChE activity
Snake-head fish (Channa striata)	Diazinon at 0.016 mg/L 0.079 mg/L 0.35 mg/L	Tissue: brain. 33% 81% 91%
Rainbow trout (O. mykiss)	Diazinon at 0.1 and 0.2 ppm	"Significant" (see references)
Nile tilapia (Oreochromis niloticus)	Chlorpyrifos at 5-15 ppb	Tissue: gonads. 45-77%

Apart from these examples, inhibition of AChE in the freshwater gastropod Chilina gibbosa and in juvenile tilapia were also found to be good biomarkers for the exposure to azinphos methyl.

Cytochrome P450 and Glutathione S-transferases, as well as other detoxification enzymes, were reported as biomarkers for rainbow trout exposed to endosulfan. The response to this toxicant was the induced activity of two microsomal monooxygenases: 7-ethoxyresorufin-O-demethylase and aldrin epoxidase. Individual isozymes of Glutathione S-tranferase were induced by dieldrin and malathion in the gilthead seabream (Sparus aurata).

Catalase, Glutathione Reductase, Glutathione peroxidase and Superoxide dismutase are important antioxidant enzymes. Toxicants may affect the balance between oxidants and antioxidants, which may get reactive oxygen species to react with lipids, membranes, DNA and proteins. This is known as oxidative stress. Exposure of freshwater characid fish (Brycon cephalus) to methyl parathion induced the activities of Superoxide dismutase, Catalase and Glutathione S-tranferase. Catalase activity also gets increased by the effect of azinphos methyl in C. gibbosa, and of methomyl in tilapia,

Table 2: Other examples of Pesticides and the biomarkers used to monitor their effects (data from Ghisi)

Pesticide (or group)	Effect	Biomarker alterations
2,4-dinitrophenol	Decouples oxidative phosphorylation	Onset of rigor mortis in poisoned individuals
2,4-D and Pentachlorophenol	DNA damage	Mutation in fish erythrocytes
Endosulfan	DNA damage	DNA breakdown in liver and gills
Organophosphates, Carbamates, Endosulfan	Interference in the nervous system	AChE activity

A broader compilation of studied biomarkers is given in Table 45. Their specificity is variable.

Table 3: Biomarker responses on animals commonly used in aquatic toxicology (adapted from Nikinmaa)

Type	Biomarker	Behaviour/Responses
Exposure	Complementary DNA microarrays	Different contaminants result in different signatures
	Bile composition	Presence and products of toxicants in the bile
	Retinol profile	Organochlorine pesticides alter the levels of retinol (vitamin A)
	Porphyrin profiles	Organochlorine contamination affects porphyrin in heme
	Mixed-function oxidases	Enzymes that contain cytochrome P450, involved in biotransformation processes
	Metallothionein levels	Affected by cadmium, copper, zinc, mercury and other metals
	DNA adducts	Formed by aromatic compounds
Exposure and Effect	ALA-D activity	Enzymatic reaction is inhibited by lead contamination

Effect	Spiggin	Androgenic effects in fish
	EROD (ethoxyreso-rufin-O-deethylase) activity	Enzyme that acts in biotransformation processes
	Lipid peroxidation	The measured compound is malonyl dial-dehyde (MDA), which belongs to the group of thiobarbituric acid reactive substances (TBARS)
	Comet assay	Genotoxicants fragment DNA
	Micronucleus test	Genotoxicants disturb mitotic spindle, so cells end up with many small micronucleus along with a main nucleus
	TUNEL assay	DNA fragments from apoptosis
	Stress protein levels	Heat shock protein amount is increased by toxicants

Some pesticides exhibit estrogen activity. These compounds, due to their physical and chemical properties, have a tendency to bioaccumulate and biomagnify in the food web. Some of the pesticides considered xenobiotic estrogens are (from Bengston):

- DDT and metabolites (o,p'-DDT, p,p'-DDT, o,p'-DDD)
- Methoxychlor (mono- and didemethylated methoxychlor)
- Chlordecone
- Dieldrin
- Alpha and beta Endosulfan
- g-HCH
- Toxaphene

These substances are known to disturb reproductive cycles and development of several species. The documented effects that have been observed include alterations in the development of reproductive system (Gray), sex changes and feminization (Fry and Toone, Davis and Bortone, Gray, Guilette), altered hormone ratios (Fox, Patnode and Curtis) and reproductive failure. Most of these observations come from studies over wildlife, although effects in humans have also been documented. o,p'-DDT has been found to increase uterine weights, as well as Methoxychlor. According to McLachlan (1993), however, an estrogenic compound is hardly identified by its structure. The actual function is what defines it, since the molecules known to bind to estrogen receptors do not share much in terms of chemical structure similarities.

Vitellogenin, a protein produced by nonmammalian vertebrate females and which composes egg yolk, is a useful effect biomarker for estrogenic substances. Production of vitellogenin is stimulated by estrogenic substances, including the pesticides listed above, so increasing levels of it in eggs, as well as its presence in the blood of males, is a clear sign of exposure to such contaminants.

Plants may show physiological alterations as result of pesticide exposure, thus allowing the selection of biomarkers. Table describes the parameters and results of a study with estuarine plants exposed to Atrazine and Metolachlor. The results indicate clear effects at sublethal dosages.

Table 4: Exposure Bioassay conditions (reproduced from Lytle, Bergston)

Pesticide	Atrazine	Metolachlor	Metolachlor	Metolachlor
Levels (ppm)	1, 10, 100	0.1, 1.0, 10	0.01, 0.1, 1	0.01, 0.1, 1
Duration (days)	24	28	16	16
Medium	Estuarine sediment	5 mixtures of sands and clay	3 mixtures of sands and clay	Seawater with Hoagland solution
Method of toxicant addition	Coated quartz sand mixed in sediment	Distilled water atop sediment	Hoagland solution atop sediment	In medium
Culture	Rhizomes	Rhizomes	Rhizome plants	Rhizome plants
Measured responses	Growth (*) Peroxidases Peroxidation Chlorophyll	Growth (*) Chlorophyll	Growth (*) Peroxidases Chlorophyll	Growth (*) Peroxidases Chlorophyll

(*) growth observed in total biomass, shoot biomass, shoot length (mean and total), root length

These two particular herbicides have different modes of action. Atrazine affects photosynthesis by inhibiting electron transport through photosystem II (Moreland, Boger and Sandmann). Metholachlor inhibits protein and lipid synthesis (Jaworski).

It is important to note that many variables are in the way of assessing the response an exposed organism (host) expresses to a certain chemical. These factors are depicted in Figure X. This is called source-chemical-host context. All the possible variations in the routes from the source of exposure to the response must be taken into account when selecting biomarkers of exposure, effect and susceptibility.

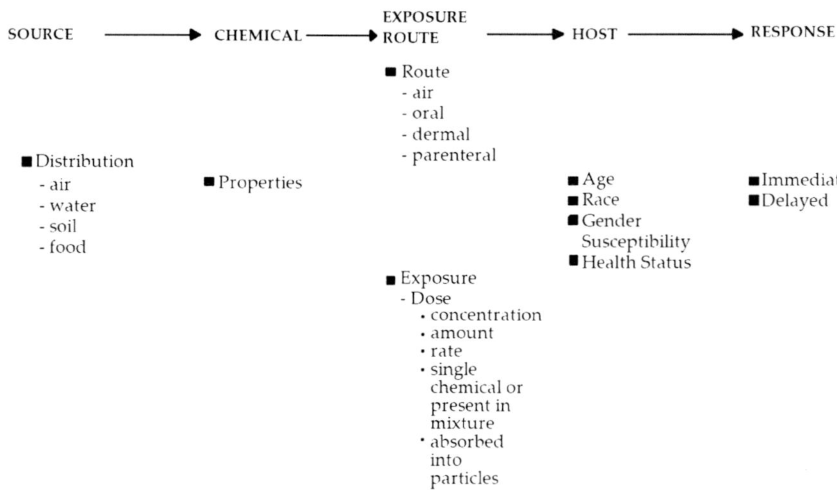

Figure 2: Some critical factors influencing interaction between host and chemical (reproduced from WHO)

The WHO still presents ten contributing factors to be considered in the light of this context:

- the identification and definition of the end-point of interest;
- the assembly of the data base to document the relationship between the chemical exposure, the possible biomarkers and the end-point. This will include data from *in vitro,* mammalian and human studies, with assessment of the validity of data and the study protocols;
- selection of biomarker(s) specific to the outcome of interest with careful consideration of the biomarker to identify what is being quantified, to assess the sensitivity and specificity of the marker in relation to exposure, and the significance with respect to health outcome or pathological change over time;

- consideration of specimens potentially available for analysis, with emphasis on protecting the integrity of the specimen between collection and analysis, and a preference for non-invasive techniques;
- review of the analytical procedures available for quantification of biomarkers and their limitations with respect to detection limit, sensitivity, precision and accuracy;
- establishment of an appropriate analytical protocol with provision for quality assurance and quality control;
- evaluation of intra- and inter-individual variation for non-exposed population;
- analysis of the data base to establish dose-effect and dose-response relationships and their variation, with special emphasis on susceptible individuals;
- calculation or prediction of risk to human health either for the general population or a sub-group;
- review of ethical and social considerations.

BIOINDICATORS

Bioindicator is a term applied to species that are used to gather information about an environment and the exposure to a pollutant, toxicant or contaminant. The characteristics of good bioindicators are:

- The species is common to a variety of environments;
- The species does not emigrate from the system under study;
- The species is tolerant to the toxicants studied;
- Measurement of the toxicant to which the species is exposed is facile.

Commercial availability (for laboratory testing) and short life span are also desirable characteristics on an indicator species.

Plants, molluscs and anemones may be good bioindicators for local contaminations, including bioaccumulation investigation. All possible exposure routes must be evaluated for a proper assessment, including food sources such as sediment. Since it is not possible to test all of the species, experts choose the most suitable and feasible ones carrying in the assumption that the results may be extended and generalized for other species and groups. The selected species in a risk assessment is called surrogate species, as summarized in Table 47.

Table 5: Surrogate species used to determine environmental impact and ecological risks (adapted from Whitford)

Studied system	**Surrogate species**
Waterfowl	Mallard Duck
Amphibian	Bullfrog
Estuarine/Marine fish	Sheepshead Minnow Silverside Minnow
Estuarine/Marine crustacean	Mysid Shrimp
Marine mollusc	Eastern Oyster
Freshwater fish	Cold water: Rainbow trout Warm water: Bluegill sunfish, Fathead Minnow
Freshwater crustacean	Daphnia (water flea)
Aquatic plants	Green algae, Diatoms, Duckweed
Sediment dwelling invertebrates	Fresh water: Amphipod (Hyalella), Midge (Chironomus) Estuarine/Marine: Amphipod (Eohaustorius and Leptocheirus)

Toxicity aspects are investigated in laboratory tests with these species. Comparison of treated groups with control groups allows for the determination of levels like NOEC (No Observed Effects Concentration) and LC50. These parameters in turn, are obtained from dose-response approaches.

In Table below is an example from recent literature of a tiered approach to risk assessment to non-target species. The investigation is carried out in a sequence, in which each stage is more complex than the previous.

Table 6: The three level tests to assess pesticide toxicity no non-target organisms (adapted from Matthews, 2006)

Species		**Tier 1** **Acute toxicity**	**Tier 2** **Reproduction test**	**Tier 3** **Field test**
Birds	Bobwhite quail or mallard ducks	LD50 (8-14 days)		
Freshwater fish	Rainbow trout or minnows	LC50 (96 h)	Effects on spawning	Fish life cycle study
Aquatic invertebrate	Daphnia, shrimp	LC50 (48 h)	Full life cycle	
Non-target invertebrate	Honey bee	LD50 (48 h)	Effects of residues on foliage	Pollination field test
	earthworms	LC50 (14 days)	Effects of residues on foliage	
Aquatic plants	algae	LC50 (96 h)	Plant vigour	
Other beneficial species		LD50 (48 h)		

RISK ASSESSMENT

The important considerations for assessment of risk are the dose rate, route, duration and frequency of exposure. These components are outlined by the WHO as:

- *hazard identification:* to confirm that the chemical is capable, subject to appropriate circumstances, of causing an adverse effect in humans;
- *dose-response assessment:* to establish the quantitative relationship between dose and effect in humans;
- *exposure assessment:* to identify and define the exposures that occur, or are anticipated to occur, in human populations.

Ecological Risk Assessment by the EPA

The definition of ecological risk assessment established by the United States Environmental Protection Agency is *"the process for evaluating how likely it is that the environment may be impacted as a result of exposure to one or more environmental stressors."* Chemicals derived of anthropogenic activity, i.e., industrial waste or pesticides used in agricultural lands, are considered some of those environmental stressors.

The EPA divides a risk assessment in 3 phases:

- Problem Formulation
- Analysis
- Risk Characterization

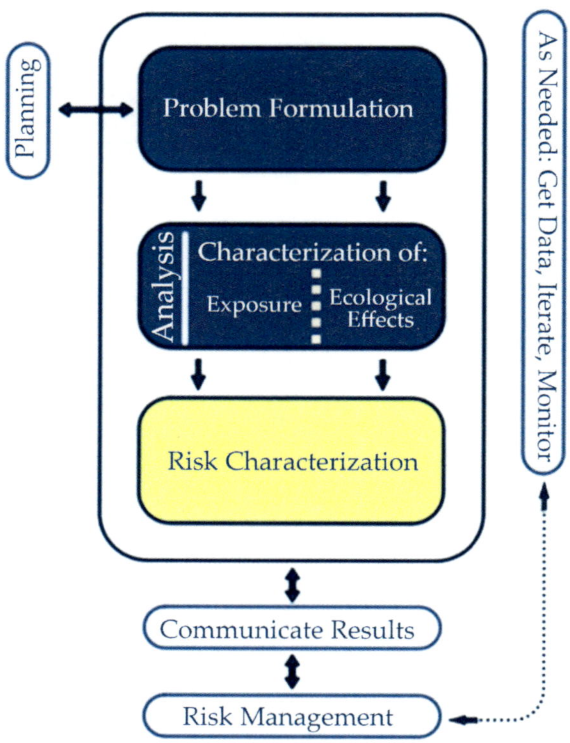

Figure 3: Risk assessment phases - EPA approach (reproduced from EPA)

Even before putting the risk assessment into practice, there is a **planning** step that may be considered a "phase 0." Basically, during the planning phase, several information are gathered in the style of an overview: agents will identify populations at risk, their susceptibility to different exposures, the actual hazards of concern, the sources of such hazards, exposure pathways (air, water, food etc), routes (ingestion, inhalation, contact etc). More information is sought at complex biological level, such as metabolism and excretion and impacts in individuals, exposure time (acute, subchronic, chronic or intermittent), and greater ecological consequences, such as mortality or behavioural alterations.

Problem formulation - an entity (or system) must be determined to be investigated. It can be a particular species, a community, or even an ecosystem. The particular properties or attributes of such system that is effectively at risk must be identified. Then, an assessment of the ecological relevance is done considering:

- Nature and intensity of effects
- Spatial and temporal scales of effects
- Potential for recovery
- Level of organization potentially affected
- The entity's role in the ecosystem

The problem of pesticide contamination is a case of ecological relevance, and the systems selected in the past decade had to be assessed with the proper scientific rigour. It is a subject that now has a major public acceptance than other environmental issues, and that is due to a diligent work done by scientists, regulators and risk managers, as well as food producers.

Analysis - the analysis phase is where data is transformed into objective, accessible information for a posteriori management. Exposure is subject to calculations such as the hazard quotients - the ratio of a pesticide level to a certain threshold. Several parameters are taken into account in that phase, as the variability from system to system is subject to a number of factors like:

- Food ingestion rate
- Bioaccumulation rate
- Bioavailability

- Life stage

Determining toxic levels of a pesticide may be done by means of a sequential process, called the tiered approach. Each tier pushes the limits of intoxication a bit further, until an end point is reached. Generally speaking:

- Tier 1 - Comprises primary acute toxicity studies done in laboratory. The objective is to determine levels that are lethal or present acute sublethal effects on a number of individuals. Ex: LC50 for fish and invertebrates.
- Tier 2 - Longer term studies, involving life cycles, reproduction and factors that impact in a species during its whole lifespan. This is one of the reasons why good bioindicator species should have a short lifespan. Ex: aquatic plant growth, full life cycles of fish, invertebrates.
- Tier 3 - Testing for the pesticide's impact on a larger scale, such as a simulated environment, or actual field condition investigation. Ex: ecosystemic alterations due to pesticide exposure, predator-prey relationships.

Risk Characterization - the outputs of the analysis phase are used to estimate risk. The information comes from parallel approaches, like field observational survey, models, exposure and effects data.

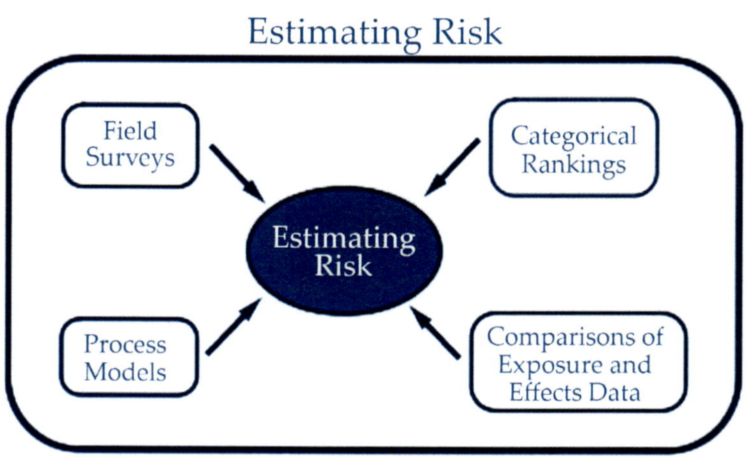

Figure 4: Approaches for ultimately estimating risk (reproduced from EPA)

The studies that a regulatory body such as the EPA requires for pesticide registration are a rich source of information for risk assessment. Some of that data are:

- The effects of said pesticide in wildlife and aquatic organisms;
- The effects of the pesticide on different plants;
- The effects on non-target species;
- The environmental fate, or what happens to the pesticide when in contact with soil, water and air;
- The half-life of the pesticide in the environment;
- The level of drift when the pesticide is sprayed.

An estimation of ecological risk includes answers to questions like:

- Is the risk acute or chronic?
- What is the severity of effects?
- What is the time over which they occur?
- Is the risk to one species or many species?
- How many organisms are at risk?

EPA issued a case study for the impact of endosulfan in benthic (sediment) invertebrates. Figure X illustrates the complete model in a visual manner, and even though this is a simplified system, it becomes obvious the large number of variables to be considered.

After application, there are at least five main routes for the pesticide to reach aquatic ecosystems: wind transport of droplets of the active ingredient from the application site to other areas; direct deposition in waterbodies at the vicinities of the application site or field; spray drift directly inserting it in waterbodies; inflow of contaminated run-off; penetration in the soil until it reaches groundwater. Endosulfan then adsorbs to sediment (soil particles and organic matter), where it will be exposed to invertebrate species. From then on, it will bioaccumulate, biomagnify and climb up the food web. The goal of the model is to address each of the ecologically relevant steps in order to objectively define a risk level for this pesticide.

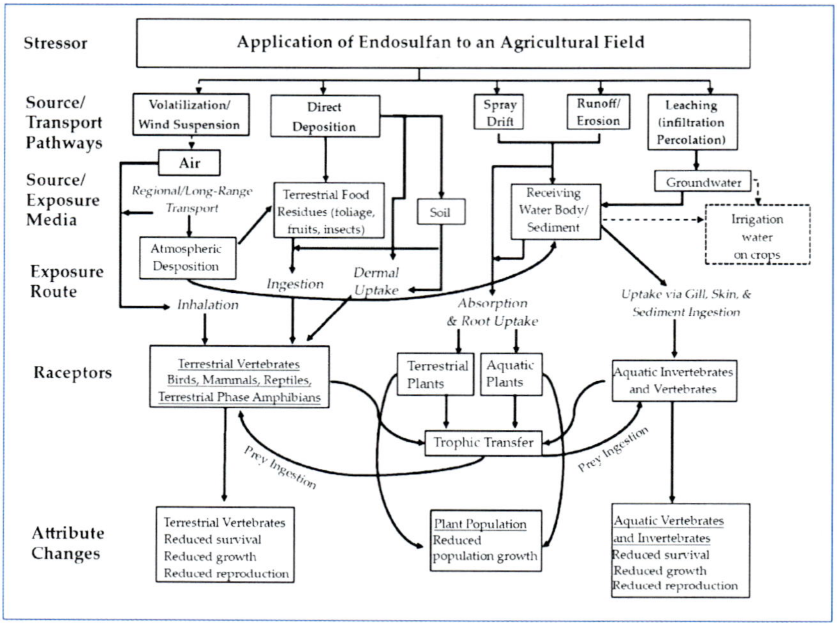

Figure 5: Conceptual model for the agricultural use of endosulfan (reproduced from EPA 2013)

Figure X also illustrates the toxicity responses that are sought in this assessment scheme. Acute, subchronic and chronic exposure tests, lethal and sublethal effects, are the responses obtained for a risk characterization.

Table 2. Assessment endpoints, measures of effect, and measures of exposure used for assessing risk to pelagic and benthic aquatic invertebrates

Taxa	Assessment Endpoint	Type of RQ	Measures of Ecological Effect[1]	Measures of Exposure[2]
Freshwater invertebrates **(pelagic)**	Survival	Acute	Acute (48-96 h) LC_{50} or EC_{50}	Peak water column EEC
	Survival, growth, reproduction and development	Chronic	Chronic (21-d) NOAEC	21-day average water column EEC
Freshwater invertebrates **(benthic)**	Survival	Acute	Acute (48-96 h) LC_{50} or EC_{50} (from water column test)	Peak EEC for pore water
	Survival, growth, reproduction and development	Chronic	Subchronic (10-d) or chronic (28-65-d) NOAEC	21-day average EEC for porewater and/or bulk sediment
Estuarine/marine invertebrates **(pelagic)**	Survival	Acute	Acute (48-96 h) LC_{50} or EC_{50}	Peak water column EEC
	Survival, growth, reproduction and development	Chronic	Chronic (28-d) NOAEC	21-day average water column EEC
Estuarine/marine invertebrates **(benthic)**	Survival	Acute	Acute (48-96 h) LC_{50} or EC_{50} (from water column test)	Peak EEC for pore water
	Survival, growth, reproduction and development	Chronic	Subchronic (10-d) or chronic (28-d) NOAEC	21-day average EEC for porewater and/or bulk sediment

Figure 6: Assessment endopoints, measures of effect and exposure used for pelagic and benthic aquatic invertebrates (reproduced from EPA)

THE TIERED APPROACH TO RISK ASSESSMENT - THE ATRAZINE CASE

A risk assessment for atrazine in water in the US based on a tiered strategy was an official recommendation of ECOFRAM (the Ecological Committee on FIFRA Risk Assessment Methods). In this scheme, a sequence of tiers is established, each with a different set of methods and tools, gradually becoming more complex and complete. In other words, the realism of the modeling in each tier is higher than the previous - the first tier is the one that carries most assumptions and conservative views. Each tier has its own problem formulation, and the data output obtained leads to a decision whether there is enough certainty of the risks associated to this particular hazard, or more investigation is necessary.

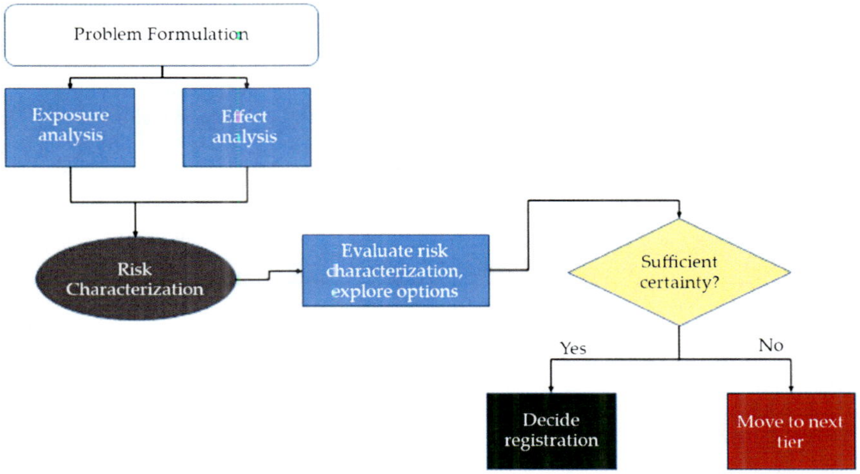

Figure 7: Risk evaluation process applied in each tier of the risk assessment, following recommendations of ECOFRAM (adapted from Giddings J et al., 2005)

A summary of the findings and results of the tiered risk assessment conducted for atrazine in water was published by SETAC (Society of Environmental Toxicology and Chemistry). We present the highlights of the results reported, in a way to help illustrate how the tiered strategy works.

Tier 1	
Problem Formulation:	For each major use pattern of atrazine, could the concentrations in water (instantaneous or time averaged) exceed toxic levels to aquatic plants and animals? Consider an extreme high exposure scenario and highly sensitive receptor species.
Effect Analysis:	Acute toxicity values (LC50 and EC50), Chronic toxicity values (NOECs) with most sensitive animal and plant set.
Exposure analysis:	Generic EECs for standard USEPA pond scenario; Corn, sorghum, turf, conifer, guava, macadamia nut, roadside uses; Environmental fate data (soil Kd, soil metabolism, aquatic metabolism)
Risk characterization:	Acute and chronic hazard quotients (HQ) for each use pattern based on peak EEC, LC50 and NOEC for the most sensitive animal and plant species
Variability and uncertainties considered:	The only exposure variability considered was among different use patterns. Spatial and temporal variability neglected. Conservative estimate of chemical persistence. Predictive ability of models based on limited validation. Variability among species was addressed by use of the most sensitive species in data set. Levels of Concern (LOCs) for hazard quotients account for some uncertainty.
Conclusion:	Possible acute and chronic hazards to aquatic plants for all modeled use patterns. No acute hazard to animals but possible chronic hazards for sugarcane and some minor uses.
Tier 2	
Problem Formulation:	For atrazine use on corn and sorghum, in what percentage of application sites or years will the annual maximum concentration of atrazine exceed the levels that cause X% effect (acute or chronic) on sensitive aquatic plants and animals? Consider a standard high exposure scenario, and up to 25 sites over 36 years.
Effect Analysis:	Acute and chronic toxicity concentration-effect relationships, for most sensitive plant and animal species
Exposure analysis:	Concentration distributions using models (PRZM/EXAMS) for ponds, annual maxima; 11 regions, up to 25 sites per region, 36 years at each site; corn, sorghum uses; Environmental fate data: Koc, aerobic soil metabolism, aerobic and anaerobic aquatic metabolism.

Risk characterization:	Acute risk curves relating mortality (%, animals) or growth inhibition (%, plants) for most sensitive species, to area-weighted probability of occurrence in each region. Chronic risk curves relating inhibition of growth or reproduction for most sensitive species, to area-weighted probability of occurrence in each region.
Variability and uncertainties considered:	Variability in exposure concentrations considered geographical location, rainfall, and use pattern. Uncertainty of exposure model input values considered in selection of conservative parameters from base data set. For concentration-response curve, a variability in individual sensitivity is considered.
Conclusion:	Low likelihood of acute or chronic effects on animals. Possible risk of acute or chronic effects on plants.

Tier 3	
Problem Formulation:	For atrazine use on corn, sorghum and sugarcane, in what percentage of sites or years will the concentration of atrazine exceed the levels that cause 50% reduction (acute) or chronic effects in animals and plants growth and survival? Consider standard high exposure, and up to 25 sites over 36 years.
Effect Analysis:	Species sensitivity distributions for acute toxicity (LC50 or EC50) to 37 plant species, 27 invertebrate species, and 19 fish species, and for chronic toxicity (NOEC) to 21 plant species and 15 animal species. Results for most sensitive life stages.
Exposure analysis:	Same as Tier 2, plus: Concentration distributions (model PRZM/EXAMS) for ponds, sugarcane uses in 10 sites and 36 years, for flowing water, corn and sorghum in 6 regions, 1 or 3 sites in each region and 19 to 36 years. Typical and maximum atrazine use rates.
Risk characterization:	Risk curves relating percentage of affected species (plant and animal, acute and chronic) to area-weighted probability of occurrence in each region; ponds (corn, sorghum, and sugarcane uses) and streams (corn and sorghum).
Variability and uncertainties considered:	Variability in exposure concentrations in ponds and flowing water considered geographical location, rainfall and application rates and patterns for corn, sorghum, sugarcane. Uncertainty in exposure models by validation of 3 runoff studies and 1 watershed monitoring study. Variability of sensitivity among species considered in distributions.
Conclusion:	Little or no risk to animals under all scenarios examined. Acute risk to plants is low except in streams in areas of greatest runoff potential. Possible chronic risk to plants in ponds.

Tier 4 (was composed of six problem formulation with respective parameters, methods, and variabilities. Only the higher level problem was taken for illustration purposes).	
Problem Formulation:	What are the relative contributions of atrazine, its degradations products, and other triazines to the toxicity of surface waters?
Effect Analysis:	Comparative toxicity data for atrazine, degradation products and other triazines.
Exposure analysis:	Measured concentrations of atrazine, degradation products and other triazines in surface waters.
Risk characterization:	Contribution of atrazine metabolites and other triazines to ambient toxicity.
Variability and uncertainties considered:	Variability in coincidental exposures to other triazines considered. Variability in contribution of total triazine exposure to effects considered. Greater uncertainty of effects of other triazines as a result of smaller data sets than atrazine but some uncertainty addressed through extrapolation of sensitive species distribution below observed data.
Conclusion:	Monitoring data indicated that exposure models generally overestimated atrazine concentrations. Whole aquatic communities (in mesocosms and microcosms) are less sensitive to atrazine than the most sensitive species of plants and animals. Risk is extremely low for animals and is low for plants except in certain high-exposure situations (high use, high runoff areas). Effects in the high exposure areas are likely to be transient, and quick recovery of the ecological system is generally expected.

Although valuable for this case of risk assessment, and following a logic and reasonable decision tree, the group recognized how artificial that approach might become, so carefulness is advised.

MODELS

A great part of risk assessment and analysis of exposure is done with the help of models. These models are complex tools that allow one to estimate, for instance, pesticide residue levels in food, in water, and in non-target organisms. The Aquatic models established by the EPA are listed below, and detailed information can be found at EPA's website https://www.epa.gov/pesticide-science-and-assessing-pesticide-risks/models-pesticide-risk-assessment

The amount of data input required for these models can be pretty extensive. Whitford et al. (2002) assembled a very comprehensive list of some of the more than 100 parameters that are considered by PRZM in order to arrive at a pesticide concentration in water:

• Snow melt factor (cm/oC) • Average storm duration (hours) • Monthly daylight hours • Minimum evaporation extraction depth (cm) • Total hectares • Soil erodibility • Topography • Canopy area coverage (%) • Canopy height at maturation (cm) • Soil loss cover management • Number of pesticides • Chemical application method • Number of pesticide applications • Target application day, month, year • Pesticide application rate (kg/ha) • Application efficiency • Spray drift fraction • Plant uptake factor • Core depth (cm) • Percent organic carbon	• Vaporization for pesticide (kcal/mol) • Diffusion coefficient for pesticide (cm2/day) • Henry's Law constant for pesticide • Pesticide solubility • Pesticide partition coefficient • Pesticide solution dispersion coefficient • Number of soil horizons • Horizon thickness (cm) • Horizon bulk density (g/cm3) • Horizon wilting point • Soil water content in horizon • Horizon field capacity • Dissolved phase decay rate (day) • Absorbed phase decay rate (day) • Vapor phase decay rate (day) • Number of crops • Crop interception storage (cm) • Crop rooting depth (cm) • Crop surface condition after harvest • Crop dry weight at full canopy (kg/m2) • Day, month, year of crop emergence • Day, month, year of crop maturation • Number of cropping periods

Below are the models currently in use by the USEPA and which are available to the public on their website.

- Pesticide in Water Calculator (PWC)

Calculates concentration of pesticides in surface and groundwater as a consequence of application in land.

- KABAM (**K**OW (based) **A**quatic **B**io**A**ccumulation **M**odel)

Estimates pesticide bioaccumulation in aquatic organisms, especially organic lipophilic ones, as well as potential risks to mammals and birds that feed on such organisms.

- Pesticide in Flooded Applications Model (PFAM)

Estimates pesticide exposure in flooded agricultural areas, such as rice paddies and cranberry bogs.

- Pesticide Root Zone Model (PRZM)

Calculates the fate of a pesticide applied to a field in a daily basis. Evapotranspiration of plants, climatic conditions, pesticide application practices and other parameters are the inputs to estimate the amount of pesticide that was effectively taken up by plants, and what ended up leaching through soil, lost by volatilization to atmosphere or in run-off.

7
CHAPTER

LABORATORY METHODS

Erick Soares Lins

The analysis of environmental samples for monitoring or assessment purposes can pose an analytical challenge. At a certain scope, we are talking about a large number of pesticides to determine, in matrices that differ from one another quite substantially, their physical and chemical properties. The analytical approach applied to that problem must bear the following:

- Sample preparation
- Separation
- Detection/Quantification
- Quality control and quality assurance

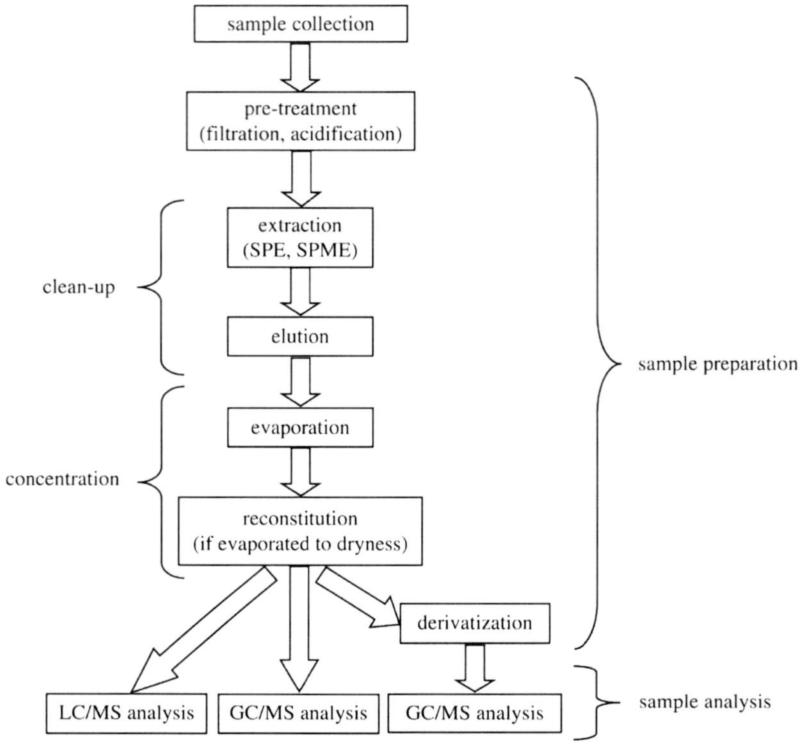

Figure 1: Overview of the main steps in sample preparation and analysis (reproduced from Comerton et al., 2009)

The purpose of the various sample preparation methods is to eliminate interferences, i.e., anything in the matrix that may be problematic to the analyte detection. When necessary, convert the analyte to a product that is more amenable to detection (derivatization), or concentrate it in the actual analytical sample. Like all other steps in the analytical procedure, it must be reproducible, robust, and applicable to varied matrix compositions. The idea is to minimize uncertainty as much as possible, since certain systems may be highly heterogeneous, and whatever result one gets in the end of the analysis, must be representative of the sample universe or population. That is a real challenge.

The steps in sample preparation may be divided in:

- Pre-treatment

Environmental samples usually contain reasonable amounts of natural organic matter, which may impair the detection of the analytes of interest. Some procedures in the pre-treatment phase include filtering with 0.2 to 0.5mm membranes, pH adjust to help in the following extraction steps. For water samples, it is important to keep them in amber flasks, as the analytes might be photosensitive, or even componentsof the sample matrix may photodegrade giving further byproducts which may also become of concern in the quality of analysis. Preservatives such as formic acid and formaldehyde may be added to the sample to help maintain the integrity of the sample against microbial activity.

- Figura
- Figure X - sample pre-treatment
- Clean Up

The most popular extraction method for multi-residue analyses in recent years is undoubtedly solid phase extraction (SPE). This extraction is very practical, as the SPE is commercially available in cartridges. The SPE cartridge is composed of a stationary phase, which is the sorbent and has to be activated before using in the real sample. The sample is percolated through the sorbent, where the analytes are retained by different methods, depending on the physical interaction of the functional groups and the molecules of interest. The analytes are then eluted with an organic solvent for the determination.

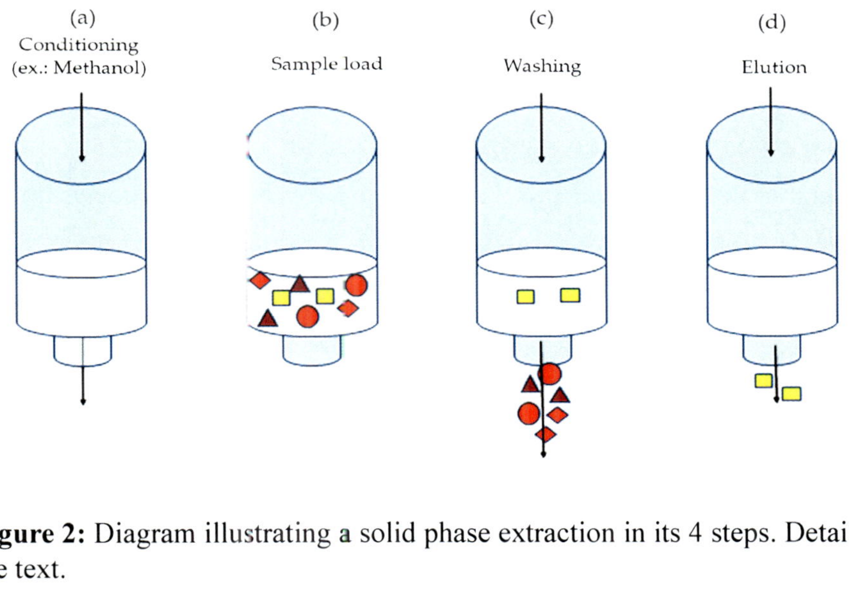

Figure 2: Diagram illustrating a solid phase extraction in its 4 steps. Details in the text.

The scheme presented in Figure 101 is as follows: First the stationary phase is conditioned with the appropriate organic solvent (a). The species of interest, along with other matrix components, is retained on the stationary phase (b). Those interferent species are washed out (c), and another solvent is used to elute the analytes (d).

SPE is widely used as an offline extraction step, but it can also be used online, i.e., coupled to the chromatographic system that will ultimately detect and quantify the analyte. Advantages of online SPE include: minimized sample manipulation, automation, increased sample throughput, ease of operation and cost efficiency.

Types of sorbent include:

Reversed phase: works by hydrophobic or nonpolar interactions

Polymeric ends on solid phase

Ex: Octadecyl (C18) and octyl (C8) bonded silica

Interactions: Van der Waals or dispersion forces

Normal phase: works by hydrophilic or polar interactions

Polar functional groups on solid phase

Ex: Diol bonded silica

Interactions: Hydrogen-bonding, dipole-dipole, dipole-induced dipole, pi-pi

Ionic exchange: works by electrostatic interaction

Cation or anion as counterions in silica

Ex: quaternary ammonium, carboxylic acid

Interactions: charged groups on the molecule with opposite charge group

Adsorption: materials with no surface modification to which the analytes adsorb when in contact. Ex: Florisil, carbon, alumina.

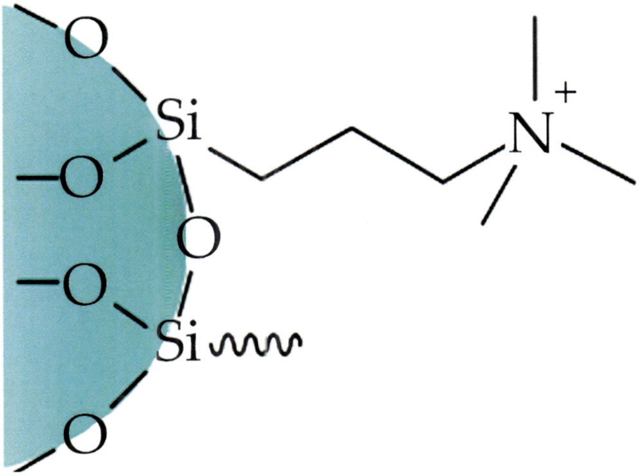

Figure 3: Representation of a Silica surface modified with quaternary ammonium for ion exchange extraction. (Reproduced from http://www.star-labsci.com/spe-cartridge/ion-exchangers/sax-spe-cartridge.html)

- Concentration and Derivatization

After elution from the extraction phase, the sample aliquot may be submitted to a gentle stream of nitrogen gas for some time to allow it to concentrate. We are talking about extracts of up to 500uL, from which portions will be injected into the chromatographic system, either gas (GC) or liquid (LC). A change in solvents may be carried out at this point, if the one used for elution is not suitable for analysis. Basically, the polarity of the pesticides will determine that. In broad-scope multiresidue methods, it is ideal to use organic solvents that are able to dissolve most of the pesticide molecules, in their varied polarity range, thus the mixtures Methanol/Water or Acetonitrile/Water.

Derivatization is the process of promoting changes in the molecular structure of the analyte in such a way that the analysis is facilitated, or the signal is enhanced. In practical terms, derivatization may be used to decrease the polarity of certain pesticides, making them more amenable to GC, i.e., broadening the scope of gas chromatography.

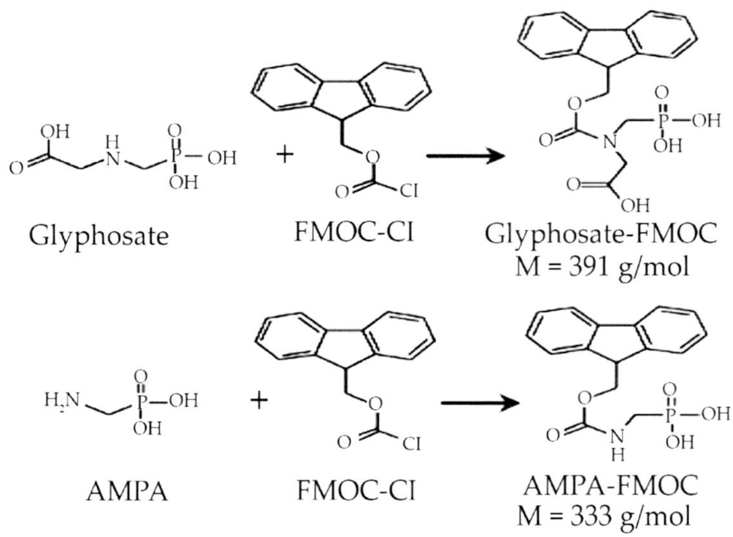

Figure 4: Use of 9-fluorenylmethyl chloroformate (FMOC-Cl) for the derivatization of Glyphosate (Reproduced from Gerstel Inc.)

Derivatization, although practically helpful, is something that naturally adds one more uncertainty factor to the analytical process. In order to address that, regulatory bodies may issue performance and reproducibility requirements for derivatization methods. The European Commission states the following guidelines governing the use of derivatizations in monitoring analysis:

Where a derivatisation method is used, this must be fully reported and justified.

The derivative must be stable and must be formed reproducibly.

Where quantification is based on the determination of a derivative, the calibration should preferably be carried out using standard solutions of that derivative, unless the derivatisation step is an on-line part of the detection system.

The mean yield and precision of the derivatisation step must be demonstrated, where possible.

Sample analysis, for a multi residue strategy, requires a separation performed by chromatographic methods.

- Chromatography

Chromatography is a separation technique that was born in the first decade of the 20th century, and has undergone amazing improvements through the decades. It uses the properties and interactions of the substances of interest to help separate them from a mixture. A sample is transported through a column (the stationary phase), being either dissolved in a solvent or vaporized to a gas (mobile phase).

Depending on the materials used for the stationary phase and the conditioning applied, some molecules may travel faster through the pores of the column than others. Polar and nonpolar interactions, as well as size exclusion, are the driving forces in the process of separating a complex mixture. Different substances will travel at different speeds and reach the detector at different times, and given constant conditions (column material, eluent, pressure, temperature and other setup) this becomes a direct parameters for the identity of the compound (the retention time).

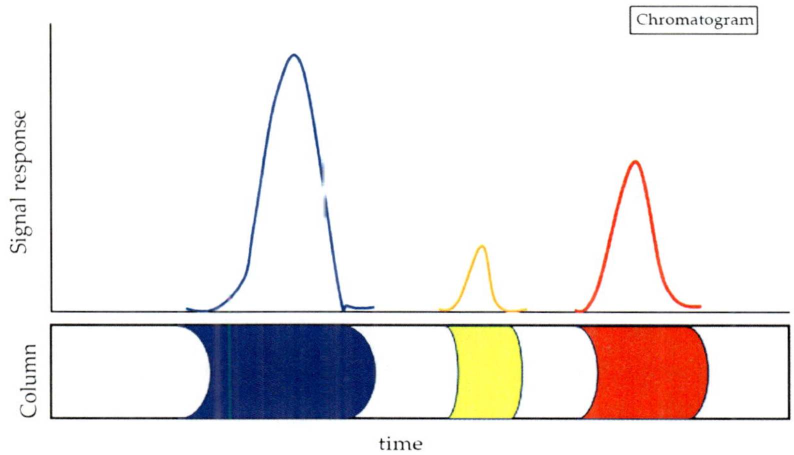

Figure 5: Schematic representation of the separation of a mixture in a chromatographic column and the signal response that is registered over time.

For pesticide analysis, two types of chromatography became the most useful in recent years, as they turned out be complementary to each other when it comes to broadening the scope of analytes. They are Gas Chromatography (GC) and Liquid Chromatography (LC).

In GC, the eluent that carries the vaporized sample through the column is an inert gas, usually Helium, because it is necessary to make sure that no side reaction would occur during the analysis. The column is a long capillary column, up to a few meters long and with an internal diameter on the order of tenths of a millimeter.

In LC, the eluent is a liquid, in which the analytes are dissolved. In modern LC, a high pressure is applied during the elution, hence the acronym HPLC - High Pressure Liquid Chromatography, sometimes also referred to as High Performance LC. Pressures of up to 350 bar are applied in HPLC. The stationary phase is a porous monolithic column, to which the stationary phase is bound.

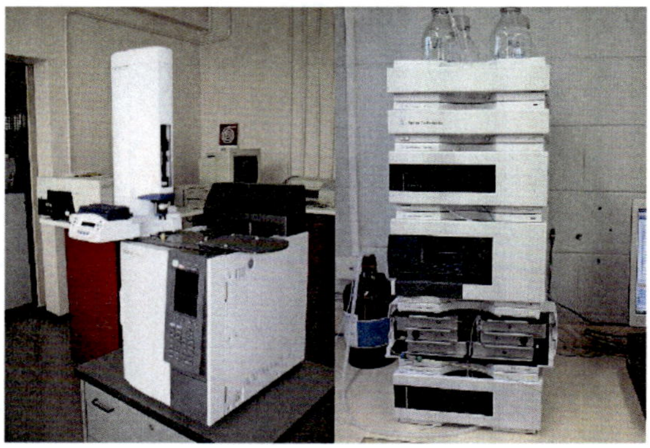

Figure 6: A modern GC (left) and a HPLC (right) instruments

There are many parameters to adjusted and fine tune in a chromatographic run that will not be detailed in this book, but which can be obtained in regular instrumental analysis materials. Also, several other types of chromatography exist and are commercially available, such as Size Exclusion, Ion Exchange, Supercritical Fluid etc., each with their own analytical applicabilities.

Every modern chromatographic system has a detector, usually UV-Vis and fluorescence, but also FPD (flame photometric detector) or ECD (electron capture detector) may be used for the determination of pesticide residues in environmental and food samples. Nonetheless, coupling mass spectrometry detector to those systems, which is also referred as hyphenated techniques, added one more dimension to the identification and confirmation of the analytes. Unlike other techniques, this is not restricted to the analytes having chromophoric moieties in their structures, for instance, which again increases the scope.

The concept of mass spectrometry consist on i) ionizing the molecules by various means, which results in charged molecules, and then ii) using the mass to charge ratios to arrive at specific values for molecules or its fragments. A mass spectrometer is composed of a source, where the ionization will be induced, a quadrupole, where species with different mass to charge ratios will travel in an electric field and where fragmentation is usually induced, and a detector. Ions and fragments that reach the detector are registered in a mass spectrum, and the values of m/z are used to elucidate or "reconstruct" an unknown molecule from comparison with a library of spectra.

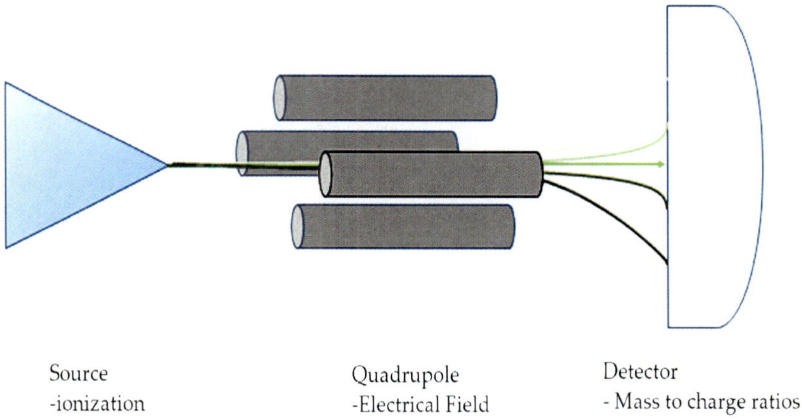

Figure 7: Schematic depiction of a single quadrupole mass spectrometer.

There are several types of ionization, being the most common for pesticide analysis: EI - Electron Ionization, CI - Chemical Ionization, APCI - Atmospheric Pressure Chemical Ionization and ESI - Electrospray Ionization. In ionization methods considered harder, such as EI, some fragmentation might already occur. Others considered softer, such as APCI, mostly only ionize maintaining the integrity of the analyte.

Thus in hyphenated techniques such as LC-MS and GC-MS, each peak representing a pesticide that was already chromatographically separated from the rest, can now provide a very specific mass spectrum, as illustrated by Figure 107.

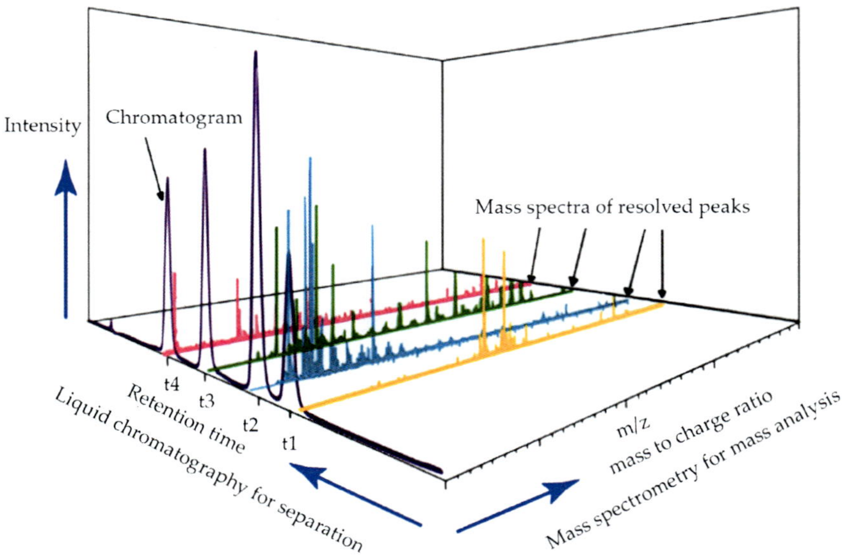

Figure 8: Substances separated in the chromatography step are individually analyzed for mass to charge ratio, enhancing the identification certainty. (credit Daniel Norena-Caro)

The USGS issued a report with an analytical method for the determination of 119 pesticides in sediment, in two ranges of organic matter. The table below presents the main information considered for compound identification in chromatography (the retention time) and mass spectrometry (the molecular ions, fragments or adducts in mass-to-charge ratio).

Table 1: Retention times, quantitation ions and confirmation ions for pesticides analyzed by GC-MS (Reproduced from Hladik M. and McWayne M., 2012 https://pubs.usgs.gov/tm/tm5c3/pdf/tm5-C3.pdf)

Herbicides/Insecticides	Retention time (minute)	Quantitation ion (m/z)	Confirmation ions (m/z)
EPTC	9.2	128	189,86
3,5-Dichloroaniline	10.4	161	163
Butylate	11.0	146	156
3,4-Dichloroaniline	11.1	161	163
Pebulate	11.9	128	161
Molinate	14.1	126	187,83
Cycloate	17.1	83	154
Ethalfluralin	17.8	276	333,316,292
Trifluralin	18.4	306	316,264
Benfluralin (Benefin)	18.5	292	276,264
2-Chloro-2,6-Diethylacetanilide	19.3	176	225,147
Pentachloroanisole	19.5	265	280,237
Fenpyroximate	20.4	213	198,142
Carbofuran	20.5	164	149

Herbicides/Insecticides	Retention time (minute)	Quantitation ion (m/z)	Confirmation ions (m/z)
Simazine	20.6	201	186,173
Clomazone	20.9	125	204
Atrazine	21.0	200	215,173
Prometon	21.1	210	225,125
Propyzamide	21.9	173	255,175
Pronamide (Propyzamide)	22.0	173	255,240,145
Diazinon	22.3	179	304,137
Triallate	23.1	169	268,142,128
Tefluthrin	23.4	177	197,141
Tebupirimfos	23.7	261	318,234,152
Propanil	24.9	161	217,163
Methyl parathion	25.2	125	263,109
Alachlor	25.3	160	188,146
Carbaryl	25.6	144	115
Fipronil desulfinyl	25.6	388	390,333

Herbicides/Insecticides	Retention time (minute)	Quantitation ion (m/z)	Confirmation ions (m/z)
Dithiopyr	26.4	354	306,286
Prometryn	26.4	241	226,184
Prodiamine	27.3	321	333,279
Metolachlor	27.6	162	238
Thiazopyr	27.8	60	363,327,306
Chlorpyrifos	27.9	197	314,199
Malathion	27.9	125	173
Thiobencarb	27.9	100	125
DCPA (Dacthal)	28.0	301	332,229
Fenthion	28.2	278	169,125,109
Flufenacet	28.7	151	211,136,123
Butralin	29.1	266	295,224
Pendimethalin	29.9	252	NA
Fipronil sulfide	30.0	351	420,255
Fipronil	30.6	367	369,351
Allethrin	31.2	123	136
Methidathion	31.6	145	85
Methoprene	32.0	73	111
Flumetralin	32.6	143	404,157
Napropamide	33.1	72	128,100
p,p'-DDE	34.2	256	318,316,248
Oxadiazon	34.8	175	344,302,258
Fipronil desulfinyl amide	34.9	406	390,308
Fipronil sulfone	34.9	383	452,255,213
Oxyflurofen	35.4	252	300
p,p'-DDD	37.8	235	237,165
p,p'-DDT	40.9	235	237,165
Hexazinone	41.4	171	NA
Propargite	42.7	135	350,201,173
Piperonyl butoxide	43.4	176	177
Resemethrin	43.7	123	171,143
Phosmet	44.2	160	133,93
Azinphos-methyl	44.2	77	160,132
Bifenthrin	45.0	181	123
Tetramethrin	45.1	164	123
Tetradifon	45.7	159	356,227,111
Phenothrin	46.2	123	183.0
Cyhalofop-butyl	46.8	256	357,229,120
Cyhalothrin	47.1	181	208,197
Fenpropathrin	47.1	181	208
Permethrin	48.3	183	181,163
Coumaphos	48.5	362	226,210,109
Pyridaben	48.5	147	309,132,117
Cyfluthrin	49.2	206	226,165,163
Cypermethrin	49.6	181	208.163
Etofenprox	50.0	163	NA
tau-Fluvalinate	50.9	250	253,181
Esfenvalerate	51.0	125	227,181,167
Deltamethrin	51.6	181	253

Fungicides	Retention time (minute)	Quantitation ion (m/z)	Confirmation ions (m/z)
Novaluron	6.5	168	335,140
Pentachloronitrobenzene	12.7	237	295,214
Tebupirimfos oxon*	13.0	218	302,260,245
Pyrimethanil	13.2	198	199
Chlorothalonil	13.3	266	264
Vinclozolin	14.5	212	198,187
Metalaxyl	14.9	206	220,160,132
Triadimefon	16.2	57	208,128
Tetraconazole	16.2	336	338,101
Fluazinam	16.9	418	372,337
Cyprodinil	16.9	224	225
Captan	17.5	79	149,117
Triadimenol	17.5	112	168,128
Triflumizole	17.6	278	206,179
Flutriafol	18.6	123	219,164
Imazalil	18.7	215	217,173
Flutolanil	18.8	173	323,281,145
Fludioxinil	18.8	248	154,127
Tribuphos*	19.3	169	258,202
Myclobutanil	19.3	179	206,150
Flusilazole	19.3	233	205
Kresoxim-methyl	19.5	116	206,131
Cyproconazole	19.9	222	139,125
Trifloxystrobin	22.0	116	222,131
Fenhexamide	22.2	97	177
Propiconazole	22.2	173	259,175
Tebuconazole	22.8	125	250,127
Zoxamide	24.0	187	258,189
Iprodione	24.0	314	316,187
Metconazole	25.0	125	250
Triticonazole	25.6	235	237,217
Fenarimol	26.8	139	251,107
Fenbuconazole	29.6	129	198
Boscalid	30.5	140	342,112
Pyraclostrobin	32.9	132	164
Difenoconazole	33.8	323	267,265
Indoxacarb*	34.3	150	264,218,203
Azoxystrobin	34.9	344	387,372
Famoxadone	35.3	330	197
Dimethomorph	36.0	301	387,265
Fluoxastrobin	37.9	188	219

Table X below is very comprehensive collection of the advantages and disadvantages of each of the most common analytical technique regarding pesticide determination, put together by Andreu and Pico (2004)

Table.2: Comparative study of different analytical technics to determine pesticides residues

Extraction technique	Characteristics	Advantages	Disadvantages
BE by Sondes	Time: 3-113 h Sample: 1-30 g	• Sample fresh solvent contact during the whole extraction step • No filtration required after extraction	• Long emraction time • Large consumption of solvent
Discontinuous LSE	Extractant 100-500 in! Only organic solvents Timm 10 min-10 h Sample: 1-30 g Dammam 30-2W ml Organic solvents, water and mixtures Ultrasonic or mechanical shaking	• Well-known procedure • Large experience in the extraction field for more than a century • Multiple extractions	• Degradation of some pesticides, such as N-methylcarbamates • Clean-up required • Filtration or centrifugation required • Large consumption of solvent • Clean-up required
MAE	Time 3-30 min Sample: 1-10 g Dammam 10-40 ml Organic solvents, water and ur	• Performed at room tempera ture to avoid degradation of some pesticides • Rapid extraction • Low solvent volumes • Extract multiple samples simultarreously	• Extraction solvent must be able to absorb microwams • Clean-up required • Degradation of some pesticides
MASE	Time 10-60 min Sample: 1-30 g Extractant: 10-150 ml Only organic solvent	• Fast extraction • Low solvent volume	• Extraction solvent must be able to absorb microwaves • Clean-up required • Degradation of some pesticides
PLE	Time 5-60 min Sample 0.4-30 g Extractant: 1-50 ml Organic solvent, water and mixtures	• Rapid extraction • Low solvent volumes • No filtration required • Automated system • On-line coupling to LC/MS by a mrbent trap when water is used as extractanq	• Clean-up required • Degradation of some pesticides • When it is coupled on-line the sorbent nap has a short life (so it should be replaced after 7-8 hi

Other instruments and detectors that have gained importance in recent years are: Triple quadrupole, Orbitrap, Ion trap, Time-of-Flight (TOF) and Direct Analysis in Real Time (DART), and soon enough these will be more commonly found in routine laboratories worldwide.

VALIDATION PROTOCOL

An important feature of laboratory methods is their validation. Even though this is more often used for enforcement and regulatory purposes, it is important to bear in mind that monitoring of environmental impact may raise information for decision making in the scope of risk assessment, so a reliable laboratorial branch is of the essence.

As defined in the ISO standard 17025: Validation is "*the confirmation by examination and the provision of objective evidence that the particular requirements for a specific intended use are fulfilled*".

In that sense, a method be considered adequate to the application it is intended. Fitness-for-purpose is defined as *"The degree to which data produced by a measurement process enables the user to make technically and administratively correct decisions for a stated purpose"*.

There are a number of guideline documents covering pesticide analysis in several matrices, mostly in food products. Most of them are government issued protocols, so publicly accessible. One can get definitions of validation studies and objectives from protocols such as the one elaborated by OECD (Organization for Economic Cooperation and Development), the US EPA, the European Commission (DG-SANCO) and so on:

Selectivity	It refers to the extent to which the method can be used to determine particular analytes in mixtures or matrices without interferences from other components of similar behaviour.
Linearity	Also referred as Working Range, determined by means of a calibration curve. It refers to the ability of a detection system to produce an acceptable, well defined, correlation between the instrumental response and the concentration of the analyte in the sample. The analyte concentration to be measured should be within the defined dynamic range of the instrument.

Precision	It is addressed in studies of Repeatability and Reproducibility. Repeatability refers to the closeness of agreement between mutually independent test results obtained with the same method on identical test material, in the same laboratory by the same operator using the same equipment within short intervals of time. The repeatability (within-run effect) includes contributions from any part of the procedure that varies within a run, including contributions from normal gravimetric and volumetric errors, heterogeneity of the test material, and other procedural errors during the analysis. Reproducibility refers to the closeness of agreement between independent results obtained with the same method on identical test material obtained but under different conditions. Within-laboratory or intra-laboratory reproducibility or single-laboratory reproducibility (run effect) contributes to day-to-day variations in the analytical system due to changes of analyst, batches of reagents, recalibration of instruments and laboratory environment (e.g. temperature changes). Between-laboratory or interlaboratory or multiple-laboratory reproducibility (laboratory effect) contributes to additional variations such as variations in calibration standards, differences between local interpretations of a protocol, differences in equipment or reagent source, or environmental factors, such as differences in average climatic conditions.
Trueness	It is the amount measured as a percentage of the amount of analyte(s) (active substance and relevant metabolites) originally added to a sample of the appropriate matrix, which contains either no detectable level of the analyte or a known detectable level. Practical data for this assessment is taken from recovery experiments.
Limit of detection (LOD)	The limit of detection of an analytical procedure is the lowest amount of an analyte in a sample that can be detected but not necessarily quantified as an exact value. At the limit of detection, a positive identification can be achieved with reasonable and/or previously determined confidence in a defined matrix using a specific analytical method.

Limit of Quantification (LOQ)	Defined from a regulatory perspective as the lowest concentration tested at which an unambiguous identification of the analyte can be proven and at which an acceptable mean recovery with an acceptable relative standard deviation (RSD) is obtained. The LOQ should be low enough to achieve the intended purpose of the method. From an analytical perspective, 6-10 times the standard deviation of the noise provides an estimate of the LOQ, which must be verified by the fortification experiments. The Lowest Calibrate Level (LCL) is also experimentally valid for this purpose.

SECTION 2
LATEST RESARCH

8

CHAPTER

CONTAMINATION OF THE AQUATIC ENVIRONMENT WITH NEONICOTINOIDS AND ITS IMPLICATION FOR ECOSYSTEMS

Francisco Sánchez-Bayo[1], Koichi Goka[2] and Daisuke Hayasaka[3]

[1]Faculty of Agriculture and Environment, The University of Sydney, Eveleigh, NSW, Australia

[2]National Institute for Environmental Studies, Tsukuba, Japan

[3]Faculty of Agriculture, Kindai University, Nara, Japan

The widespread use of systemic neonicotinoid insecticides in agriculture results first in contamination of the soil of the treated crops, and secondly in the transfer of residues to the aquatic environment. The high toxicity of these insecticides to aquatic insects and other arthropods has been recognized, but there is little awareness of the impacts these chemicals have on aquatic environments and the ecosystem at large. Recent monitoring studies in several countries, however, have revealed a world-wide contamination of creeks, rivers and lakes with these insecticides, with residue levels in the low μg/L (ppb) range. The current extent of aquatic contamination by neonicotinoids is reviewed first, and the findings contrasted with the known acute and chronic toxicity of neonicotinoids to various aquatic organisms. Impacts on populations and aquatic communities, mostly using mesocosms, are reviewed next to identify the communities most at risk from those that undergo little or no impact. Finally, the ecological links between aquatic and terrestrial organisms are considered. The consequences for terrestrial vertebrate species that depend mainly on this food source are discussed together with impacts on ecosystem function. Gaps in knowledge stem from difficulties in obtaining long-term experimental data that relates the effects on individual organisms to impacts on populations and ecosystems. The paper concludes with a summary of findings and the implications they have for the larger ecosystem.

INTRODUCTION

Neonicotinoids are a novel class of chemical insecticides derived from the natural toxin nicotine. The first compound that was launched to the market in the early 1990s, imidacloprid, was hailed as the solution to the environmental problems caused by older insecticides such as organochlorines and organophosphates, e.g., spray drift onto non-target areas (Siebers et al., 2003), broad-spectrum toxicity to most organisms (Brown, 1978), fish kills (Fox and Matthiessen, 1982), bioaccumulation in fatty tissues (Matthiessen et al., 1982) and poisoning effects on the applicators (Cataño et al., 2008), among others.

Because they are systemic, neonicotinoids are typically applied to the roots of the crop plant, avoiding thus the need for spraying and contaminating nearby land by drift, although they can also be applied as foliar sprays (Elbert et al., 2008). In addition, the selectivity of neonicotinoids toward arthropods, and insects in particular, was an achievement only paralleled by the pyrethroids (derived from the natural toxin pyrethrum). However, unlike the latter insecticides, neonicotinoids are not toxic to either fish or zooplankton species, a great advantage for using them in environmental programs. Thirdly, their hydrophilic properties avoided any chance of bioaccumulation in organisms, and furthermore they were harmless to mammals (Tomizawa and Casida, 2005). These features provided safety to both environment and operators, mainly farmers, and became the key selling points in marketing (Jeschke and Nauen, 2008). Within a decade, imidacloprid was the top selling insecticide in the world, having displaced older chemistries (Jeschke et al., 2011). Newly developed neonicotinoids followed suit, to the point that they now constitute the largest group of insecticides in the global market (Simon-Delso et al., 2015).

Their marketing success, however, was tarnished by their association with honey bee failures in France (van der Sluijs et al., 2013). Many people now have become aware of the existence of neonicotinoids through reports in the media and the internet about bee declines. Hundreds of research papers have been written on this topic in recent years (Osborne, 2012), and the evidence suggests that neonicotinoids impacts on bees and other pollinators cannot be ignored (EFSA, 2013). Consequently, authorities have started to impose measures in Europe to reduce their use in crops that attract bees, like rapeseed (canola), sunflower, and maize (European Commission, 2013), and some countries (i.e., France and Germany) have recently banned the use of seeds treated with neonicotinoids (Garric and Hir, 2016).

As the debate about neonicotinoids has been focused on bees not many people are aware of their impacts on aquatic ecosystems. Yet they pose threats to this environment (Sánchez-Bayo, 2014), more subtle perhaps but broader in scope when we analyse the consequences for the larger aquatic ecosystem.

This paper is a review of current knowledge about the toxicity of neonicotinoids to aquatic species, starting with their effects at the individual level, discussing their impacts on populations and aquatic communities, and concluding with the consequences that these impacts have on ecosystems. In addition, a meta-analysis of the contamination of freshwater systems with neonicotinoids to date adds a global perspective to the issues. There is urgency in assessing the advantages and disadvantages of the widespread use of this class of insecticides, so as not to repeat the mistakes of the past (Krebs et al., 1999).

EFFECTS AT THE ORGANISMAL LEVEL

The first neonicotinoid launched to the market, imidacloprid, had very low acute toxicity to the standard aquatic species used in regulatory testing of chemicals. The 48-h median lethal concentration (LC50) for the waterflea *Daphnia magna* was between 10 and 85 mg/L (ppm), which is several orders of magnitude higher than the LC50 of pyrethroids, organophosphorus and carbamate insecticides to the same species (Song et al., 1997; Tomlin, 2009). Similarly, the 96-h LC50 for fish species were in the range 83–281 mg/L for rainbow trout (*Oncorhynchus mykiss*) and zebrafish (*Danio rerio*) (Ding et al., 2004; Tomlin, 2009). These data suggested that neonicotinoids would not have major impacts, if any, on aquatic ecosystems.

It was later found that other aquatic taxa were much more sensitive to imidacloprid than the standard test species. Thus, freshwater ostracods have 48-h LC50s in the range 185–719 μg/L (ppb), which are 50–120 times lower than that of *D. magna* (Sánchez-Bayo and Goka, 2006b). Moreover, the LC50s for the amphipod *Hyalella azteca* are between 115 and 7 μg/L (ppb) depending on whether the exposure is for 2 or 28 days, and for midge larvae (*Chironomus tentants*) LC50s are in the range from 5 to 0.9 μg/L for the respective 4 and 28-day exposures (Stoughton et al., 2008). The latter values are four orders of magnitude lower than those for *D. magna*, indicating that somehow the initial toxicity assessment of this insecticide was flawed. Survival of midge larvae (*Chironomus riparius*) is also reduced in waters contaminated with mixtures of neonicotinoids (imidacloprid, thiacloprid) and pyrethroid insecticides (deltamethrin and esfenvalerate) (Kunce et al., 2015).

One particular aspect of neonicotinoids became apparent only after years of testing: median toxicity values varied significantly depending on the time of exposure. As mentioned above, the estimated LC50s for amphipods and midge larvae were one or two orders of magnitude lower for exposures of 28 days compared to standard exposures of 2 or 4 days. This translates in a large acute/chronic toxicity ratio, which for the mayfly *Cloeon dipterum* is 800 times when exposed for 28 days to imidacloprid (van den Brink et al., 2016). For the freshwater ostracod *Cypridopsis vidua*, the difference in LC50 between 2- and 5-day exposures is three orders of magnitude! (Sánchez-Bayo, 2009). This trend toward lower LC50s with increasing exposure time has been confirmed for several other species, including *D. magna, Gammarus*amphipods, black fly larvae, alderflies, mayfly and dragonfly nymphs (Beketov and Liess, 2008a; Roessink et al., 2013) when exposed to imidacloprid, thiamethoxam or thiacloprid (Figure 1). The consequence is an apparent "delayed mortality" (Beketov and Liess, 2008a), which can be observed in mesocosm trials that use a single pulse exposure: most of the organisms do not die immediately but start dying in large numbers after a week, and their populations disappear completely after a few weeks (Sánchez-Bayo and Goka, 2006a; Hayasaka et al., 2012a).

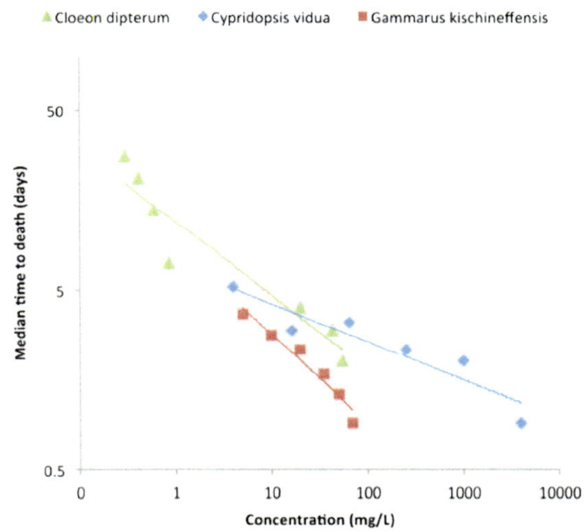

Figure 1: Time-cumulative mortality of three aquatic species exposed to neonicotinoids: mayfly *Cloeon dipterum* **to thiacloprid (green triangles), ostracod** *Cypridopsis vidua* **to imidacloprid (blue diamonds) and amphipod** *Gammarus kischineffensis* **to thiamethoxam (red squares)**. Data sources: mayfly (van den Brink et al., 2016); ostracod (Sánchez-Bayo, 2009); amphipod (Uğurlu et al., 2015).

The physiological mechanism responsible for such unusual toxicological response is based on the agonistic mode of action of this class of chemicals, and was deduced by Tennekes (2010a) based on the model of Druckrey and Küpfmüller (1949). Neonicotinoids bind irreversibly to the nicotinic acetylcholine receptors (nAChR) embedded in the synaptic membranes of neurons, and their activation elicits a continuous electric impulse that eventually leads to the death of the neuron. The neuronal death toll accumulates as more and more chemical molecules bind to other nAChRs until the organism cannot cope with the damage and dies (Rondeau et al., 2014). Although an antagonistic mode of action on the same receptors has been reported for thiacloprid and thiamethoxam in *Lymnaea stagnalis* snails (Vehovszky et al., 2015), this inhibition of neurotransmission results in a similar outcome.

Aquatic organisms are constantly being exposed to residues of chemicals present in water, a medium from which they cannot escape. The time to reach the organism›s death threshold depends on the internal concentration of insecticide, which in turn depends on its external concentration and the kinetics and detoxification ability of each species (Escher et al., 2011). The latter explains the enormous differences in susceptibility to this class of insecticides by various aquatic and terrestrial taxa, which range several orders of magnitude among insects and crustaceans alone (Morrissey et al., 2015).

The main difference between this mode of action and that of other pesticides is that effects are cumulative with time, because neurons do not regenerate. It has been termed time-cumulative toxicity (Tennekes and Sánchez-Bayo, 2013) to distinguish it from the more common toxicological response of insecticide inhibitors (e.g., organophosphorus, pyrethroids), which may bind irreversibly to specific receptors or enzymes but whose effects are temporary and can be reversed once the target receptors or enzymes are regenerated (Matsumura, 1985).

Aside from mortality, exposure to neonicotinoids causes a number of sublethal effects on aquatic organisms, such as feeding inhibition (Alexander et al., 2007; Kreutzweiser et al., 2007; Nyman et al., 2013), impaired movement (Motobayashi et al., 2012), reduced fecundity (Böttger et al., 2013), reduced body size in mayflies (Alexander et al., 2008) and fish (Hayasaka et al., 2012a) and immune-suppression in fish (Sánchez-Bayo and Goka, 2005). Downstream drift also occurs probably as an avoidance response to toxic conditions (Beketov and Liess, 2008b). All these effects were ignored for years, as the focus of neonicotinoid research was on bees, not on aquatic organisms. Obviously, some of these sublethal effects can be reversed if they do not rely directly upon the nervous system.

Impacts at the Population Level

Measurements of acute toxicity such as LC50s are useful to determine the potency of a chemical. Equally useful are the estimations of lowest effect concentrations (LOECs) based on observations of chronic exposure, although they are less accurate and reliable. What matters is to protect the populations of as many species as possible so as to maintain the integrity of the aquatic ecosystem services. To achieve that goal, it is imperative to know the range of sensitivities amongst species in different taxonomic groups, so that an evaluation of risks can be made. For aquatic species, toxicity data are scarce for all neonicotinoids except for imidacloprid and, to a lesser extent, thiacloprid. Therefore, assessments of risks and water quality thresholds for neonicotinoids are currently based on the acute toxicity of imidacloprid, mostly derived from short-term exposures of 2 or 4 days (Morrissey et al., 2015) and some chronic data (Smit et al., 2015).

The ECOTOX database of the US Environmental Protection Agency (EPA) provides LC50s of imidacloprid for 57 aquatic species belonging to several taxonomic groups (9 classes in 4 phyla). Sensitivities span six orders of magnitude, from the most susceptible mayflies (LC50 ~ 1 ppb) to the most tolerant fish (LC50 ~ 650 ppm) (Figure 2). It is apparent that the most susceptible species are aquatic insects, followed by crustaceans such as amphipods, ostracods and shrimps, then tubicifid worms and mussels. All cladoceran crustaceans (waterfleas) are very tolerant except perhaps *Ceriodaphnia dubia*, which is as sensitive as ostracods. Waterfleas are, therefore, not representative of other invertebrate taxa for imidacloprid nor any other neonicotinoid compound (Beketov and Liess, 2008a; Daam et al., 2013; Hayasaka et al., 2013).

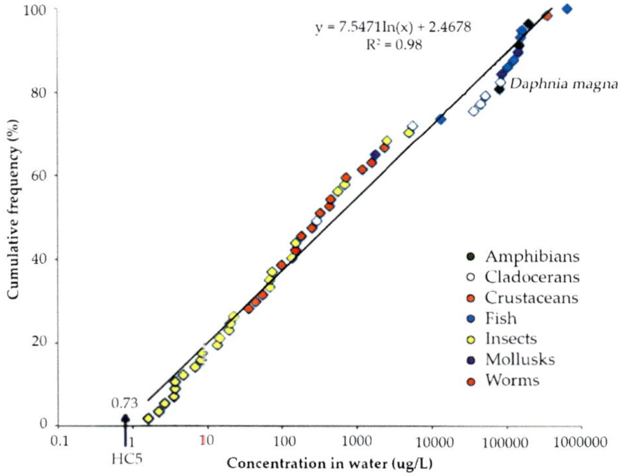

Figure 2: Species sensitivity distribution (SSD) for acute toxicity of imidaclo-

prid (LC50) to aquatic organisms. The hazard concentration affecting 5% of species (HC5, μg/L) is indicated. Source: ECOTOX database https://cfpub.epa.gov/ecotox/.

Species sensitivity distributions (SSD, Figure 2) have been used by government agencies in some countries to derive water quality thresholds that protect their aquatic environment. In the Netherlands and other European countries the protective level for short-term peak concentrations of imidacloprid is 0.2 μg/L, whereas for long-term exposures the threshold is 8.3 ng/L (Smit et al., 2015). In the United States, the chronic invertebrate Aquatic Life Benchmark is 1.05 μg/L, in Canada 0.23 μg/L (CCME, 2007; Anderson et al., 2015) and in Sweden 13 ng/L (Kreuger et al., 2010). Thresholds for other neonicotinoids are about the same order of magnitude in the US, but most countries have not established yet any regulation concerning neonicotinoids, while many still base their ecological assessments on the misleading toxicity data for *Daphnia* and fish.

The above regulatory thresholds are only a guide. Unlike with previous pesticides, protective levels for neonicotinoids cannot be achieved by setting a concentration benchmark because, as already explained, the effects of neonicotinoids increase with exposure time. An alternative is to assess the impact on populations using the predicted affected fraction (PAF) of species, which is determined by comparing waterborne residue levels from monitoring surveys with the SSD. Data on water residues for these compounds have been gathered in the past decade; prior to 2005 only a few surveys found some imidacloprid in a watershed and two streams of New York State (USGS, 2002; Phillips and Bode, 2004) and in drains from potato fields in Canada (Denning et al., 2004; CCME, 2007). A meta-analysis of all residue data from 11 countries available to date (Table S1) revealed the following:

(i) up to six neonicotinoids are currently present in water bodies all over the world. Average concentrations were similar for all compounds, ranging from 0.08 μg/L (dinotefuran) to 0.73 μg/L (imidacloprid); the highest concentrations detected so far were for imidacloprid and thiamethoxam (320 and 225 μg/L, respectively) (Figure 3).

(ii) average residue levels have increased over the past 15

years, with highest rates of increase for clothianidin and thiamethoxam reflecting the worldwide trend in usage of these two compounds (Simon-Delso et al., 2015; Figure 4A).

(iii) the frequency of detection varies widely from country to country, with 100% detections for some compounds in several regions. On average neonicotinoid detections were found in 13% (acetamiprid) to 57% (dinotefuran) of all waters, and they also showed an increasing trend with time; again, the highest increases were for clothianidin and thiamethoxam (Figure 4B).

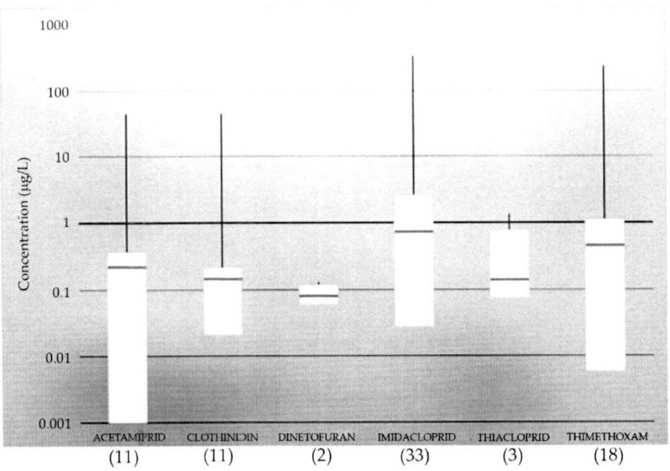

Figure 3: Worldwide survey of neonicotinoid residues in water. The number of surveys reporting each chemical is in brackets. Boxes contain the residues between the first and third quartile; blue lines indicate the geometric mean; vertical lines show the outliers. Sources: see Table S1.

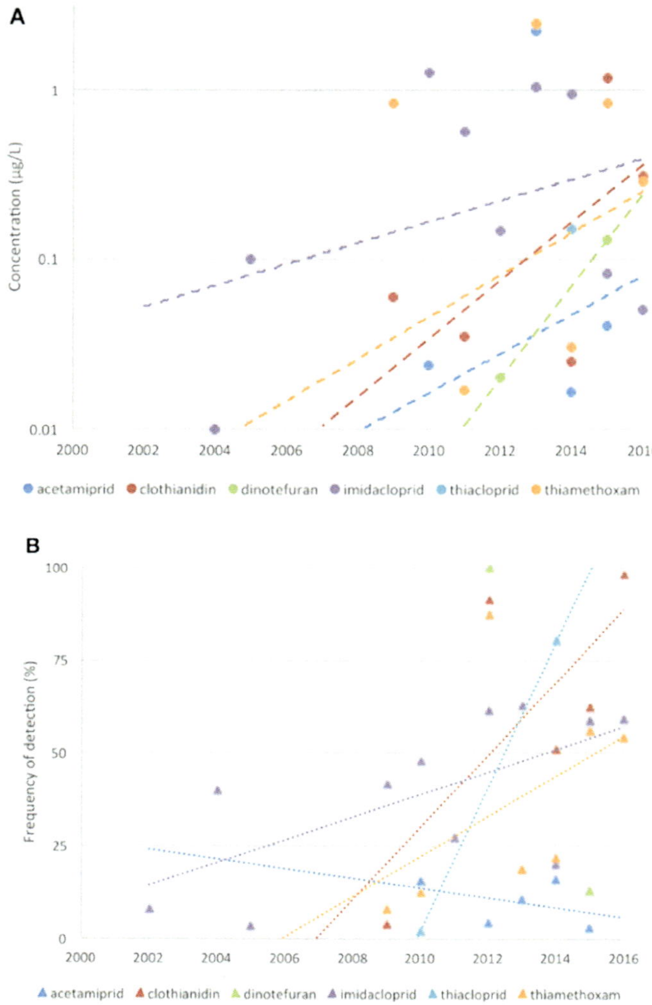

Figure 4: Temporal trends of (A) concentrations of neonicotinoids in waters of the world, and (B) their frequency of detection. Sources: see Table S1.

Contrasting these data with the protective levels established in some countries, the average concentration of all neonicotinoids in water exceeded the European guidelines 27% of the time, and the Canadian and United States guidelines 66 and 79% of the time, respectively, whereas maximum concentrations can exceed the European guidelines 35% of the time (Morrissey et al., 2015). These findings are of concern. The increasing trend in detections is obviously due to two factors: (i) a major effort in looking for these compounds in recent times, which contrasts with the absence of data in previous years; and (ii) better analytical capabilities, with current limits of detection around 1 ng/L or less using either HPLC (Sánchez-Bayo and Hyne, 2014) or LC-MS/MS instrumentation (Hladik and Calhoun, 2012; Yamamoto et al., 2012). However, the increasing residue levels are of great concern, as they indicate that residues in soil, where most of these insecticides are applied, are accumulating over the years. Indeed, there is evidence that such accumulation is happening in countries with a long history of using seeds treated with imidacloprid (Jones et al., 2014; Douglas and Tooker, 2015), although residues of thiamethoxam or clothianidin may plateau if crop rotation is used over a few years (Schaafsma et al., 2016). This accumulation results primarily from the fact that 80–90% of the insecticide in the coated-seeds and granules remains in the soil at the end of the cropping season (Sur and Stork, 2003; Goulson, 2013), and dissipation from soil is slower than in water: half-lives of neonicotinoids in soil are between 50 and 600 days for the four most commonly used compounds (Bonmatin et al., 2015), whereas photolytic hydrolysis in water can dissipate waterborne residues in a few days (Phong et al., 2009; Thuyet et al., 2011). Degradation in sediments is faster for newly developed compounds like cycloxaprid (Liu et al., 2015). Up to 6% of imidacloprid residues in soil can be transported in runoff after storm events (Thuyet et al., 2012), but most of the residual chemical would remain in the applied field, from where it moves readily into ground waters, particularly thiamethoxam, imidacloprid, clothianidin (González-Pradas et al., 2002; Miranda et al., 2011; Bajeer et al., 2012) and dinotefuran (Kurwadkar et al., 2014). The increasing use of products containing neonicotinoids and their repeated application as coated seeds in agricultural fields (Douglas and Tooker, 2015) adds every year a new layer of residues to the soil, and hence to the waters, where residue levels are a reflection of those present in soil at any time (Hladik et al., 2014; Schaafsma et al., 2015).

By comparing the distribution of waterborne residues of imidacloprid to its SSD, estimations of the PAF are made to assess its current impact on aquatic organisms, i.e., the loss of half the populations exposed (Figure 5). Surveys to date indicate that up to 40% of species are being seriously affected in streams of Maryland, where average residue levels are 5.4 μg/L and can reach 131 μg/L (Johnson and Pettis, 2014), and a similar proportion in draining ditches from greenhouses in Sweden (average 3.2 μg/L and highest 89 μg/L) (Kreuger et al., 2010). In three agricultural valleys of California, imidacloprid is currently affecting up to 11% of aquatic species (Starner and Goh, 2012), and in the Sydney basin up to 14% of species are being affected in streams that receive water from turf farms (Sánchez-Bayo and Hyne, 2014). Only streams and estuaries contaminated mostly with urban runoff, e.g., San Francisco (Weston et al., 2015), have minimal number of species affected. One can expect similar impacts for the other neonicotinoid residues, although it is not possible to assess them at this stage so long as the data available are insufficient.

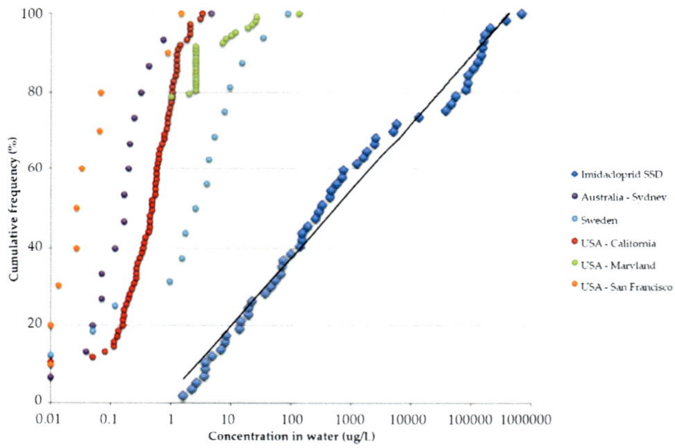

Figure 5: Distribution of waterborne residues of imidacloprid in several countries contrasted with the acute SSD for this compound. Data sources: Australia (Sánchez-Bayo and Hyne, 2014); Sweden (Kreuger et al., 2010); USA California (Starner and Goh, 2012); USA Maryland (Johnson and Pettis, 2014); USA San Francisco (Weston et al., 2015).

This preliminary assessment is only based on the acute toxicity data (Figure 2) as determined in laboratories. For more realistic assessments of the long-term impacts, field and mesocosm studies are required, as explained in the next section.

IMPACTS ON AQUATIC COMMUNITIES

Some 22 studies on the impacts of neonicotinoids on aquatic communities have been conducted to date. Most of them comprise mesocosms that used imidacloprid, with five studies using thiacloprid and one acetamiprid in addition to those two compounds (Table S2). These studies were carried out in Japan (rice mesocoms), Portugal (field trials), Canada, and Germany (streams and microcosms). The most striking feature of these studies is their consistency in reporting population and community effects at levels well below the LC50s of the aquatic species tested. This is unusual, since field or mesocosms studies under realistic scenarios typically report fewer impacts of pesticides and other toxicants than in closed laboratory conditions (Cleveland et al., 2002). Reduced exposures, due mainly to chemical losses by microbial degradation, hydrolysis and other environmental factors, are usually responsible for the lesser impacts under field conditions (Maund et al., 1997).

A reduced abundance in aquatic insects is apparent when concentrations of imidacloprid in water are above 1 or 2 μg/L (Sánchez-Bayo and Goka, 2006a; Pestana et al., 2009; Hayasaka et al., 2012a; Colombo et al., 2013). Population reductions in the short-term are caused by direct toxicity, but in mesocosms such reductions affect the structure of the macroinvertebrate communities when residues are one or two orders of magnitude lower than the LC50s for most species, as more tolerant taxa tend to increase in numbers to fill the niche vacuum thus created in the ecosystem. Some of these changes are predictable. For instance, waterfleas increased in numbers when grazers such as ostracods were eliminated in rice mesocosms treated with imidacloprid (Sánchez-Bayo et al., 2007), whereas the disappearance of chironomid larvae brought about increases in *Radix* sp. snails (Colombo et al., 2013). In other cases, opportunistic predators (e.g., Heteroptera, Coleoptera, Odonata) and scavengers (e.g., Amphipoda) that feed on the moribund insects or their corpses have a temporary increase in food availability (Sánchez-Bayo and Goka, 2006a). Interestingly, the negative impacts on predatory copepod populations in rice fields are followed by upsurges of mosquitoes but not of chironomids. Consequently, the overall biodiversity of the aquatic communities is negatively affected (Pestana et al., 2009). Similar impacts are observed in mesocosms treated with thiacloprid at 3.2 μg/L or above (Kattwinkel et al., 2016). This is not surprising, as the HC5 for thiacloprid derived from outdoor stream mesocosms is 0.72 μg/L (Beketov et al., 2008), the same as that calculated for imidacloprid (HC5

= 0.73 μg/L, Figure 2). Treatments of rice mesocosms with imidacloprid at different rates over 4 years in Japan resulted in community impacts that were related to the initial concentrations of this insecticide in water, from 240 to 40 μg/L. Average reductions of 46–62% were recorded among the plankton, neuston, nekton or benthos communities (Figure 6A), but the specific taxa groups and species affected differed from 1 year to another (Sánchez-Bayo et al., 2007; Hayasaka et al., 2012b). The greatest impacts (>45% reductions) occurred in ostracods, mayflies and snails, followed by chironomids, dragonflies, damselflies and some Hemiptera predators such as Corixidae, Mesoveliidae and *Anthocoris* sp. Emergence of dragonflies is also reduced by more than 80% (Jinguji et al., 2013). By contrast, waterfleas increased by 75% and Diptera larvae (excluding chironomids) by 15%. A similar trial in a rice field in Portugal measured initial concentrations of 52 μg/L imidacloprid in water, and average weighted concentrations of 8 μg/L, which were estimated to affect 40–63% of the aquatic species (Daam et al., 2013). These observations are consistent with the impacts in Canadian stream mesocosms (Figure 6B), where weekly imidacloprid pulses at 2 or 20 μg/L had the highest reductions on worms (75%), caddisflies (70%), mayflies and stoneflies (both 68%) (Pestana et al., 2009).

A

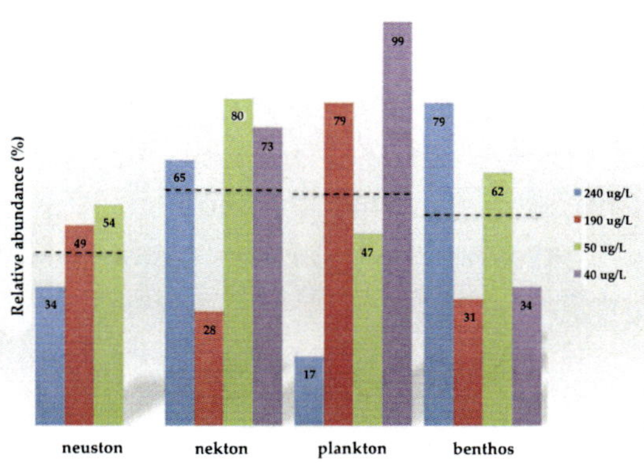

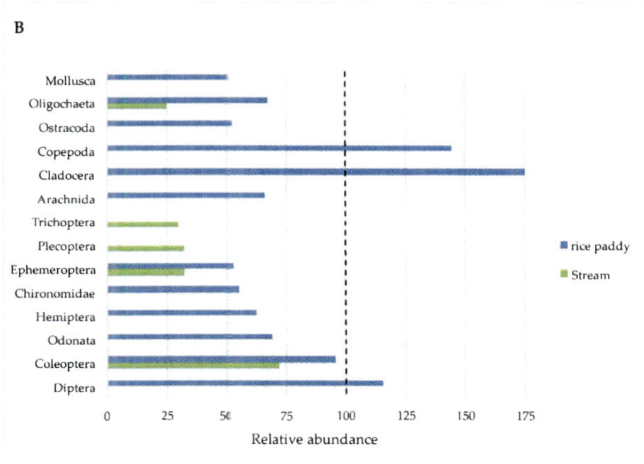

Figure 6: Relative abundance with respect to controls of aquatic invertebrates in imidacloprid-treated mesocosms. (A) Communities in rice paddies for different initial concentrations of imidacloprid; dashed lines indicate average reductions. (B) Invertebrate taxa in rice paddies and streams; vertical dashed line indicates the control. Data sources: rice paddies (Sánchez-Bayo and Goka, 2006a; Sánchez-Bayo et al., 2007; Hayasaka et al., 2012a,b); streams (Pestana et al., 2009).

Moreover, many of these populations are decimated and their recovery is either slow or, if there is competition with other species, it does not take place (Liess et al., 2013). Recovery of most populations only occurs when the neonicotinoid concentrations in water or sediment are below 1 μg/L (Hayasaka et al., 2012b), whereas many univoltine or semivoltine species do not recover at all. Nor does the structure of the communities revert to the original situation, because some species disappear while others take over and increase in numbers (Beketov et al., 2008; Hayasaka et al., 2012a). These impacts contrast with those caused by other pesticides, which tend to produce a large initial mortality upon target and non-target populations alike but allow the recovery of the species affected within a few weeks (van den Brink et al., 1996; Brock et al., 2010). Thus, aquatic communities of rice paddies recovered completely in a week after foliar application of the insecticide etofenprox took place (Sánchez-Bayo et al., 2007). The reason for this difference lies in the delayed mortality after chronic exposure to very low concentrations of neonicotinoids in water or sediments and, for more tolerant species like *Gammarus roeseli*, in their reduced fecundity under similar conditions (Böttger et al., 2013). By contrast, many pyrethroids and organophosphates (with the exception of persistent compounds like chlorpyrifos) do not produce time-cumulative mortality (Parsons and Surgeoner, 1991) since their exposure is limited in time (Lahr et al., 2000; Medina et al., 2004).

Apart from the lethal effects on communities, blackfly larvae (*Simulium latigonium*), mayfly nymphs (*Baetis rhodani*) and amphipods (*Gammarus pulex*) drift downstream when concentrations of various neonicotinoids in water are lower than 1/10 of the species LC50 (Beketov and Liess, 2008b), indicating that these organisms respond to the adverse effects of these neurotoxicants at sublethal levels, i.e., below 0.37, 0.46, and 27 μg/L of either thiacloprid, acetamiprid or imidacloprid for the respective species above.

Finally, when mayfly nymphs (*Baetis rhodani*) and *Gammarus fossarum* are exposed together to sublethal levels of thiacloprid, the amphipod increases its predation on the nymphs but reduces its shredding of litter at concentrations as low as 0.5–1 μg/L (Englert et al., 2012). Imidacloprid also reduces the litter decomposition carried out by stoneflies (*Pteronarcys dorsata*) and crane flies (*Tipula* sp.) at concentrations below 12 μg/L (Kreutzweiser et al., 2008b), probably due to a feeding inhibition effect of this insecticide, which has also been observed with other detritivores such as *Gammarus pulex* (Nyman et al., 2013) and *Eisenia fetida* earthworms (Kreutzweiser et al., 2008a). The implications of these impacts for the larger ecosystem are discussed next.

Impacts on the Ecosystem

The consequences of all the above for the larger ecosystem have not been studied in detail yet. Difficulties in obtaining long-term experimental data that relates the effects on individual organisms to impacts on ecosystems prevent carrying out such studies. However, it is clear that some predictions can be made from the limited set of observations about the effects on aquatic communities reported so far. At least two main areas of concern can be identified: reduced capacity for decomposition of organic debris by aquatic organisms and starvation of insectivores and other vertebrate fauna that depend on invertebrates as a major or only food source (Figure 7).

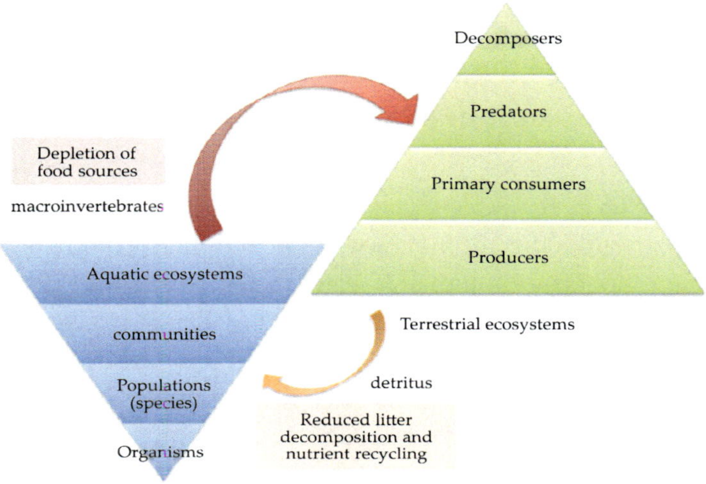

Figure 7: Ecosystem impacts of water-borne neonicotinoid residues.

REDUCED DECOMPOSITION CAPACITY

The recycling of organic matter that falls into water bodies is an essential ecosystem function that not only provides food for a wide range of aquatic and benthic organisms but also ensures the water quality is adequate for all other organisms that use it, including ourselves.

It is well established that mayfly (Ephemeroptera), caddisfly (Trichoptera), and stonefly (Plecoptera) nymphs are the most sensitive aquatic organisms to most pollutants, so they are considered bioindicators of water quality (Morse et al., 1993; von der Ohe et al., 2007). They are shredders of leaves and other debris found at the bottom of creeks and streams that run through forested and agricultural areas, although not the only ones: larvae of crane flies (Tipulidae), black flies (Simulidae) and other Diptera taxa perform the same function, together with amphipods, ostracods and aquatic isopods. The fact that litter decomposition by stoneflies, crane flies, mayflies and amphipods is significantly reduced by concentrations of neonicotinoids that are currently found in many aquatic environments is of concern (Kreutzweiser et al., 2007). Even if some individuals may survive in depleted populations, they still will be unable to carry out the decomposition function properly due to the feeding inhibition caused by these neurotoxicants, which will render those individuals unfit to do their job. Naturally, insufficient feeding leads to reduced ability for reproduction (Böttger et al., 2013), so the long term prospects are poor for the detritivore populations affected.

To many regulators of chemicals, whether mayflies or other macroinvertebrates are depleted is not important, or at least not as much as the increase in productivity that farmers may obtain from using products like neonicotinoids, although the latter benefits are questionable—see (Seagraves and Lundgren, 2012; Macfadyen et al., 2014). Just because macroinvertebrates are not seen, since they are small and live at the bottom of ponds and streams, this does not mean they can be dispensed with. As Suter and Cormier have argued, these small creatures are present in ecosystems for an important reason (Suter and Cormier, 2015).

Given that more than half of the waters are contaminated (Figure 4) with neonicotinoid levels that impair this important ecosystem function, higher organic and inorganic pollution can be expected wherever these insecticides are present. Microbial degradation of the debris may still occur, but it would be slower and produce undesirable by-products such as methane and sulfides (Sorrell and Boon, 1992; Kwok et al., 2005). The combined impacts by neonicotinoids and other pollutants could gradually poison the surface waters in many parts of the world.

Starvation of Insectivores and Invertebrate Feeders

Indirect impacts are a common feature of many pesticides, one that has no relation to the toxicity on the species ultimately affected (Sánchez-Bayo, 2011). Thus, neonicotinoids do not cause fish mortality directly, but because aquatic invertebrates are a rich food source for many species of fish, depletion and disappearance of this source in waters contaminated with neonicotinoids could affect fish stocks in freshwater ecosystems. In the Netherlands, where residues of imidacloprid in water are the highest in the world (Table S1), correlations between such residues and the decline of arthropod taxa such as Ephemeroptera, Odonata, Diptera and some crustaceans have been found (van Dijk et al., 2013). Although not all the declines can be blamed on neonicotinoids, because other pesticide residues are also found and can have similar impacts (Beketov et al., 2013; Vijver and van den Brink, 2014), it would seem that neonicotinoids are clearly involved.

As already mentioned, populations of aquatic species exposed to neonicotinoids often do not recover. Indeed, the overall abundance of macroinvertebrates in mesocosms treated with imidacloprid or thiacloprid was lower than that in the controls after a few months, while some species disappeared completely (Sánchez-Bayo and Goka, 2006a; Beketov et al., 2008). This suggests that recovery of the extinct populations in the following year must require re-colonization from nearby areas. The elimination of predatory species results in the increase of prey species, with some of them, like mosquitoes (Figure 6B), being a nuisance and a health hazard. In agricultural areas treated extensively with seeds containing neonicotinoids the chances of re-colonization are less frequent for species that are not very mobile. Aquatic insects and invertebrate species are being removed from many land and water areas and heading toward extinction. This dire prediction is not far from the reality in some places. An entomological survey carried out in a region of Germany comprising agricultural land and a nature reserve reported a decline of 75% in flying insect abundance between 1989 and 2013 (Sorg et al., 2013). Many of these flying insects have aquatic life cycles, and their disappearance is probably due to their larvae not having survived in water. This astonishing reduction in entomofauna parallels the decline of wild bee species in North America and the British Isles (Fitzpatrick et al., 2007; Cameron et al., 2011; Woodcock et al., 2016) and butterflies in California and England (Gilburn et al., 2015; Forister et al., 2016) in the same period. It must be remembered that neonicotinoids were introduced in the early 1990s.

Riverine ecosystems are notorious for the rich biodiversity they encompass (Sánchez-Bayo, 1991). Many of the vertebrates living around rivers, lakes and ponds are insectivorous species that depend almost exclusively on aquatic invertebrates as their food source: frogs, newts, skinks and lizards, a large array of birds including passerines and waders, bats and shrews. All these animals, whether terrestrial or amphibian, draw their food from flying insects, their aquatic larvae, crustaceans and worms that live in the water environment. Consequently, the depletion of this food source must necessarily affect them (Tennekes, 2010b). To date, the only study available that makes a connection between bird declines and neonicotinoids in water was carried out in the Netherlands (Hallmann et al., 2014). The authors of that study collected information on 15 species of passerine birds in that country over 20 years since 1993, and correlated their abundance with residue concentrations of imidacloprid and other pesticide residues in water during the same period. All bird species studied were either insectivorous or fed insects and larvae to their offspring during the breeding season. The only pesticide that explained the declining trends of 14 bird species was imidacloprid, whereas other factors that were taken into consideration, such as urban or agricultural area, availability of some cereal crops, fertilizer use and others, were discarded by the statistical analysis. For the 6 species that showed a significant decline with imidacloprid residues, the average bird population decline was 3.5% per year in areas with residue levels above 20 ng/L (ppt). These levels are below the HC5 for imidacloprid (0.73 μg/L, Figure 2) and well below the LC50s of all aquatic insects tested. However, as demonstrated in the microcosm and mesocosm studies, they are sufficient to cause sublethal effects and delayed mortality, all of which can eliminate entire populations of invertebrates, without recovery in many cases.

Starvation by depletion of food sources due to pesticides was demonstrated for gray partridges (*Perdix perdix*) in England (Potts, 1986). Also, applications of fipronil insecticide for locust control in Madagascar reduced the abundance of two species of tenrec, a skink and iguanian lizards that depended on termites as their main food source (Peveling et al., 2003). Evidence of similar impacts by neonicotinoids on vertebrate taxa other than birds does not exist because of difficulties in obtaining relevant long-term experimental data. However, if terrestrial birds, lizards and mammals can be taken as examples of what occurs in nature when pesticides reduce the food source, it is reasonable to think that other taxa that are experiencing worldwide declines, such as frogs and bats can be affected by indirect neonicotinoid impacts on the aquatic environment (Mason et al., 2013). Establishing the link between food depletion and population declines in some species is not difficult, but linking food depletion to individual chemical causes is a more challenging task.

CONCLUSION

Negative impacts of neonicotinoids in aquatic environments are a reality.

Initial assessments that considered these insecticides harmless to aquatic organisms may have led to a relaxation of monitoring efforts, resulting in the worldwide contamination of many aquatic ecosystems with neonicotinoids.

The decline of many populations of invertebrates, due mostly to the widespread presence of waterborne residues and the extreme chronic toxicity of neonicotinoids, is affecting the structure and function of aquatic ecosystems. Consequently, vertebrates that depend on insects and other aquatic invertebrates as their sole or main food resource are being affected. Declines of insectivore bird species are quite evident so far, but many other terrestrial and amphibian species may be at risk.

Solutions must be found soon if we are to save the biodiversity not only of aquatic ecosystems, but all other ecosystems linked by the food web. Since the prophylactic use of seeds treated with neonicotinoids is responsible for most of the soil and aquatic contamination, while there is evidence of little productivity gain, one obvious solution is to stop the marketing of seeds coated with these insecticides (van der Sluijs et al., 2015) and use alternative and carefully targeted methods for pest control in agriculture (Douglas and Tooker, 2015; Furlan and Kreutzweiser, 2015), such as integrated pest management (IPM). At the same time, remediation systems based on photolytic processes (Malato et al., 2001) and wetlands phyto-remediation (Beketov and Liess, 2008c) should be implemented to reduce as much as possible the current contamination by these and other pollutants.

AUTHOR CONTRIBUTIONS

FS wrote the first draft and KG and DH added and corrected the original version. All three authors contributed equally to the experimental data reviewed here.

CONFLICT OF INTEREST STATEMENT

The authors declare that the research was conducted in the absence of any commercial or financial relationships that could be construed as a potential conflict of interest.

REFERENCES

1. Alexander, A. C., Culp, J. M., Liber, K., and Cessna, A. J. (2007). Effects of insecticide exposure on feeding inhibition in mayflies and oligochaetes. *Environ. Toxicol. Chem.* 26, 1726–1732. doi: 10.1897/07-015R.1
2. Alexander, A. C., Heard, K. S., and Culp, J. M. (2008). Emergent body size of mayfly survivors. *Freshw. Biol.* 53, 171–180. doi: 10.1111/j.1365-2427.2007.01880.x
3. Anderson, J. C., Dubetz, C., and Palace, V. P. (2015). Neonicotinoids in the Canadian aquatic environment: a literature review on current use products with a focus on fate, exposure, and biological effects. *Sci. Total Environ.* 505, 409–422. doi: 10.1016/j.scitotenv.2014.09.090
4. Bajeer, M. A., Nizamani, S. M., Sherazi, S. T. H., and Bhanger, M. I. (2012). Adsorption and leaching potential of imidacloprid pesticide through alluvial soil. *Am. J. Anal. Chem.* 3, 604–611. doi: 10.4236/ajac.2012.38079
5. Beketov, M. A., Kefford, B. J., Schäfer, R. B., and Liess, M. (2013). Pesticides reduce regional biodiversity of stream invertebrates. *Proc. Natl. Acad. Sci. U.S.A.* 110, 11039–11043. doi: 10.1073/pnas.1305618110
6. Beketov, M. A., and Liess, M. (2008a). Acute and delayed effects of the neonicotinoid insecticide thiacloprid on seven freshwater arthropods. *Environ. Toxicol. Chem.* 27, 461–470. doi: 10.1897/07-322R.1
7. Beketov, M. A., and Liess, M. (2008b). Potential of 11 pesticides to initiate downstream drift of stream macroinvertebrates. *Arch. Environ. Contam. Toxicol.* 55, 247–253. doi: 10.1007/s00244-007-9104-3
8. Beketov, M. A., and Liess, M. (2008c). Variability of pesticide exposure in a stream mesocosm system: macrophyte-dominated vs. non-vegetated sections. *Environ. Pollut.* 156, 1364–1367. doi: 10.1016/j.envpol.2008.08.014
9. Beketov, M., Schäfer, R. B., Marwitz, A., Paschke, A., and Liess, M. (2008). Long-term stream invertebrate community alterations induced by the insecticide thiacloprid: effect concentrations and

recovery dynamics. *Sci. Total Environ.* 405, 96–108. doi: 10.1016/j.scitotenv.2008.07.001

10. Bonmatin, J. M., Giorio, C., Girolami, V., Goulson, D., Kreutzweiser, D. P., Krupke, C., et al. (2015). Environmental fate and exposure; neonicotinoids and fipronil. *Environ. Sci. Pollut. Res.* 22, 35–67. doi: 10.1007/s11356-014-3332-7
11. Böttger, R., Feibicke, M., Schaller, J., and Dudel, G. (2013). Effects of low-dosed imidacloprid pulses on the functional role of the caged amphipod *Gammarus roeseli* in stream mesocosms. *Ecotoxicol. Environ. Saf.* 93, 93–100. doi: 10.1016/j.ecoenv.2013.04.006
12. Brock, T. C. M., Belgers, J. D. M., Roessink, I., Cuppen, J. G. M., and Maund, S. J. (2010). Macroinvertebrate responses to insecticide application between sprayed and adjacent nonsprayed ditch sections of different sizes. *Environ. Toxicol. Chem.* 29, 1994–2008. doi: 10.1002/etc.238
13. Brown, A. W. A. (1978). *Ecology of Pesticides.* New York, NY: John Wiley & Sons, Inc.
14. Cameron, S. A., Lozier, J. D., Strange, J. P., Koch, J. B., Cordes, N., Solter, L. F., et al. (2011). Patterns of widespread decline in North American bumble bees. *Proc. Natl. Acad. Sci. U.S.A.* 108, 662–667. doi: 10.1073/pnas.1014743108
15. Cataño, H. C., Carranza, E., Huamaní, C., and Hernández, A. F. (2008). Plasma cholinesterase levels and health symptoms in Peruvian farm workers exposed to organophosphate pesticides. *Arch. Environ. Contam. Toxicol.* 55, 153–159. doi: 10.1007/s00244-007-9095-0
16. CCME (2007). *Canadian Water Quality Guidelines: Imidacloprid. Scientific Supporting Document.* Winnipeg: Canadian Council of Ministers of the Environment.
17. Cleveland, C. B., Mayes, M. A., and Cryer, S. A. (2002). An ecological risk assessment for spinosad use on cotton. *Pest Manage. Sci.* 58, 70–84. doi: 10.1002/ps.424
18. Colombo, V., Mohr, S., Berghahn, R., and Pettigrove, V. J. (2013). Structural changes in a macrozoobenthos assemblage after imidacloprid pulses in aquatic field-based microcosms. *Arch. Environ. Contam. Toxicol.* 65, 683–692. doi: 10.1007/s00244-013-

9940-2

19. Daam, M. A., Santos Pereira, A. C., Silva, E., Caetano, L., and Cerejeira, M. J. (2013). Preliminary aquatic risk assessment of imidacloprid after application in an experimental rice plot. *Ecotoxicol. Environ. Saf.* 97, 78–85. doi: 10.1016/j.ecoenv.2013.07.011
20. Denning, A., Ernst, W. R., Julien, G. R., Doe, K. G., Cook, A., Bernier, M., et al. (2004). *An Assessment of Buffer Zone Effectiveness in Reducing Pesticide Runoff from Potato Fields in Prince Edward Island (2001–2002).* Atlantic Region: Environment Canada, Environmental Protection Branch.
21. Ding, Z., Yang, Y., Jin, H., Yu, H., Feng, J., Zhang, X., et al. (2004). Acute toxicity and bioconcentration factor of three pesticides on *Brachydanio rerio. Ying Yong Sheng Tai Xue Bao* 15, 888–890.
22. Douglas, M., and Tooker, J. F. (2015). Large-scale deployment of seed treatments has driven rapid increase in use of neonicotinoid insecticides and preemptive pest management in U.S. field crops. *Environ. Sci. Technol.* 49, 5088−5097. doi: 10.1021/es506141g
23. Druckrey, H., and Küpfmüller, K. (1949). *Dosis und Wirkung - Beiträge zur theoretischen Pharmakologie.* Freiburg im Breisgau: Cantor GmbH.
24. EFSA (2013). Conclusion on the peer review of the pesticide risk assessment for bees for the active substance imidacloprid. *EFSA J.*11:3068. doi: 10.2903/j.efsa.2013.3068
25. Elbert, A., Haas, M., Springer, B., Thielert, W., and Nauen, R. (2008). Applied aspects of neonicotinoid uses in crop protection. *Pest Manag. Sci.* 64, 1099–1105. doi: 10.1002/ps.1616
26. Englert, D., Bundschuh, M., and Schulz, R. (2012). Thiacloprid affects trophic interaction between gammarids and mayflies. *Environ. Pollut.* 167, 41–46. doi: 10.1016/j.envpol.2012.03.024
27. Escher, B. I., Ashauer, R., Dyer, S., Hermens, J. L. M., Lee, J.-H., Leslie, H. A., et al. (2011). Crucial role of mechanisms and modes of toxic action for understanding tissue residue toxicity and internal effect concentrations of organic chemicals. *Integr. Environ. Assess. Manag.* 7, 28–49. doi: 10.1002/ieam.100
28. European Commission (2013). *Commission Implementing*

Regulation (EU) No 485/2013. Brussels: Official Journal of the European Union.

29. Fitzpatrick, Ú., Murray, T. E., Paxton, R. J., Breen, J., Cotton, D., Santorum, V., et al. (2007). Rarity and decline in bumblebees - a test of causes and correlates in the Irish fauna. *Biol. Conserv.* 136, 185–194. doi: 10.1016/j.biocon.2006.11.012
30. Forister, M. L., Cousens, B., Harrison, J. G., Anderson, K., Thorne, J. H., Waetjen, D., et al. (2016). Increasing neonicotinoid use and the declining butterfly fauna of lowland California. *Biol. Lett.* 12:20160475. doi: 10.1098/rsbl.2016.0475
31. Fox, P. J., and Matthiessen, P. (1982). Acute toxicity to fish of low-dose aerosol applications of endosulfan to control tsetse fly in the Okavango delta, Botswana. *Environ. Polluti. Ser. A* 27, 129–142. doi: 10.1016/0143-1471(82)90105-2
32. Furlan, L., and Kreutzweiser, D. (2015). Alternatives to neonicotinoid insecticides for pest control: case studies in agriculture and forestry. *Environ. Sci. Pollut. Res.* 22, 135–147. doi: 10.1007/s11356-014-3628-7
33. Garric, A., and Hir, P. L. (2016). Loi sur la biodiversité: la France bannit les pesticides tueurs d›abeilles. *Le Monde*.
34. Gilburn, A. S., Bunnefeld, N., Wilson, J. M., Botham, M. S., Brereton, T. M., Fox, R., et al. (2015). Are neonicotinoid insecticides driving declines of widespread butterflies? *PeerJ* 3:e1402. doi: 10.7717/peerj.1402
35. González-Pradas, E., Ureía-Amate, M. D., Flores-Céspedes, F., Fernández-Pérez, M., Garratt, J., and Wilkins, R. (2002). Leaching of imidacloprid and procymidone in a greenhouse of southeast of Spain. *Soil Sci. Soc. Am. J.* 66, 1821–1828. doi: 10.2136/sssaj2002.1821
36. Goulson, D. (2013). An overview of the environmental risks posed by neonicotinoid insecticides. *J. Appl. Ecol.* 50, 977–987. doi: 10.1111/1365-2664.12111
37. Hallmann, C. A., Foppen, R. P. B., van Turnhout, C. A. M., de Kroon, H., and Jongejans, E. (2014). Declines in insectivorous birds are associated with high neonicotinoid concentrations. *Nature* 511, 341–343. doi: 10.1038/nature13531

38. Hayasaka, D., Korenaga, T., Sánchez-Bayo, F., and Goka, K. (2012a). Differences in ecological impacts of systemic insecticides with different physicochemical properties on biocenosis of experimental paddy fields. *Ecotoxicology* 21, 191–201. doi: 10.1007/s10646-011-0778-y
39. Hayasaka, D., Korenaga, T., Suzuki, K., Saito, F., Sánchez-Bayo, F., and Goka, K. (2012b). Cumulative ecological impacts of two successive annual treatments of imidacloprid and fipronil on aquatic communities of paddy mesocosms. *Ecotoxicol. Environ. Saf.* 80, 355–362. doi: 10.1016/j.ecoenv.2012.04.004
40. Hayasaka, D., Suzuki, K., Nomura, T., Nishiyama, M., Nagai, T., Sánchez-Bayo, F., et al. (2013). Comparison of acute toxicity of two neonicotinoid insecticides, imidacloprid and clothianidin, to five cladoceran species. *J. Pestic. Sci.* 38, 44–47. doi: 10.1584/jpestics.D12-061
41. Hladik, M. L., and Calhoun, D. L. (2012). *Analysis of the Herbicide Diuron, Three Diuron Degradates, and Six Neonicotinoid Insecticides in Water—Method Details and Application to Two Georgia streams*. Reston, VA: U.S. Geological Survey.
42. Hladik, M. L., Kolpin, D. W., and Kuivila, K. M. (2014). Widespread occurrence of neonicotinoid insecticides in streams in a high corn and soybean producing region, USA. *Environ. Pollut.* 193, 189–196. doi: 10.1016/j.envpol.2014.06.033
43. Jeschke, P., and Nauen, R. (2008). Neonicotinoids—from zero to hero in insecticide chemistry. *Pest Manag. Sci.* 64, 1084–1098. doi: 10.1002/ps.1631
44. Jeschke, P., Nauen, R., Schindler, M., and Elbert, A. (2011). Overview of the status and global strategy for neonicotinoids. *J. Agric. Food Chem.* 59, 2897–2908. doi: 10.1021/jf101303g
45. Jinguji, H., Thuyet, D., Ueda, T., and Watanabe, H. (2013). Effect of imidacloprid and fipronil pesticide application on *Sympetrum infuscatum* (Libellulidae: Odonata) larvae and adults. *Paddy Water Environ.* 11, 277–284. doi: 10.1007/s10333-012-0317-3
46. Johnson, J. D., and Pettis, J. S. (2014). A survey of imidacloprid levels in water sources potentially frequented by honeybees (*Apis mellifera*) in the Eastern USA. *Water Air Soil Pollut.* 225, 1–6. doi:

10.1007/s11270-014-2127-2

47. Jones, A., Harrington, P., and Turnbull, G. (2014). Neonicotinoid concentrations in arable soils after seed treatment applications in preceding years. *Pest Manag. Sci.* 70, 1780–1784. doi: 10.1002/ps.3836
48. Kattwinkel, M., Reichert, P., Rüegg, J., Liess, M., and Schuwirth, N. (2016). Modeling macroinvertebrate community dynamics in stream mesocosms contaminated with a pesticide. *Environ. Sci. Technol.* 50, 3165–3173. doi: 10.1021/acs.est.5b04068
49. Krebs, J. R., Wilson, J. D., Bradbury, R. B., and Siriwardena, G. M. (1999). The second Silent Spring? *Nature* 400, 611–612. doi: 10.1038/23127
50. Kreuger, J., Graaf, S., Patring, J., and Adielsson, S. (2010). *Pesticides in Surface Water in Areas with Open Ground and Greenhouse Horticultural Crops in Sweden 2008*. Uppsala: Swedish University of Agricultural Sciences.
51. Kreutzweiser, D., Good, K., Chartrand, D., Scarr, T., and Thompson, D. (2007). Non-target effects on aquatic decomposer organisms of imidacloprid as a systemic insecticide to control emerald ash borer in riparian trees. *Ecotoxicol. Environ. Saf.* 68, 315–325. doi: 10.1016/j.ecoenv.2007.04.011
52. Kreutzweiser, D. P., Good, K. P., Chartrand, D. T., Scarr, T. A., Holmes, S. B., and Thompson, D. G. (2008a). Effects on litter-dwelling earthworms and microbial decomposition of soil-applied imidacloprid for control of wood-boring insects. *Pest Manag. Sci.* 64, 112–118. doi: 10.1002/ps.1478
53. Kreutzweiser, D. P., Good, K. P., Chartrand, D. T., Scarr, T. A., and Thompson, D. G. (2008b). Toxicity of the systemic insecticide, imidacloprid, to forest stream insects and microbial communities. *Bull. Environ. Contam. Toxicol.* 80, 211–214. doi: 10.1007/s00128-007-9347-8
54. Kunce, W., Josefsson, S., Örberg, J., and Johansson, F. (2015). Combination effects of pyrethroids and neonicotinoids on development and survival of *Chironomus riparius*. *Ecotoxicol. Environ. Saf.* 122, 426–431. doi: 10.1016/j.ecoenv.2015.09.008
55. Kurwadkar, S., Wheat, R., McGahan, D. G., and Mitchell, F. (2014).

Evaluation of leaching potential of three systemic neonicotinoid insecticides in vineyard soil. *J. Contam. Hydrol.* 170, 86–94. doi: 10.1016/j.jconhyd.2014.09.009

56. Kwok, Y. C., Hsieh, D. P. H., and Wong, P. K. (2005). Toxicity identification evaluation (TIE) of pore water of contaminated marine sediments collected from Hong Kong waters. *Mar. Pollut. Bull.* 51, 1085–1091. doi: 10.1016/j.marpolbul.2005.06.009
57. Lahr, J., Diallo, A. O., Gadji, B., Diouf, P. S., Bedeaux, J. J. M., Badji, A., et al. (2000). Ecological effects of experimental insecticide applications on invertebrates in Sahelian temporary ponds. *Environ. Toxicol. Chem.* 19, 1278–1289. doi: 10.1002/etc.5620190509
58. Liess, M., Foit, K., Becker, A., Hassold, E., Dolciotti, I., Kattwinkel, M., et al. (2013). Culmination of low-dose pesticide effects. *Environ. Sci. Technol.* 47, 8862–8868. doi: 10.1021/es401346d
59. Liu, X., Xu, X., Li, C., Zhang, H., Fu, Q., Shao, X., et al. (2015). Degradation of chiral neonicotinoid insecticide cycloxaprid in flooded and anoxic soil. *Chemosphere* 119, 334–341. doi: 10.1016/j.chemosphere.2014.06.016
60. Macfadyen, S., Hardie, D. C., Fagan, L., Stefanova, K., Perry, K. D., DeGraaf, H. E., et al. (2014). Reducing insecticide use in broad-acre grains production: an Australian study. *PLoS ONE* 9:e89119. doi: 10.1371/journal.pone.0089119
61. Malato, S., Caceres, J., Agüera, A., Mezcua, M., Hernando, D., Vial, J., et al. (2001). Degradation of imidacloprid in water by photo-Fenton and TiO_2 photocatalysis at a solar pilot plant: a comparative study. *Environ. Sci. Technol.* 35, 4359–4366. doi: 10.1021/es000289k
62. Mason, R., Tennekes, H., Sánchez-Bayo, F., and Jepsen, P. U. (2013). Immune suppression by neonicotinoid insecticides at the root of global wildlife declines. *J. Environ. Immunol. Toxicol.* 1, 3–12. doi: 10.7178/jeit.1
63. Matsumura, F. (1985). *Toxicology of Pesticides*. New York, NY: Plenum Press.
64. Matthiessen, P., Fox, P. J., Douthwaite, R. J., and Wood, A. B. (1982). Accumulation of endosulfan residues in fish and their predators after aerial spraying for the control of tsetse fly in Botswana. *Pestic.*

Sci. 13, 39–48. doi: 10.1002/ps.2780130107

65. Maund, S. J., Sherratt, T. N., Stickland, T., Biggs, J., Williams, P., Shillabeer, N., et al. (1997). Ecological considerations in pesticide risk assessment for aquatic ecosystems. *Pestic. Sci.* 49, 185–190.
66. Medina, M., Barata, C., Telfer, T., and Baird, D. J. (2004). Effects of cypermethrin on marine plankton communities: a simulated field study using mesocosms. *Ecotoxicol. Environ. Saf.* 58, 236–245. doi: 10.1016/j.ecoenv.2003.07.001
67. Miranda, G. R. B., Raetano, C. G., Silva, E., Daam, M. A., and Cerejeira, M. J. (2011). Environmental fate of neonicotinoids and classification of their potential risks to hypogean, epygean, and surface water ecosystems in Brazil. *Hum. Ecol. Risk Assess.* 17, 981–995. doi: 10.1080/10807039.2011.588159
68. Morrissey, C. A., Mineau, P., Devries, J. H., Sánchez-Bayo, F., Liess, M., Cavallaro, M. C., et al. (2015). Neonicotinoid contamination of global surface waters and associated risk to aquatic invertebrates: a review. *Environ. Int.* 74, 291–303. doi: 10.1016/j.envint.2014.10.024
69. Morse, J. C., Stark, B. P., and McCafferty, P. W. (1993). Southern Appalachian streams at risk: implications for mayflies, stoneflies, caddisflies, and other aquatic biota. *Aquat. Conserv. Mar. Freshw. Ecosyst.* 3, 293–303. doi: 10.1002/aqc.3270030404
70. Motobayashi, T., Genka, M., Khanh Phong, T., and Watanabe, H. (2012). Effects of formulation and treatment method of imidacloprid in nursery boxes on aquatic insects inhabiting rice paddy fields. *Jpn. J. Appl. Entomol. Zool.* 56, 169–172. doi: 10.1303/jjaez.2012.169
71. Phong, T. K., Nhung, D. T. T., Motobayashi, T., Thuyet, D. Q., and Watanabe, H. (2009). Fate and transport of nursery-box-applied tricyclazole and imidacloprid in paddy fields. *Water Air Soil Pollut.* 202, 3–12. doi: 10.1007/s11270-008-9953-z
72. Nyman, A.-M., Hintermeister, A., Schirmer, K., and Ashauer, R. (2013). The insecticide imidacloprid causes mortality of the freshwater amphipod *Gammarus pulex* by interfering with feeding behavior. *PLoS ONE* 8:e62472. doi: 10.1371/journal.pone.0062472
73. Osborne, J. L. (2012). Ecology: bumblebees and pesticides. *Nature* 491, 43–45. doi: 10.1038/nature11637
74. Parsons, J. T., and Surgeoner, G. A. (1991). Effect of exposure time

on the acute toxicities of permethrin, fenitrothion, carbaryl and carbofuran to mosquito larvae. *Environ. Toxicol. Chem.* 10, 1219–1227. doi: 10.1002/etc.5620100913

75. Pestana, J. L. T., Alexander, A. C., Culp, J. M., Baird, D. J., Cessna, A. J., and Soares, A. M. V. M. (2009). Structural and functional responses of benthic invertebrates to imidacloprid in outdoor stream mesocosms. *Environ. Pollut.* 157, 2328–2334. doi: 10.1016/j.envpol.2009.03.027
76. Peveling, R., McWilliam, A. N., Nagel, P., Rasolomanana, H., Raholijaona, Rakotomianina, L., et al. (2003). Impact of locust control on harvester termites and endemic vertebrate predators in Madagascar. *J. Appl. Ecol.* 40, 729–741. doi: 10.1046/j.1365-2664.2003.00833.x
77. Phillips, P. J., and Bode, R. W. (2004). Pesticides in surface water runoff in south-eastern New York State, USA: seasonal and stormflow effects on concentrations. *Pest Manag. Sci.* 60, 531–543. doi: 10.1002/ps.879
78. Potts, G. R. (1986). *The Partridge - Pesticides, Predation and Conservation*. London: Collins.
79. Roessink, I., Merga, L. B., Zweers, H. J., and van den Brink, P. J. (2013). The neonicotinoid imidacloprid shows high chronic toxicity to mayfly nymphs. *Environ. Toxicol. Chem.* 32, 1096–1100. doi: 10.1002/etc.2201
80. Rondeau, G., Sánchez-Bayo, F., Tennekes, H. A., Decourtye, A., Ramírez-Romero, R., and Desneux, N. (2014). Delayed and time-cumulative toxicity of imidacloprid in bees, ants and termites. *Sci. Rep.* 4:5566. doi: 10.1038/srep05566
81. Sánchez-Bayo, F. (1991). "Temporal diversity and structure within a bird community," in *Diversidad Biologica/Biological Diversity*, eds F. D. Pineda, M. A. Casado, J. M. D. Miguel, and J. Montalvo (Madrid: Centro de Estudios Ramon Areces S.A.), 155–159.
82. Sánchez-Bayo, F. (2009). From simple toxicological models to prediction of toxic effects in time. *Ecotoxicology* 18, 343–354. doi: 10.1007/s10646-008-0290-1
83. Sánchez-Bayo F. (2011). "Impacts of agricultural pesticides on terrestrial ecosystems," in Ecological Impacts of Toxic Chemicals,

eds F. Sánchez-Bayo, P. J. van den Brink, and R. Mann (Beijing: Bentham Science Publishers), 63–87. Available online at: http://www.benthamdirect.com/51436/volume/1

84. Sánchez-Bayo, F. (2014). The trouble with neonicotinoids. *Science* 346, 806–807. doi: 10.1126/science.1259159
85. Sánchez-Bayo, F., Ahmad, R., and Goka, K. (2007). "Evaluation of the standard quotient and EcoRR methodologies based on field monitoring from rice fields," in *Rational Environmental Management of Agrochemicals*, eds I. R. Kennedy, K. R. Solomon, S. J. Gee, A. N. Crossan, S. Wang, and F. Sánchez-Bayo (Washington, DC: American Chemical Society), 66–86.
86. Sánchez-Bayo, F., and Goka, K. (2005). Unexpected effects of zinc pyrithione and imidacloprid on Japanese medaka fish (*Oryzias latipes*). *Aquat. Toxicol.* 74, 285–293. doi: 10.1016/j.aquatox.2005.06.003
87. Sánchez-Bayo, F., and Goka, K. (2006a). Ecological effects of the insecticide imidacloprid and a pollutant from antidandruff shampoo in experimental rice fields. *Environ. Toxicol. Chem.* 25, 1677–1687. doi: 10.1897/05-404R.1
88. Sánchez-Bayo, F., and Goka, K. (2006b). Influence of light in acute toxicity bioassays of imidacloprid and zinc pyrithione to zooplankton crustaceans *Aquat. Toxicol.* 78, 262–271. doi: 10.1016/j.aquatox.2006.03.009
89. Sánchez-Bayo, F., and Hyne, R. V. (2014). Detection and analysis of neonicotinoids in river waters - development of a passive sampler for three commonly used insecticides. *Chemosphere* 99, 143–151. doi: 10.1016/j.chemosphere.2013.10.051
90. Schaafsma, A., Limay-Rios, V., Baute, T., Smith, J., and Xue, Y. (2015). Neonicotinoid insecticide residues in surface water and soil associated with commercial maize (corn) fields in southwestern Ontario. *PLoS ONE* 10:e0118139. doi: 10.1371/journal.pone.0118139
91. Schaafsma, A., Limay-Rios, V., Xue, Y., Smith, J., and Baute, T. (2016). Field-scale examination of neonicotinoid insecticide persistence in soil as a result of seed treatment use in commercial maize (corn) fields in southwestern Ontario. *Environ. Toxicol.*

Chem. 35, 295–302. doi: 10.1002/etc.3231

92. Seagraves, M. P., and Lundgren, J. G. (2012). Effects of neonicitinoid seed treatments on soybean aphid and its natural enemies. *J. Pest Sci.* 85, 125–132. doi: 10.1007/s10340-011-0374-1
93. Siebers, J., Binner, R., and Wittich, K.-P. (2003). Investigation on downwind short-range transport of pesticides after application in agricultural crops. *Chemosphere* 51, 397–407. doi: 10.1016/S0045-6535(02)00820-2
94. Simon-Delso, N., Amaral-Rogers, V., Belzunces, L. P., Bonmatin, J. M., Chagnon, M., Downs, C., et al. (2015). Systemic insecticides (neonicotinoids and fipronil): trends, uses, mode of action and metabolites. *Environ. Sci. Pollut. Res.* 22, 5–34. doi: 10.1007/s11356-014-3470-y
95. Smit, C. E., Posthuma-Doodeman, C. J. A. M., van Vlaardingen, P. L. A., and de Jong, F. M. W. (2015). Ecotoxicity of imidacloprid to aquatic organisms: derivation of water quality standards for peak and long-term exposure. *Hum. Ecol. Risk Assess.* 21, 1608–1630. doi: 10.1080/10807039.2014.964071
96. Song, M. Y., Stark, J. D., and Brown, J. J. (1997). Comparative toxicity of four insecticides, including imidacloprid and tebufenozide, to four aquatic arthropods. *Environ. Toxicol. Chem.* 16, 2494–2500. doi: 10.1002/etc.5620161209
97. Sorg, M., Schwan, H., Stenmans, W., and MüLler, A. (2013). Ermittlung der Biomassen flugaktiver Insekten im Naturschutzgebiet Orbroicher Bruch mit Malaise Fallen in den Jahren 1989 und 2013. *Proc. Krefeld Entomol. Soc.* 1, 1–5.
98. Sorrell, B. K., and Boon, P. I. (1992). Biogeochemistry of billabong sediments. II Seasonal variations in methane production. *Freshw. Biol.* 27, 435–445. doi: 10.1111/j.1365-2427.1992.tb00552.x
99. Starner, K., and Goh, K. (2012). Detections of the neonicotinoid insecticide imidacloprid in surface waters of three agricultural regions of California, USA, 2010–2011. *Bull. Environ. Contam. Toxicol.* 88, 316–321. doi: 10.1007/s00128-011-0515-5
100. Stoughton, S. J., Liber, K., Culp, J., and Cessna, A. (2008). Acute and chronic toxicity of imidacloprid to the aquatic invertebrates *Chironomus tentans* and *Hyalella azteca* under

constant- and pulse-exposure conditions. *Arch. Environ. Contam. Toxicol.* 54, 662–673. doi: 10.1007/s00244-007-9073-6

101. Sur, R., and Stork, A. (2003). Uptake, translocation and metabolism of imidacloprid in plants. *Bull. Insectol.* 56, 35–40.

102. Suter, G. W. II, and Cormier, S. M. (2015). Why care about aquatic insects: uses, benefits, and services. *Integr. Environ. Assess. Manag.* 11, 188–194. doi: 10.1002/ieam.1600

103. Tennekes, H. A. (2010a). The significance of the Druckrey-Küpfmüller equation for risk assessment - the toxicity of neonicotinoid insecticides to arthropods is reinforced by exposure time. *Toxicology* 276, 1–4. doi: 10.1016/j.tox.2010.07.005

104. Tennekes, H. A. (2010b). *The Systemic Insecticides: A Disaster in the Making*. Zutphen: ETS Nederland BV.

105. Tennekes, H. A., and Sánchez-Bayo, F. (2013). The molecular basis of simple relationships between exposure concentration and toxic effects with time. *Toxicology* 309, 39–51. doi: 10.1016/j.tox.2013.04.007

106. Thuyet, D. Q., Jorgenson, B. C., Wissel-Tyson, C., Watanabe, H., and Young, T. M. (2012). Wash off of imidacloprid and fipronil from turf and concrete surfaces using simulated rainfall. *Sci. Total Environ.* 414, 515–524. doi: 10.1016/j.scitotenv.2011.10.051

107. Thuyet, D. Q., Watanabe, H., Yamazaki, K., and Takagi, K. (2011). Photodegradation of imidacloprid and fipronil in rice–paddy water. *Bull. Environ. Contam. Toxicol.* 86, 548–553. doi: 10.1007/s00128-011-0243-x

108. Tomizawa, M., and Casida, J. (2005). Neonicotinoid insecticide toxicology: mechanisms of selective action. *Annu. Rev. Pharmacol. Toxicol.* 45, 247–268. doi: 10.1146/annurev.pharmtox.45.120403.095930

109. Tomlin, C. D. S. (ed.). (2009). *The e-Pesticide Manual, 12th Edn*. Surrey: British Crop Protection Council.

110. Uğurlu, P., Ünlü, E., and Satar, E. I. (2015). The toxicological effects of thiamethoxam on *Gammarus kischineffensis* (Schellenberg 1937) (Crustacea: Amphipoda). *Environ. Toxicol. Pharmacol.* 39, 720–726. doi: 10.1016/j.etap.2015.01.013

111. USGS (2002). *Concentrations of Pesticides and Pesticide Degradates in the Croton River Water- Shed in Southeastern New York, July–September 2000*. New York, NY: United States Geological Survey.

112. van den Brink, P. J., Smeden, J. M. V., Bekele, R. S., Dierick, W., De Gelder, D. D., Noteboom, M., et al. (2016). Acute and chronic toxicity of neonicotinoids to nymphs of a mayfly species and some notes on seasonal differences. *Environ. Toxicol. Chem.* 35, 128–133. doi: 10.1002/etc.3152

113. van den Brink, P. J., Wijngaarden, R. P. A. V., Lucassen, W. G. H., Brock, T. C. M., and Leeuwangh, P. (1996). Effects of the insecticide Dursban 4E (active ingredient chlorpyrifos) in outdoor experimental ditches: II. Invertebrate community responses and recovery. *Environ. Toxicol. Chem.* 15, 1143–1153.

114. van der Sluijs, J. P., Amaral-Rogers, V., Belzunces, L. P., Lexmond, M. F. I. J. B. V., Bonmatin, J.-M., Chagnon, M., et al. (2015). Conclusions of the Worldwide Integrated Assessment on the risks of neonicotinoids and fipronil to biodiversity and ecosystem functioning. *Environ. Sci. Pollut. Res.* 22, 148–154. doi: 10.1007/s11356-014-3229-5

115. van der Sluijs, J. P., Simon-Delso, N., Goulson, D., Maxim, L., Bonmatin, J.-M., and Belzunces, L. P. (2013). Neonicotinoids, bee disorders and the sustainability of pollinator services. *Curr. Opin. Environ. Sustain.* 5, 293–305. doi: 10.1016/j.cosust.2013.05.007

116. van Dijk, T. C., van Staalduinen, M. A., and van der Sluijs, J. P. (2013). Macro-invertebrate decline in surface water polluted with imidacloprid. *PLoS ONE* 8:e62374. doi: 10.1371/journal.pone.0062374

117. Vehovszky, Á., Farkas, A., Ács, A., Stoliar, O., Székács, A., Mörtl, M., et al. (2015). Neonicotinoid insecticides inhibit cholinergic neurotransmission in a molluscan (*Lymnaea stagnalis*) nervous system. *Aquat. Toxicol.* 167, 172–179. doi: 10.1016/j.aquatox.2015.08.009

118. Vijver, M. G., and van den Brink, P. J. (2014). Macro-invertebrate decline in surface water polluted with imidacloprid: a rebuttal and some new analyses. *PLoS ONE* 9:e89837. doi: 10.1371/journal.

pone.0089837

119. von der Ohe, P. C., Prüss, A., Schäfer, R. B., Liess, M., de Deckere, E., and Brack, W. (2007). Water quality indices across Europe - a comparison of the good ecological status of five river basins. *J. Environ. Monit.* 9, 970–978. doi: 10.1039/b704699p

120. Weston, D. P., Chen, D., and Lydy, M. J. (2015). Stormwater-related transport of the insecticides bifenthrin, fipronil, imidacloprid, and chlorpyrifos into a tidal wetland, San Francisco Bay, California. *Sci. Total Environ.* 527–528, 18–25. doi: 10.1016/j.scitotenv.2015.04.095

121. Woodcock, B. A., Isaac, N. J. B., Bullock, J. M., Roy, D. B., Garthwaite, D. G., Crowe, A., et al. (2016). Impacts of neonicotinoid use on long-term population changes in wild bees in England. *Nat. Commun.* 7:12459. doi: 10.1038/ncomms12459

122. Yamamoto, A., Terao, T., Hisatomi, H., Kawasaki, H., and Arakawa, R. (2012). Evaluation of river pollution of neonicotinoids in Osaka City (Japan) by LC/MS with dopant-assisted photoionisation. *J. Environ. Monit.* 14, 2189–2194. doi: 10.1039/c2em30296a

9
CHAPTER

RISKS OF USING ANTIFOULING BIOCIDES IN AQUACULTURE

Francisco Antonio Guardiola, Alberto Cuesta, José Meseguer and Maria Angeles Esteban

Department of Cell Biology and Histology, Faculty of Biology, University of Murcia, Murcia 30100, Spain

ABSTRACT

Biocides are chemical substances that can deter or kill the microorganisms responsible for biofouling. The rapid expansion of the aquaculture industry is having a significant impact on the marine ecosystems. As the industry expands, it requires the use of more drugs, disinfectants and antifoulant compounds (biocides) to eliminate the microorganisms in the aquaculture facilities. The use of biocides in the aquatic environment, however, has proved to be harmful as it has toxic effects on the marine environment. Organic booster biocides were recently introduced as alternatives to the organotin compounds found in antifouling products after restrictions were imposed on the use of tributyltin (TBT). The replacement products are generally based on copper metal oxides and organic biocides. The biocides that are most commonly used in antifouling paints include chlorothalonil, dichlofluanid, DCOIT (4,5-dichloro-2-n-octyl-4-isothiazolin-3-one, Sea-nine 211®), Diuron, Irgarol 1051, TCMS pyridine (2,3,3,6-tetrachloro-4-methylsulfonyl pyridine), zinc pyrithione and Zineb. There are two types of risks associated with the use of biocides in aquaculture: (i) predators and humans may ingest the fish and shellfish that have accumulated in these contaminants and (ii) the development of antibiotic resistance in bacteria. This paper provides an overview of the effects of antifouling (AF) biocides on aquatic organisms. It also provides some insights into the effects and risks of these compounds on non-target organisms.

Keywords: biocides; antifouling; environment; aquaculture; resistance; antibiotics

INTRODUCTION

According to the Biocides Directive (98/8/EC) [1], biocides are active substances or preparations that are intended to destroy, deter, render harmless and exercise control or prevent the action of any other harmful organism through chemical or biological means. Biocides are classified into 23 different product types, each of which is comprised of multiple subgroups. Biocides are used because of their potential to destroy a wide range of organisms and for their relatively easy applicability to vessels [2] and aquaculture systems. The settlement of microorganisms, plants and animals is a natural phenomenon that occurs continuously and vigorously on immersed surfaces. This process is called biofouling [3] (Figure 1). Biofouling is a problem for any structure placed in the aquatic environment. It can be controlled through the use of both chemical biocides and non-biocidal technologies [4].

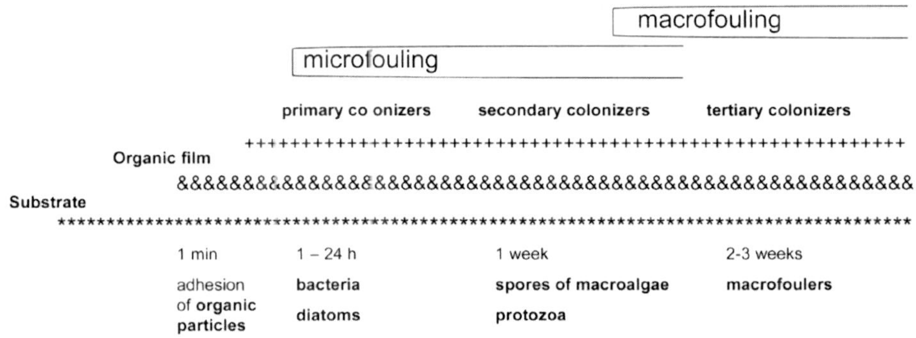

Figure 1: Temporal organization of biofouling (modified from [5]).

Aquaculture in general, and the fish farming industry in particular, suffer significantly from the effects of biofouling [6]. The aquaculture industry makes periodic discharges of wastes from farm activities. These waste products include detergents, effluent from net washing, antifoulants, heavy metals and even chemicals, such as drugs [7]. The chemicals are essential for aquaculture as they help increase and control the production of seeds in hatcheries, increase feeding efficiency, improve survival rates, control pathogens and diseases and reduce transport stress [8,9].

Nevertheless, despite the beneficial effects of the chemicals to aquaculture, they may also cause potential harm to aquatic organisms [10–12] and even to humans. The chemicals may be ingested by farmed fish and shellfish, which are, in turn, consumed by humans. Ingestion of the contaminated fish and shellfish can pose a great risk to human health [9,13]. The conditions and locations of the aquaculture farms play a significant role on the spread of these chemicals and heavy metals into the environment [14].

Marine pollution caused by the chemicals utilised in aquaculture activities, however, is not yet well documented. In addition, available information indicates low concentrations (low ng L^{-1} level) [15] of these compounds in the environment. This is due to factors, such as the complexity of the matrix, the high dilution factor, and degradation phenomena [9,16]. Nevertheless, the health risk in animals and humans may increase when bacterial resistance to antibiotics and heavy metals caused by the use of biocides occurs [17–19]. The aim of this study is to review the main effects and risks of using antifouling biocides in aquaculture on aquatic systems, shellfish, fish and humans.

PRESENT SITUATION OF BIOCIDE USE IN AQUACULTURE

Aquaculture is the farming of aquatic organisms, including fish, molluscs, crustaceans and aquatic plants, using techniques designed to increase the production and productivity of these organisms beyond the natural capacity of the environment [9]. Since wild fish stocks are reaching the limits of exploitation, we have to rely to a far greater extent on products produced from aquaculture [20]. However, the practice of aquaculture has become so widespread that it has begun to have significant impact on the environment and on natural resources. A number of concerns have been expressed by both environmental activists and scientists regarding this practice [21–25].

With the rapid expansion of the aquaculture industry and with the tightening of the legislation on the use of antifouling (AF) biocides, the problems of aquaculture biofouling have increased [5,6,26,27]. The herbicides or fungicides currently used in aquaculture were originally developed for use in agriculture or as additives for boat anti-fouling paints. As such, the published data regarding their occurrence in marine waters are mainly related to such activities [15,28,29]. Accordingly, many studies have investigated and demonstrated the presence of pesticides and biocides in surface waters [30–34].

With the gradual elimination of triorganotin-based formulations (e.g., tributyltin (TBT)), copper has become the principal biocidal component of most AF paints. It usually comes in the form of copper oxide (Cu_2O) [5]. Inorganic zinc is often used in combination with copper to increase the overall toxicity of the formulation or to facilitate the leaching process [35]. Organic booster biocides, such as Irgarol 1051®, Sea Nine 211®, dichlofluanid, chlorothalonil, zinc pyrithione, and Zineb are also added to the paint to enhance its effectiveness [36]. The main AFs used in aquaculture and their effect are shown in Table 1. Nevertheless, these alternatives to TBT are also toxic and their contamination of the aquatic environment has been a topic of increasing importance in recent years [29]. Several studies have evaluated the toxicity of booster biocides on non-target species and have found most of them to be growth inhibitors for freshwater and marine autotrophs [37], influencing key species, such as sea grasses [38] and corals [39]. Therefore, there is increasing interest in the impact of these compounds on the aquatic ecosystems [40].

Table 1: The main antifouling biocides used in aquaculture and their effect on aquatic organisms

Common Names [a]	Application [a]	Mode of Action [a]	Species	Effects
Chorothalonil	Fungicide	Inhibition of mitochondrial electron transport	**Crustaceans**	
			Cancer magister	Behaviour Larval mortality [43]
			Penaeus duoramun	Mortality [44]
			Molluscs	
			Crassostrea virginica	Growth [45] Embryotoxicity [46]
			Mytilus edulis	Embryotoxicity [47] Mortality [48]
			Tunicates	
			Ciona intestinalis	Embryotoxicity Inhibition of larval settlement [47]
			Teleosts	
			Anguilla japonica	Mortality [49]
			Cyprinodon variegates	Mortality [45]
			Galaxias auratus	Mortality [50]
			Galaxias maculates	Mortality [50]
			Galaxias truttaceus	Mortality [50]
			Leiostomus xanthurus	Mortality [44]
			Pseudaphritis urvillii	Juvenile mortality [51]
Copper pyrithione (CuPT)	Microbicide	Multi-site inhibitor (metabolic processes)	**Crustaceans**	
			Artemia salina	Inhibition of Na/K ATPase and Mg^{2+} ATPase enzyme activities [52]
			Molluscs	
			Mytilus galloprovincialis	Alterations in Na/K ATPase and Ca^{2+} ATPase activities [53]
			Teleosts	
			Fundulus heteroclitus	Alterations in gill osmoregulation [54]
			Oncorhynchus kisutch	Alterations in gill osmoregulation [55]
Dichlofuanid	Fungicide	Inhibitor of PS II electron transport	**Echinoderms**	
			Glyptocidaris crenularis	Embryotoxicity [56]

Common Names [a]	Application [a]	Mode of Action [a]	Specie	Effects
DCOIT (Sea-Nine 211®)	Herbicide	Inhibitor of PS II electron transport	**Crustaceans**	
			Balanus amphitrite	Larva mortality [57,58]
			Molluscs	
			Crassostrea virginica	Embryo-larva immobility [46]
			Mytilus edulis	Embryo-larva immobility and embryotoxicity [46,47]
			Echinoderms	
			Hemicentrotus pulcherrimus	Embryotoxicity [59]
			Anthocidaris crassispina	Embryotoxicity [59]
			Tunicates	
			Ciona intestinalis	Embryotoxicity and inhibition of larval settlement [47]
			Teleosts	
			Cyprinodon variegatus	Mortality [46]
Diuron	Herbicide	Inhibitor of PS II electron transport	**Algae**	
			Scenedesmus vacuolatus	Toxic for the reproduction [60]
			Microalgae	
			planktonic periphytic	Reduction of chlorophyll *a* levels [61–63]
			Teleosts	
			Carassius auratus	Acetylcholinesterase inhibition [64]
Irgarol-1051	Herbicide	Inhibitor of PS II electron transport	**Algae**	
			Dunaliella tertiolecta *Synechococcus sp* *Emiliania huxleyi* *Zostera marina*	Decreasing in growth, inhibition of cell number and decrease in the
			Fucus vesiculosus *Enteromorpha intestinalis* *Ulva intestinalis*	photosynthetic activity [38,65–70]
TCMS pyridine (2,3,3,6-tetrachloro-4-methylsulfonylpyridine)	Fungicide	Inhibitor of mitochondrial electron transport	**Tunicates**	
			Botryllus schlosseri	Immunotoxic [71,72]
Zinc pyrithione (ZnPT)	Microbicide	Multi-site inhibitor (metabolic processes)	**Teleosts**	
			Oryzias latipes	Embryotoxicity [73]
			Danio rerio	Embryo-larva [74]
Zineb	Fungicide	Multi-site inhibitor (metabolic processes)	**Teleosts**	
			Salmo gairdneri	Embryotoxicity [75]

[a] From [76].

In the aquatic environment, fishes have been found appropriate to be used as a model for the immunotoxicity testing because they are representatives of aquatic organisms and, therefore, bioindicators of aquatic animal health. As vertebrates that have immune systems strikingly similar to those of mammals, they can also be used to identify potential threats to terrestrial wildlife and humans [41,42]. The risk to predators and humans through the consumption of fish is very low, especially for humans, since the latter are less exposed to the dangers of contamination due to the fact that fish constitutes only a small part of their diet [9]. However, the risk may be increased by mechanisms of resistance.

THE MAIN TYPE OF ANTIFOULING USED IN AQUACULTURE AND ITS EFFECT ON AQUATIC ORGANISMS

Chorothalonil

Chlorothalonil (2,4,5,6-tetrachloroisophthalonitrile) is a pesticide used widely in agriculture, silviculture and urban settings. This pesticide can enter surface waters through rainfall runoff, spray drift or atmospheric deposition, subsequently impacting aquatic biota [77]. It is used as a booster biocide in marine paints as one of the chemicals replacing the widely banned organotin fungicides, such as tributyltin, resulting in greater potential for chlorothalonil contamination of marine waters and sediments [78,79]. Chlorothalonil is a broad-spectrum fungicide with a K^{ow} of 2.64–4.28 and a water solubility of 0.9 mg L^{-1} [80].

Chlorothalonil can be acutely toxic (50% lethal concentration, LC_{50}) to fish following 96 h exposures ranging from 8.2 to 76 μg L^{-1}, depending on the species and the exposure conditions [48,51]. Chlorothalonil can accumulate in the tissue of fish. Bioaccumulation factors have been reported to be 18 for willow shiner (*Gnathopogon caerulescens*) and 25 for carp (*Cyprinus carpio*) following sublethal exposures (1.1–1.4 μg L^{-1}) [81]. It has been suggested that leukocytes may be a potential target of toxicity because significant decreases in leukocyte values were found in the Australian freshwater fish *Pseudaphritis urvulii,* which was exposed for 10 d to 4.4 μg L^{-1} chlorothalonil [51]. *In vitro* studies have demonstrated that the exposure of fish (*Morone saxatilus*) macrophages and oyster hemocytes to chlorothalonil (10 ± 500 µg L^{-1}) suppressed immunostimulated ROS (reactive oxygen species) and baseline NADPH (nicotinamide adenine dinucleotide phosphate) concentration but did not inhibit phagocytosis [82,83]. There are numerous toxicity studies for chlorothalonil on marine animals, such as crustaceans [43–46], molluscs [44–48], tunicates [47] and teleosts [44–46,49–51].

Copper Oxide

Copper is an essential metal. However, although it is an effective biocide, it may also affect non-target organisms and cause environmental concerns [84]. The toxicity of copper in water is greatly affected by the chemical form or speciation of the copper and to what degree it is bound to various ligands that may be in the water, making the copper unavailable to organisms [85]. The speciation is essential for understanding the copper's bioavailability and subsequent toxicity to aquatic organisms [4]. Copper oxide leaches from the boat surfaces and enters the water as a free copper ion (Cu^{+}), which is immediately oxidised to Cu^{2+} and forms complexes with inorganic and organic ligands [4].

Copper is a trace element needed at miniscule levels for the proper functioning of all organisms [84]. However, it can be toxic at higher concentrations [85]. Copper is generally toxic to aquatic organisms, with a lethal concentration 50 (LC_{50}) value varying from 5 to 100,000 µg L^{-1} [86,87]. However, organisms have different mechanisms by which they cope with and process copper [84]. Generally, copper is actively regulated in fish, decapod crustaceans and algae. It is stored in bivalves, barnacles and aquatic insects [84,88].

The bioavailability, biodistribution to various parts of the organism and bioaccumulation of copper are dramatically influenced by water chemistry. Therefore, water pH, hardness, organic content and salinity play important roles in copper-induced toxicity [84,85]. Thus, increased pH accentuates copper toxicity because of the reduced competition between copper and hydrogen ions at the cell surface [84,89]. In a similar manner, cations that are involved in water hardness also compete with Cu^{2+} for biological binding sites [84,90].

Copper bound to organic matter is widely thought to be non-bioavailable and, therefore, non-toxic [4,91,92]. Dissolved organic carbon (DOC) content is among the most important factors in reducing copper toxicity in both fresh- and salt-water species [84]. DOC forms organic complexes with copper, thereby reducing copper's bioavailability [84]. The effects of DOC on reducing the toxicity of copper have been reported in fish [93,94], bivalves [92], echinoderms [95], macroalgae [96], unicellular algae [97], estuarine copepod [98] and planktonic crustaceans [99]. Some authors confirm that water salinity influences the biodistribution and bioaccumulation of copper, affecting its toxicity [54,98,100–102]. Therefore, in oysters, copper accumulation was inversely related to salinity [100].

Copper causes toxicity by impairing the osmoregulation and ion regulation in the gill of numerous aquatic animals [54,55]. In brine shrimp, copper inhibited the Na/K ATPase and Mg^{2+} ATPase enzyme activity [52]. In mussel, *Mytilus galloprovincialis*, copper interfered with Ca^{2+} homeostasis in the gill, causing alterations in the Na/K ATPase and Ca^{2+} ATPase [53]. Copper depresses the transcription of key genes within the olfactory signal transduction pathway [103]. Additionally, copper toxicity can be induced by generating reactive oxygen species (ROS) [53,104].

It seems remarkable that phytoplankton species have different sensitivities to copper toxicity: resistant (diatoms), intermediate sensitivity (coccolithophores and dinoflagellates) and most sensitive (cyanobacteria) [105,106].

Dichlofluanid

Dichlofluanid(*N*-dichlorofluoromethylthio-*N*0-dimethyl-*N*-phenylsulphamide) has been commonly used as a herbicide on crops (Lee *et al.*, 2010). Dichlofluanid has a lower toxicity compared with other AF agents, although some studies have identified its toxic effects [107–109], such as embryotoxicity in sea urchin, *Glyptocidaris crenularis* [56].

DCOIT (Sea Nine 211®)

One of the new alternative biocides is 4,5-dichloro-2-*n*-octyl-4-isothiazolin-3-one (DCOIT), the active ingredient of the Sea Nine 211® AF Agent manufactured by Rohm and Haas Company [110].

Aquatic microcosm and marine sediment studies demonstrate that the predominant route of DCOIT dissipation in the marine environment is its rapid biodegradation [110]. DCOIT predominantly undergoes biotic degradation under both aerobic and anaerobic conditions with biological degradation over 200 times faster than hydrolysis or photolysis [4,58,111]. Biodegradation is a very effective mechanism for the detoxification of the compound since the resulting metabolites are five orders of magnitude less toxic than the parent compound [112,113]. However, Sea-Nine antifoulant is acutely toxic to a wide range of aquatic organisms although no chronic toxicological effects have been observed in the extensive toxicology tests conducted on it [114]. DCOIT has a log K_{OW} of 2.8 and an aqueous solubility of 14 mg L^{-1} [4].

There are numerous studies that have investigated the toxicity and effects of DCOIT on marine animals. These studies demonstrated the following: larval mortality in crustaceans [57,58]: embryo-larva immobility and embryotoxicity in molluscs [46,47], embryotoxicity in echinoderms [59], embryotoxicity and inhibition of larval settlement in tunicates [47] and mortality in teleosts [46,115].

Diuron

Diuron (1-(3,4-dichlorophenyl)-3,3-dimethylurea) also persists in seawater, but it is less persistent in marine sediments with a half-life of 14 days [116,117]. Diuron is relatively soluble in water (35 mg L^{-1}) and has a reported log K_{OW} of 2.8 [4]. Diuron is present at high concentrations in marine surface waters but it has only been detected at low concentrations in sediments [118,119]. Diuron is persistent in the marine environment and partitions poorly between water and sediments. It can remain suspended and available for uptake by marine organisms [120].

While the toxic effect of the antifoulant herbicide diuron to the photosynthetic aquatic biota has been widely studied, its sublethal effects on the different life stages of fish have been under-reported [121]. Diuron has been proven to be very toxic for the reproduction of the green freshwater alga *Scenedesmus vacuolatus* [60]. It has also been proven to affect planktonic and periphytic microalgae by reducing the chlorophyll *a* levels [61–63]. Moreover, it has been proven to be toxic to certain bacterial species [122–124].

Irgarol-1051®

Irgarol-1051 (2-methylthio-4-terbutylamino-6-cyclopropylamino-s-triazine) is a slightly soluble and moderately lipophilic triazine herbicide used in concert with copper to control fouling on boat hulls [125]. Irgarol inhibits electron transport in the photosystem II (PSII) [126] by binding to the D1 protein [127]. Irgarol may affect non-target photosynthetic organisms, such as phytoplankton, periphyton and aquatic macrophytes [128] when leaching into the marine environment [129].

Only a few studies have addressed the possible effect of Irgarol on marine non-target algae [130]. The effect of Irgarol on green alga *Dunaliella tertiolecta* [65], *Synechococcus sp* and *Emiliania huxleyi* [66], natural phytoplankton communities [131], periphyton colonization [129] and phytoplankton species [130,132,133] has been investigated and the results showed a decrease in growth, inhibition in cell number and a decrease in the photosynthetic activity of these organisms. These effects have been seen in many different marine plants and algae, such as the eelgrass *Zostera marina* [38,67], the brown macroalga *Fucus serratus* [69], the green macroalga *Enteromorpha intestinalis* [70] and the green macroalga *Ulva intestinalis* [71].

TCMS Pyridine

TCMS (2,3,5,6-tetrachloro-4-methylsulphonyl pyridine), which was used in both the textile and leather industries, is one of the more recent AF compounds introduced to the market [134]. The toxicity of TCMS towards living organisms has already been evidenced [29,135,136] and substantiated in *in vitro* studies [137,138]. TCMS has been found to cause immunotoxic effects at concentrations higher than 10 μM in haemocyte cultures of the colonial ascidian *Botryllus schlosseri*, causing oxidative stress in the process [71,72]. Both diuron and TCMS pyridine exerted immunosuppressant effects on the Botryllus hemocytes when used at concentrations higher than 250 μM and 10 μM, respectively, causing (i) deep changes in the cytoskeleton that irreversibly affect cell morphology and phagocytosis; (ii) induction of DNA damage; and (iii) leakage of oxidative and hydrolytic enzymes due to membrane alteration. Unlike organotin compounds, diuron and TCMS pyridine do not inhibit cytochrome-c-oxidase and only TCMS pyridine triggers oxidative stress.

Zinc Pyrithione

Zinc pyrithione (ZnPT) (bis(1hydroxy-2(1*H*)-pyridethionato-o,s)-(T-4) zinc), one of the most popular surrogate AF biocides, has long been widely used as algaecide, bactericide and fungicide [5,139]. ZnPT was found to be highly toxic to aquatic plants and animals [140], but it was assumed to be environmentally neutral because it could easily photo-degrade to less toxic compounds [140,141]. ZnPT is toxic to Japanese medaka fish (*Oryzias latipes*) and also causes teratogenic effects, such as spinal cord deformities in embryos and on the larvae of zebra fish (*Danio rerio*) [74] at very low sublethal concentrations [73]. However, there is a lack of data on the toxicity of ZnPT [139].

Zineb

Zineb (zinc ethylenebis-(dithiocarbamate)) is a widely used foliar fungicide with prime agricultural and industrial applications [142]. Zineb has been registered for use on fruits, vegetables, field crops, ornamental plants and for the treatment of many seeds [142]. It has also been registered as a fungicide in paints and for mould control on fabrics, leather, linen, painted and wood surfaces, and so on [143]. The occurrence of the dithiocarbamates in coastal environments was not reported until 2009 [144] although it is known that these compounds exhibit teratogenicity in fish embryos at relatively low concentrations [75].

Capsaicin, Econea and Medetomidine

Capsaicin, Econea and medetomidine can be collectively termed as "emerging" biocides [4]. Capsaicin (8-methyl-*n*-vanillyl-6-nonenamide) is a compound that may emerge as an AF biocide in the future. It has even been evaluated as a marine AF [4,145]. Econea (2-(p-chlorophenyl)-3-cyano-4- bromo-5-trifluoromethyl pyrrole) is being marketed as a metal-free biocidal additive replacement for copper [4]. Medetomidine (4-[1-(2,3-dimethylphenyl)ethyl]-3Himidazole), on the other hand, is a neuroactive catemine that has been shown to be effective in preventing barnacle cyprid settlement by interfering with the regulation of cement production [4,146].

BIOACCUMULATION

The bioconcentration of pesticides and other chemicals into aquatic organisms mainly proceeds by passive diffusion through gills, epithelial tissues, or the gastrointestinal tract [147]. Bioconcentration factors (BCFs) are available for certain biocides in specific tissues. They represent the concentration of a biocide in the tissue per concentration of the biocide in water (L kg^{-1}) [4]. DCOIT has been shown to bioaccumulate in fish at very low levels following exposure to radiolabelled DCOIT [58]. There are no reports of the bioaccumulation of diuron with BCFs of 75 and 22 L kg^{-1}, suggesting that its accumulation in aquatic organisms is unlikely [4,148]. Irgarol 1051 accumulates in freshwater macrophytes [149] and marine macrophytes [67] with BCFs of up to 30,000 L kg^{-1}. It also accumulates in the green alga *Tetraselmis suecica* under laboratory conditions with BCFs of up to 150,000 mL g^{-1} [150]. In addition, the accumulation of Zineb in trout (*Salmo gairdneri*) is reported to be low with a BCF of <100 L kg^{-1} [4].

RESISTANCE

Scientific evidence from bacteriological, biochemical and genetic data indicate that the use of active molecules in the biocidal products may contribute to the increased occurrence of antibiotic resistant bacteria. The selective stress exerted by biocides may favour the existence of bacteria expressing resistance mechanisms and their dissemination. Some biocides have the capacity to maintain the presence of mobile genetic elements that carry genes involved in cross-resistance between biocides and antibiotics. The dissemination of these mobile elements, their genetic organisation and the formation of biofilms, provide conditions that could create a potential risk of development of cross-resistance between antibiotics and biocides [151].

CONCLUSIONS

Biocides are used as components in paints to coat the structures of vessels, as a means of disinfecting aquaculture facilities and cages, as well as in controlling the biofouling phenomenon (antifouling). The use of biocides is not as well-regulated as drug use in aquaculture because the information available on the effects of these agents to the marine ecosystems is still limited. Hence, it is important to know the risks associated with the existence of those biocides in the marine environment. It is also important to evaluate the effects of these compounds through the continuous monitoring of biocide concentration profiles in water, sediment and biota to provide information that could lead to concerted action to ban or regulate their use.

ACKNOWLEDGEMENTS

This work was partly supported by projects of the Ministry of Education and Science (AGL2008-05119-C02-01) and the *Fundación Séneca*(04538/GERM/06). A. Cuesta thanks the Ministry of Science and Innovation of Spain the Ramón y Cajal contract.

REFERENCES

1. European Parliament and of the Council. Directive 98/8/EC of the European Parliament and of the Council of 16 February Concerning the Placing of Biocidal Products on the Market. *Official Journal of the European Communities L 123/1 of 24.04.98* **1998**.
2. La Carbona, S.; Viitasalo-Frösen, S.; Masson, D.; Sassi, J.; Pineau, S.; Lehtiniemi, M.; Corroler, D. Efficacy and environmental acceptability of two ballast water treatment chemicals and an alkylamine based-biocide. *Sci. Total Environ* **2010**, *409*, 247–255.
3. Löschau, M.; Krätke, R. Efficacy and toxicity of self-polishing biocide-free antifouling paints. *Environ. Pollut* **2005**, *138*, 206–267.
4. Thomas, K.V.; Brooks, S. The environmental fate and effects of antifouling paint biocides. *Biofouling* **2010**, *26*, 73–88.
5. Yebra, D.M.; Kiil, S.; Dam-Johansen, K. Antifouling technology-past, present and future step towards efficient and environmentally friendly antifouling coatings. *Prog. Org. Coat* **2004**, *50*, 75–104.
6. Bazes, A.; Silkina, A.; Defer, D.; Bernède-Bauduin, C.;

Quéméner, E.; Braud, J.P.; Bourgougnon, N. Active substances from *Ceramium botryocarpum* used as antifouling products in aquaculture. *Aquaculture* **2006**, *258*, 664–674.

7. Read, P.; Fernandes, T. Management of environmental impacts of marine aquaculture in Europe. *Aquaculture* **2003**, *226*, 139–163.
8. Huntington, T.C.; Roberts, H.; Cousins, N.; Pitta, V.; Marchesi, N.; Sanmamed, A.; Hunter-Rowe, T.; Fernandes, T.F.; Tett, P.; McCue, J.; *et al. Some Aspects of the Environmental Impact of Aquaculture in Sensitive Areas*; Report to the DG Fish and Maritime Affairs of the European Commission: Brussels, Belgium, 2006.
9. Muñoz, I.; Martínez Bueno, M.J.; Agüera, A.; Fernández-Alba, A.R. Environmental and human health risk assessment of organic micro-pollutants occurring in a Spanish marine fish farm. *Environ. Pollut* **2010**, *158*, 1809–1816.
10. Liess, M.; Schulz, R. Linking insecticide contamination and population response in an agricultural stream. *Environ. Toxicol. Chem***1999**, *18*, 1948–1955.
11. Chèvre, N.; Loepfe, C.; Singer, H.; Stamm, C.; Fenner, K.; Escher, B.I. Including mixtures in the determination of water quality criteria for herbicides in surface water. *Environ. Sci. Technol* **2006**, *40*, 426–435.
12. Wittmer, I.K.; Scheidegger, R.; Bader, H.P.; Singer, H.; Stamm, C. Loss rates of urban biocides can exceed those of agricultural pesticides. *Sci. Total Environ* **2011**, *409*, 920–932.
13. Hites, R.A.; Foran, J.A.; Carpenter, D.O.; Hamilton, M.C.; Knuth, B.A.; Schwager, S.J. Global assessment of organic contaminants in farmed salmon. *Science* **2004**, *303*, 226–229.
14. Gräslund, S.; Bengtsson, B. Chemicals and biological products used in southeast Asian shrimp farming, and their potential impact on the environment—A review. *Sci. Total Environ* **2001**, *280*, 93–131.
15. Hernando, M.D.; De Vettori, S.; Bueno, M.J.M.; Fernández-Alba, A.R. Toxicity evaluation with *Vibrio fischeri* test of organic chemicals used in aquaculture. *Chemosphere* **2007**, *68*, 724–730.
16. Pouliquen, H.; Delépée, R.; Larhantec-Verdier, M.; Morvan, M.; Le Bris, H. Comparative hydrolysis and photolysis of four antibacterial agents (oxytetracycline oxolinic acid, flumequine

and florfenicol) in deionised water, freshwater and seawater under abiotic conditions. *Aquaculture* **2007**, *262*, 23–28.

17. Brözel, V.S.; Cloete, T.E. Resistance of *Pseudomonas aeruginosa* to isothiazolone. *J. Appl. Bacteriol* **1994**, *76*, 576–582.
18. Bruins, M.R.; Kapil, S.; Oehme, F.W. Microbial resistance to metals in the environment. *Ecotoxicol. Environ. Saf. Mar* **2000**, *45*, 198–207.
19. Chelossi, E.; Faimali, M. Comparative assessment of antimicrobial efficacy of new potential biocides for treatment of cooling and ballast waters. *Sci. Total Environ* **2006**, *356*, 1–10.
20. Penders, J.; Stobberingh, E.E. Antibiotic resistance of motile aeromonads in indoor catfish and eel farms in the southern part of The Netherlands. *Int. J. Antimicrob. Agents* **2008**, *31*, 261–265.
21. Dierberg, F.E.; Kiattisimukul, W. Issues, impacts, and implications of shrimp aquaculture in Thailand. *Environ. Manag* **1996**, *20*, 649–666.
22. Goldburg, R.; Triplett, T. *Murky Waters: Environmental Effects of Aquaculture in the United States*; Environmental Defense Fund: Washington, DC, USA, 1997; pp. 1–198.
23. Naylor, R.L.; Goldburg, R.J.; Mooney, H.; Beveridge, M.; Clay, J.; Folke, C.; Kautsky, N.; Lubchenco, J.; Primavera, J.; Williams, M. Nature's subsidies to shrimp and salmon farming. *Science* **1998**, *282*, 883–884.
24. Naylor, R.L.; Goldburg, R.J.; Primavera, J.H.; Kautsky, N.; Beveridge, M.C.M.; Clay, J.; Folke, C.; Lubchenco, J.; Mooney, H.; Troell, M. Effect of aquaculture on world fish supplies. *Nature* **2000**, *405*, 1017–1024.
25. Boyd, C.E. Guidelines for aquaculture effluent management at the farm-level. *Aquaculture* **2003**, *226*, 101–112.
26. Fusetani, N. Biofouling and antifouling. *Nat. Prod. Rep* **2004**, *21*, 94–104.
27. Braithwaite, R.A.; McEvoy, L.A. Marine biofouling on fish farms and its remediation. *Adv. Mar. Biol* **2005**, *47*, 215–252.
28. Martinez, K.; Ferrer, I.; Hernando, M.D.; Fernandez-Alba, A.R.; Marce, R.M.; Borrull, F.; Barcelo, D. Occurrence of antifouling

biocides in the Spanish Mediterranean marine environment. *Environ. Technol* **2001**, *22*, 543–552.

29. Konstantinou, I.K.; Albanis, T.A. Worldwide occurrence and effects of antifouling paint booster biocides in the aquatic environment. *Environ. Int* **2004**, *30*, 235–248.
30. Hoffmann, R.S.; Capel, D.C.; Larson, S.J. Comparison of pesticides in eight U.S. urban streams. *Environ. Toxicol. Chem* **2000**, *19*, 2249–2258.
31. Blanchoud, H.; Moreau-Guigon, E.; Farrugia, F.; Chevreuil, M.; Mouchel, J.M. Contribution by urban and agricultural pesticide uses to water contamination at the scale of the Marne watershed. *Sci. Total Environ* **2007**, *375*, 168–179.
32. Singer, H.; Jaus, S.; Hanke, I.; Lück, A.; Hollender, J.; Alder, A.C. Determination of biocides and pesticides by on-line solid phase extraction coupled with mass spectrometry and their behaviour in wastewater and surface water. *Environ. Pollut* **2010**, *158*, 3054–3064.
33. Wittmer, I.K.; Bader, H.P.; Scheidegger, R.; Singer, H.; Lück, A.; Hanke, I.; Carlsson, C.; Stamm, C. Significance of urban and agricultural land use for biocide and pesticide dynamics in surface waters. *Water Res* **2010**, *44*, 2850–2862.
34. Wittmer, I.K.; Scheidegger, R.; Bader, H.P.; Singer, H.; Stamm, C. Loss rates of urban biocides can exceed those of agricultural pesticides. *Sci. Total Environ* **2011**, *409*, 920–932.
35. Watermann, B.T.; Daehne, B.; Sievers, S.; Dannenberg, R.; Overbeke, J.C.; Klijnstra, J.W.; Heemken, O. Bioassays and selected chemical analysis of biocidefree antifouling coatings. *Chemosphere* **2005**, *60*, 1530–1541.
36. Parks, R.; Donnier-Marechal, M.; Frickers, P.E.; Turner, A.; Readman, J.W. Antifouling biocides in discarded marine paint particles. *Mar. Pollut. Bull* **2010**, *60*, 1226–1230.
37. Okamura, H.; Nishida, T.; Ono, Y.; Shim, W.J. Phytotoxic effects of antifouling compounds on nontarget plant species. *Bull. Environ. Contam. Toxicol* **2003**, *71*, 881–886.
38. Chesworth, J.C.; Donkin, M.E.; Brown, M.T. The interactive

effects of the antifouling herbicides Irgarol 1051 and Diuron on the seagrass *Zostera marina* (L.). *Aquat. Toxicol* **2004**, *66*, 293–305.

39. Owen, R.; Knap, A.; Toaspern, M.; Carbery, K. Inhibition of coral photosynthesis by the antifoling herbicide Irgarol 1051. *Mar. Pollut. Bull* **2002**, *44*, 623–632.
40. Sánchez-Rodríguez, A.; Sosa-Ferrera, Z.; Santana-del Pino, A.; Santana-Rodríguez, J.J. Probabilistic risk assessment of common booster biocides in surface waters of the harbours of Gran Canaria (Spain). *Mar. Pollut. Bull* **2011**, *62*, 985–991.
41. Anderson, D.P.; Zeeman, M.G. Immunotoxicology in Fish. In *Fundamentals of Aquatic Toxicology: Effects, Environmental Fate, and Risk Assessment*, 2nd ed; Rand, G.M., Ed.; Taylor & Francis: Washington, DC, USA, 1995; pp. 371–404.
42. Luebke, R.W.; Hodson, P.V.; Faisal, M.; Ross, P.S.; Grasman, K.A.; Zelikoff, J. Aquatic pollution-induced immunotoxicity in wildlife species. *Fundam. Appl. Toxicol* **1997**, *37*, 1–15.
43. Armstrong, D.A.; Buchanan, D.V.; Caldwell, R.S. A mycosis caused by *Lagneidium* sp. in laboratory-reared larvae of the dungeness crab, *Cancer magister*, and possible chemical treatments. *J. Invertebr. Pathol* **1976**, *28*, 329–336.
44. Mayer, F.L. *Acute Toxicity Handbook of Chemicals to Estuarine Organisms, Gulf Breeze, Florida*; US Environmental Protection Agency: Washington, DC, USA, 1987.
45. Montforts, M.H.M.M. *Chlorothalonil*; CSR Advisory Report 06105A00: Bilthoven, The Netherlands, 1999.
46. U.S. EPA Office of Pesticide Programs. *Pesticide Ecotoxicity Database (Formerly: Environmental Effects Database (EEDB))*; Environmental Fate and Effects Division, U.S. EPA: Washington, DC, USA, 2000.
47. Bellas, J. Comparative toxicity of alternative antifouling biocides on embryos and larvae of marine invertebrates. *Sci. Total Environ* **2006**, *367*, 573–585.
48. Ernst, W.; Doe, K.; Jonah, P.; Young, J.; Julien, G.; Hennigar, P. The toxicity of chlorothalonil to aquatic fauna and the impact of its operational use on a pond ecosystem. *Arch. Environ. Contam. Toxicol* **1991**, *21*, 1–9.

49. Yokoyama, T.; Saka, H.; Fujita, S.; Nishiuchi, Y. Sensitivity of Japanese eel, *Anguilla japonica*, to 68 kinds of agricultural chemicals. *Bull. Agric. Chem. Insp. Stn* **1988**, *28*, 26–33.

50. Davies, P.E.; White, R.W.G. The toxicology and metabolism of chlorothalonil in fish. I. Lethal levels for *Salmo gairdneri*, *Galaxias maculatus*, *G. truttaceus* and *G. auratus* and the fate of 14C-TCIN in *S. gairdneri*. *Aquat. Toxicol* **1985**, *7*, 93–105.

51. Davies, P.E.; Cook, L.S.J.; Goenarso, D. Sublethal responses to pesticides of several species of Australian freshwater fish and crustaceans and rainbow trout. *Environ. Toxicol. Chem* **1994**, *13*, 1341–1354.

52. Katranitsas, A.; Castritsi-Catharios, J.; Persoone, G. The effects of a copper-based antifouling paint on mortality and enzymatic activity of a non-target marine organism. *Mar. Pollut. Bull* **2003**, *46*, 1491–1494.

53. Viarengo, A.; Pertica, M.; Mancinelli, G.; Burlando, B.; Canesi, L.; Orunesu, M. *In vivo* effects of copper on calcium homeostasis mechanisms of mussel gill cell plasma membranes. *Comp. Biochem. Physiol. Part C* **1996**, *113*, 421–425.

54. Blanchard, J.; Grosell, M. Effects of salinity on copper accumulation in the common killifish (*Fundulus heteroclitus*). *Environ. Toxicol. Chem* **2005**, *24*, 1403–1413.

55. McIntyre, J.K.; Baldwin, D.H.; Meador, J.P.; Scholz, N.L. Chemosensory deprivation in juvenile Coho salmon exposed to dissolved copper under varying water chemistry conditions. *Environ. Sci. Technol* **2008**, *42*, 1352–1358.

56. Xu, X.; Wang, X.; Li, Y.; Wang, Y.; Wang, Y. Acute toxicity and synergism of binary mixtures of antifouling biocides with heavy metals to embryos of sea urchin *Glyptocidaris crenularis*. *Hum. Exp. Toxicol* **2011**, *30*, 1009–1021.

57. Willemsen, P.R.; Overbeke, K.; Suurmond, A. Repetitive testing of TBTO, Sea-Nine 211 and farnesol using *Balanus amphitrite*(Darwin) cypris larvae: variability in larval sensitivity. *Biofouling* **1998**, *12*, 133–147.

58. Jacobson, A.H.; Willingham, G.L. Sea-Nine antifoulant: An environmentally acceptable alternative to organotin antifoulants. *Sci.*

Total Environ **2000**, *258*, 103–110.

59. Kobayashi, N.; Okamura, H. Effects of new antifouling compounds on the development of sea urchin. *Mar. Pollut. Bull* **2002**, *44*, 748–751.
60. Backhaus, T.; Faust, M.; Scholze, M.; Gramatica, P.; Vighi, M.; Grimme, L.H. Joint algal toxicity of phenylurea herbicides is equally predictable by concentration addition and independent action. *Environ. Toxicol. Chem* **2004**, *23*, 258–264.
61. Molander, S.; Blanck, H. Detection of pollution-induced community tolerance (PICT) in marine periphyton communities established under diuron exposure. *Aquat. Toxicol* **1992**, *22*, 129–144.
62. Arrhenius, A.; Grönvall, F.; Scholze, M.; Backhaus, T.; Blanck, H. Predictability of the mixture toxicity of 12 similarly acting congeneric inhibitors of photosystem II in marine periphyton and epipsammon communities. *Aquat. Toxicol* **2004**, *68*, 351–367.
63. Perschbacher, P.W.; Ludwig, G.M. Effects of diuron and other aerially applied cotton herbicides and defoliants on the plankton communities of aquaculture ponds. *Aquaculture* **2004**, *233*, 197–203.
64. Bretaud, S.; Toutant, J.P.; Saglio, P. Effects of Carbofuran, Diuron, and Nicosulfuron on Acetylcholinesterase Activity in Goldfish (*Carassius auratus*). *Ecotoxicol. Environ. Saf* **2000**, *47*, 117–124.
65. Gatidou, G.; Kotrikla, A.; Rontogianni, V.; Thomaidis, N.S.; Lekkas, T.D. The Toxic Effects of the Antifouling Biocide Irgarol 1051 and Its Principal Metabolite on the Green Alga Dunaliella Tertiotecta. Proceedings of 8th International Conference on Environmental Science and Technology, Lemnos island, Greece, 8–10 September 2003.
66. Devilla, R.A.; Brown, M.T.; Donkin, M.; Tarran, G.A.; Aiken, J.; Readman, J.W. Impact of antifouling booster biocides on single microalgal species and on a natural marine phytoplankton community. *Mar. Ecol. Prog. Ser* **2005**, *256*, 1–12.
67. Scarlett, A.; Donkin, P.; Fileman, T.W.; Morris, R.J. Occurrence of the Antifouling Herbicide, Irgarol 1051, within Coastal-water Seagrasses from Queensland, Australia. *Mar. Pollut. Bull* **1999**, *38*,

687–691.

68. Braithwaite, R.A.; Fletcher, R.L. The toxicity of Irgarol 1051 and Sea-Nine 211 to the non-target macroalga *Fucus serratus* Linnaeus, with the aid of an image capture and analysis system. *J. Exp. Mar. Bio. Ecol* **2005**, *322*, 111–121.
69. Scarlett, A.; Donkin, M.E.; Fileman, T.W.; Donkin, P. Occurrence of the marine antifouling agent irgarol 1051 within the Plymouth Sound locality: Implications for the green macroalga *Enteromorpha intestinalis*. *Mar. Pollut. Bull* **1997**, *34*, 645–651.
70. Tolhurst, L.E.; Barry, J.; Dyer, R.A.; Thomas, K.V. The effect of resuspending sediment contaminated with antifouling paint particles containing Irgarol 1051 on the marine macrophyte *Ulva intestinalis*. *Chemosphere* **2007**, *68*, 1519–1524.
71. Menin, A.; Ballarin, L.; Marin, M.G.; Cima, F. Immunotoxicity in ascidians of antifouling compounds alternative to organotins, the case of Diuron and TCMS pyridine. *J. Environ. Sci. Health B* **2008**, *43*, 644–654.
72. Menin, A.; Ballarin, L. Immunomodulatory molecules in the compound ascidian *Botryllus schlosseri*: Evidence from conditioned media. *J. Invertebr. Pathol* **2008**, *99*, 275–280.
73. Sánchez-Bayo, F.; Goka, K. Unexpected effects of zinc pyrithione and imidacloprid on Japanese medaka fish (*Oryzias latipes*). *Aquat. Toxicol* **2005**, *74*, 285–293.
74. Goka, K. Embryotoxicity of zinc pyrithione, an antidandruff chemical, in Fish. *Environ. Res* **1999**, *81*, 81–83.
75. van Leeuwen, C.J.; Espeldoorn, A.; Mol, F. Aquatic toxicological aspects of dithiocarbamates and related compounds. III. Embryolarval studies with rainbow trout (Salmo gairdneri). *Aquat. Toxicol* **1986**, *9*, 129–145.
76. Dafforn, K.A.; Lewis, J.A.; Johnston, E.L. Antifouling strategies: History and regulation, ecological impacts and mitigation. *Mar. Pollut. Bull* **2011**, *62*, 453–465.
77. *Reregistration Eligibility Decision: Chlorothalonil*, U.S. Environmental Protection Agency (USEPA) ed.; Office of Prevention, Pesticides and Toxic Substances, EPA: Washington,

DC, USA, 1999.

78. Voulvoulis, N.; Scrimshaw, M.D.; Lester, J.N. Occurrence of four biocides utilized in antifouling paints, as alternatives to organotin compounds, in waters and sediments of a commercial estuary in the UK. *Mar. Pollut. Bull* **2000**, *40*, 938–946.

79. Voulvoulis, N.; Scrimshaw, M.D.; Lester, J.N. Partitioning of selected antifouling in the aquatic environment. *Mar. Environ. Res* **2000**, *53*, 1–16.

80. Caux, P.Y.; Kent, R.A.; Fan, G.T.; Stephenson, G.L. Environmental fate and effects of chlorothalonil: a Canadian perspective. *Crit. Rev. Environ. Sci. Technol* **1996**, *26*, 45–93.

81. Tsuda, T.; Aoki, S.; Kojima, M.; Fujita, T. Accumulation and excretion of pesticides used in golf courses by carp (*Cyprinus carpio*) and willow shiner (*Gnathopogon caerulescens*). *Comp. Biochem. Physiol. C* **1992**, *101*, 63–66.

82. Baier-Anderson, C.; Anderson, R.S. Suppression of superoxide production by chlorothalonil in striped bass (*Morone saxatilus*) macrophages: the role of cellular sulfhydryls and oxidative stress. *Aquat. Toxicol* **2000**, *50*, 85–96.

83. Baier-Anderson, C.; Anderson, R.S. The effects of chlorothalonilon oyster hemocyte activation: Phagocytosis, reduced pyridine nucleotides, and reactive oxygen species production. *Environ. Res* **2000**, *83*, 72–78.

84. Kiaune, L.; Singhasemanon, N. Pesticidal copper (I) oxide: Environmental fate and aquatic toxicity. *Rev. Environ. Contam. Toxicol***2011**, *213*, 1–26.

85. Burridge, L.; Weis, J.; Cabello, F.; Pizarro, J. Chemical use in salmon aquaculture: A review of current practices and possible environmental effects. *Aquaculture* **2010**, *306*, 7–23.

86. *Ambient Water Quality Criteria for Copper*, U.S. Environmental Protection Agency (EPA) ed.; Office of Water and Office of Science and Technology, U.S. Government Printing Office: Washington, DC, USA, 1985.

87. *Ambient Water Quality Criteria - Saltwater Copper Addendum*, U.S. Environmental Protection Agency (EPA) ed.; Office of Water and Office of Science and Technology, U.S. Government Printing

Office: Washington, DC, USA, 1995.

88. Brix, K.V.; DeForest, D.K. *Critical Review of the Use of Bioconcentration Factors for Hazard Classification of Metals and Metal Compounds*; Parametrix Inc.: Washington, DC, USA, 2000.
89. Wilde, K.L.; Stauber, J.L.; Markich, S.J.; Franklin, N.M.; Brown, P.L. The effects of pH on the uptake and toxicity of copper and zinc in a tropical freshwater alga (*Chlorella* sp). *Arch. Environ. Contam. Toxicol* **2006**, *51*, 174–185.
90. Boulanger, B.; Nikolaidis, N.P. Mobility and aquatic toxicity of copper in an urban watershed. *J. Am. Water Resour. Assoc* **2003**, *39*, 325–336.
91. Arnold, W.R.; Santore, R.C.; Cotsifas, J.S. Predicting copper toxicity in estuarine and marine waters using the biotic ligand model. *Mar. Pollut. Bull* **2005**, *50*, 1634–1640.
92. Brooks, S.J.; Bolam, T.; Tolhurst, L.; Bassett, J.; La Roche, J.; Waldock, M.; Barry, J.; Thomas, K.V. The effects of dissolved organic carbon on the toxicity of copper to the developing embryos of the Pacific oyster, *Crassostrea gigas*. *Environ. Toxicol. Chem* **2007**, *26*, 1756–1763.
93. Playle, R.C.; Dixon, D.G.; Burnison, K. Copper and cadmium binding to fish gills: Modification by dissolved organic carbon and synthetic ligands. *Can. J. Fish. Aquat. Sci* **1993**, *50*, 2667–2677.
94. Playle, R.C.; Dixon, D.G.; Burnison, K. Copper and cadmium binding to fish gills: Estimates of metal-gill stability constants and modelling of metal accumulation. *Can. J. Fish. Aquat. Sci* **1993**, *50*, 2678–2687.
95. Lorenzo, J.I.; Nieto, O.; Beiras, R. Anodic stripping voltammetry measures copper bioavailability for sea urchin larvae in the presence of fulvic acids. *Environ. Toxicol. Chem* **2006**, *25*, 73–81.
96. Brooks, S.J.; Bolam, T.; Tolhurst, L.; Bassett, J.; La Roche, J.; Waldock, M.; Barry, J.; Thomas, K.V. Dissolved organic carbon reduces the toxicity of copper to germlings of the macroalgae, *Fucus vesiculosus*. *Ecotoxicol. Environ. Saf* **2008**, *70*, 88–98.
97. Florence, T.M.; Stauber, J.L. Toxicity of copper complexes to the marine diatom *Nitzschia closterium*. *Aquat. Toxicol* **1986**, *8*, 11–26.

98. Hall, L.W., Jr; Anderson, R.D.; Lewis, B.L.; Arnold, W.R. The influence of salinity and dissolved organic carbon on the toxicity of copper to the estuarine copepod, *Eurytemora affinis*. *Arch. Environ. Contam. Toxicol.* **2008**, *54*, 44–56.

99. Kramer, K.J.M.; Jak, R.G.; van Hattum, B.; Hoftman, R.N.; Zwolsman, J.J.G. Copper toxicity in relation to surface water-dissolved organic matter: biological effects to *Daphnia magna*. *Environ. Toxicol. Chem* **2004**, *23*, 2971–2980.

100. Amiard-Triquet, C.; Berthet, B.; Martoja, R. Influence of salinity on trace metal (Cu, Zn, Ag) accumulation at the molecular, cellular and organism level in the oyster *Crassostrea gigas Thunberg*. *Biol. Met* **1991**, *4*, 144–150.

101. Grosell, M.; Blanchard, J.; Brix, K.V.; Gerdes, R. Physiology is pivotal for interactions between salinity and acute copper toxicity to fish and invertebrates. *Aquat. Toxicol* **2007**, *84*, 162–172.

102. Polo, A.D.; Scrimshaw, M. Challenges for the development of a biotic ligand model predicting copper toxicity in estuaries and seas. *Environ. Toxicol. Chem* **2011**. in press.

103. Tilton, F.; Tilton, S.C.; Bammler, T.K.; Beyer, R.; Farin, F.; Stapleton, P.L.; Gallagher, E.P. Transcription biomarkers and mechanisms of copper-induced olfactory injury in zebrafish. *Environ. Sci. Technol* **2008**, *42*, 9404–9411.

104. Bopp, S.K.; Abicht, H.K.; Knauer, K. Copper-induced oxidative stress in rainbow trout gill cells. *Aquat. Toxicol* **2008**, *86*, 197–204.

105. Brand, L.; Sunda, W.G.; Guillard, R.R.L. Reduction of marine phytoplankton reproduction rates by copper and cadmium. *J. Exp. Mar. Biol. Ecol* **1986**, *96*, 225–250.

106. Beck, N.G.; Bruland, K.W.; Rue, E.L. Short-term biogeochemical influence of a diatom bloom on the nutrient and trace metal concentrations in South San Francisco Bay microcosm experiments. *Estuaries* **2002**, *25*, 1063–1076.

107. Wang, H.; Li, Y.; Huang, H.; Xu, X.; Wang, Y. Toxicity evaluation of single and mixed antifouling biocides using the *Strongylocentrotus intermedius* sea urchin embryo test. *Environ. Toxicol. Chem* **2011**, *30*, 692–703.

108. Lee, S.E.; Won, H.S.; Lee, Y.W. Study on the new antifouling

compounds in Korean coasts. *Bull. Environ. Contam. Toxicol* **2010**, *85*, 538–543.

109. Lee, S.; Chung, J.; Won, H.; Lee, D.; Lee, Y.W. Analysis of antifouling agents after regulation of tributyltin compounds in Korea. *J. Hazard. Mater* **2011**, *185*, 1318–1325.
110. Steen, R.J.; Ariese, F.; van Hattum, B.; Jacobsen, J.; Jacobson, A. Monitoring and evaluation of the environmental dissipation of the marine antifoulant 4,5-dichloro-2-*n*-octyl-4-isothiazolin-3-one (DCOIT) in a Danish Harbor. *Chemosphere* **2004**, *57*, 513–521.
111. Callow, M.E.; Willingham, G.L. Degradation of antifouling biocides. *Biofouling* **1996**, *10*, 239–249.
112. Jacobson, A. Environmentally Preferable Alternatives to TBT and Their Impact on the Environment. Proceedings of IBC UK Conferences, Marine Environmental Regulations: The Cost to the Shipping Industry, LM 1120, London, UK, 23–24 September 1998; IBC UK Limited: London, UK, 1998.
113. Jacobson, A.; Williams, T.M. Environmental fate of isothiazolone biocides. *Chem. Oggi* **2000**, *18*, 105–108.
114. Shade, W.D.; Hurt, S.S.; Jacobson, A.H.; Reinert, K.H. Ecological Risk Assessment of a Novel Marine Antifoulant. In *Environmental Toxicology and Risk Assessment*; Gorsuch, J.W., Dwyer, F.W., Ingersoll, C.M., LaPoint, T.W., Eds.; American Society for Testing and Materials: Philadelphia, PA, USA, 1994; Volume 2, pp. 381–408.
115. Cima, F.; Bragadin, M.; Ballarin, L. Toxic effects of new antifouling compounds on tunicate haemocytes I. Sea-Nine 211™ and chlorothalonil. *Aquat. Toxicol* **2008**, *86*, 299–312.
116. Callow, M.E.; Finlay, J.A. A simple method to evaluate the potential for degradation of antifouling biocides. *Biofouling* **1995**, *9*, 153–165.
117. Thomas, K.V.; McHugh, M.; Waldock, M. Antifouling paint booster biocides in UK coastal waters: inputs, occurrence and environmental fate. *Sci. Total Environ* **2002**, *293*, 117–127.
118. Thomas, K.V.; Blake, S.J.; Waldock, M.J. Antifouling paint booster biocide contamination in UK marine sediments. *Mar. Pollut. Bull***2000**, *40*, 739–745.

119. Lamoree, M.H.; Swart, C.P.; van der Horst, A.; van Hattum, B. Determination of diuron and the antifouling paint biocide Irgarol 1051 in Dutch marinas and coastal waters. *J. Chromatogr. A* **2002**, *970*, 183–190.

120. Okamura, H.; Watanabe, T.; Aoyama, I.; Hasobe, M. Toxicity evaluation of new antifouling compounds using suspension-cultured fish. *Chemosphere* **2002**, *46*, 945–951.

121. Gagnon, M.M.; Rawson, C.A. Diuron increases spinal deformity in early-life-stage pink snapper *Pagrus auratus*. *Mar. Pollut. Bull***2009**, *58*, 1078–1095.

122. Canna-Michaelidou, S.; Nicolaou, A.S. Evaluation of the genotoxicity potential (by Mutatox™ test) of ten pesticides found as water pollutants in Cyprus. *Sci. Total Environ* **1996**, *193*, 27–35.

123. Strachan, G.; Preston, S.; Maciel, H.; Porter, A.J.R.; Paton, G.I. Use of bacterial biosensors to interpret the toxicity and mixture toxicity of herbicides in freshwater. *Water Res* **2001**, *35*, 3490–3495.

124. Tixier, C.; Sancelme, M.; Bonnemoy, F.; Cuer, A.; Veschambre, H. Degradation products of a phenylurea herbicide, diuron: synthesis, ecotoxicity, and biotransformation. *Environ. Toxicol. Chem* **2001**, *20*, 1381–1389.

125. Hall, L.W., Jr; Killen, W.D.; Anderson, R.D.; Balcomb, R.; Gardinali, P. Ecological risk of Irgarol 1051 and its major metabolite in coastal California marinas and reference areas. *Mar. Pollut. Bull.* **2009**, *58*, 702–710.

126. Holt, J.S. Mechanisms and agronomic aspects of herbicide resistance. *Ann. Rev. Plant Physiol. Plant Mol. Biol* **1993**, *44*, 203–229.

127. Ranke, J.; Jastorff, B. Multidimensional risk analysis of antifouling biocides. *Environ. Sci. Poll. Res* **2000**, *7*, 109–110.

128. Hall, L.W., Jr; Gidding, J.M.; Solomon, K.R.; Balcomb, R. An ecological risk assessment for the use of Irgarol 1051 as an algaecide for antifoulingpaints. *Crit. Rev. Toxicol.* **1999**, *29*, 367–437.

129. Dahl, B.; Blanck, H. Toxic effects of the antifouling agent Irgarol 1051 on periphyton communities in coastal water microcosms. *Mar. Poll. Bul* **1996**, *32*, 342–350.

130. Buma, A.G.J.; Sjollema, S.B.; van de Poll, W.H.; Klamer, H.J.C.; Bakker, J.F. Impact of the antifouling agent Irgarol 1051 on marine phytoplankton species. *J. Sea Res* **2009**, *61*, 133–139.

131. Readman, J.W.; Devilla, R.A.; Tarran, G.; Llewellyn, C.A.; Fileman, T.W.; Easton, A.; Burkill, P.H.; Mantoura, R.F.C. Flow cytometry and pigment analysis as tools to investigate the toxicity of herbicides to natural phytoplankton communities. *Mar. Environ. Res* **2004**, *58*, 353–358.

132. Devilla, R.A.; Brown, M.T.; Donkin, M.; Readman, J.W. The effects of a PSII inhibitor on phytoplankton community structure as assessed by HPLC pigment analyses, microscopy and flow cytometry. *Aquat. Toxicol* **2005**, *71*, 25–38.

133. Zamora-Ley, I.M.; Gardinali, P.R.; Jochem, F.J. Assessing the effects of Irgarol 1051 on marine phytoplankton populations in Key Largo Harbor, Florida. *Mar. Poll. Bull* **2006**, *52*, 935–941.

134. Bragadin, M.; Iero, A.; Cima, F.; Ballarin, L.; Manente, S. TCMS inhibits ATP synthesis in mitochondria: A systematic analysis of the inhibitory mechanism. *Toxicol. In Vitro* **2007**, *21*, 1127–1133.

135. Huh, W.K.; Masuji, Y.; Tada, J.; Arata, J.; Kaniwa, M. Allergic contact dermatitis froma pyridine derivative in polyvinyl chloride leather. *Am. J. Contact Dermat* **2001**, *12*, 35–37.

136. Voulvoulis, N.; Scrimshaw, M.D.; Lester, J.N. Partitioning of selected antifouling in the aquatic environment. *Mar. Environ. Res* **2001**, *53*, 1–16.

137. Karlsson, J.; Eklund, B. New biocide antifouling paints are toxic. *Mar. Pollut. Bull* **2004**, *49*, 456–464.

138. Sabev, H.A.; Handley, P.S.; Robson, G.D. *In situ* quantification of biocide efficacy using GFP transformed *Aureobasidium pullutans*. *J. Appl. Microbiol* **2004**, *97*, 1132–1139.

139. Bao, W.W.V.; Leung, K.M.Y.; Kwok, K.W.H.; Zhang, A.Q.; Lui, G.C.S. Synergistic toxic effects of zinc pyrithione and copper to three marine species: Implications on setting appropriate water quality criteria. *Mar. Pollut. Bull* **2008**, *57*, 616–623.

140. Turley, P.A.; Fenn, R.J.; Ritter, J.C. Pyrithiones as antifoulants: Environmental chemistry and preliminary risk assessment. *Biofouling***2000**, *15*, 175–182.

141. Turley, P.A.; Fenn, R.J.; Ritter, J.C.; Callow, M.E. Pyrithiones as antifoulants: environmental fate and loss of toxicity. *Biofouling* **2005**, *21*, 31–40.

142. Andrioli, N.B.; Soloneski, S.; Larramendy, M.L.; Mudry, M.D. Cytogenetic and microtubule array effects of the zineb-containing commercial fungicide formulation Azzurro® on meristematic root cells of *Allium cepa* L. *Mutat. Res* **2011**. in press.

143. U.S. Environmental Protection Agency (EPA). *Compendium of Registered Pesticides*; US Government Printing Office: Washington, DC, USA, 1974.

144. Mochida, K.; Fujii, K. Futher Effects of Alternative Biocides on Aquatic Organisms. In *Ecotoxicology of Antifouling Biocides*; Arai, T., Harino, H., Ohji, M., Langston, W., Eds.; Springer: Tokyo, Japan, 2009; pp. 383–396.

145. Xu, Q.; Barrios, C.A.; Cutright, T.; Newby, B.Z. Evaluation of toxicity of capsaicin and zosteric acid and their potential application as antifoulants. *Environ. Toxicol* **2005**, *20*, 467–474.

146. Dahlström, M.; Jonsson, P.R.; Lausmaa, J.; Arnebrant, T.; Sjögren, M.; Holmberg, K.; Martensson, L.G.E.; Elwing, H. Impact of polymer surface affinity of novel antifouling agents. *Biotechnol. Bioeng* **2004**, *86*, 1–8.

147. Katagi, T. Bioconcentration, bioaccumulation, and metabolism of pesticides in aquatic organisms. *Rev. Environ. Contam. Toxicol* **2010**, *204*, 1–132.

148. Kenaga, E. Predicted bioconcentration factors and soil sorption coefficients of pesticides and other chemicals. *Ecotoxicol. Environ. Saf* **1980**, *4*, 26–38.

149. Tóth, S.; Becker-van Slooten, K.; Spack, L.; de Alencastro, L.; Tarradellas, J. Irgarol 1051, an antifouling compound in freshwater, sediment, and biota of Lake Geneva. *Bull. Environ. Toxicol. Chem* **1996**, *57*, 426–433.

150. Dyer, R.A.; Tolhurst, L.E.; Hilton, M.J.; Thomas, K.V. Bioaccumulation of the antifouling paint booster biocide Irgarol 1051 by the green alga *Tetraselmis suecica*. *Bull. Environ. Contam. Toxicol* **2006**, *77*, 1519–1524.

151. Assessment of the Antibiotic Resistance Effects of Biocides;

Scientific Committee on Emerging and Newly Identified Health Risks (SCENIHR), January 2009. Available online: http://ec.europa.eu/health/ph_risk/committees/04_scenihr/docs/scenihr_o_021.pdfaccessed on 12 November 2011.

10

CHAPTER

IMMUNOTOXICOLOGICAL THREATS OF POLLUTANTS IN AQUATIC INVERTEBRATES

Sajal Ray[1], Soumalya Mukherjee[1], Niladri Sekhar Bhunia[1], Anindya Sundar Bhunia[1] and Mitali Ray[1]

[1] Aquatic Toxicology Laboratory, Department of Zoology, University of Calcutta, Kolkata, West Bengal, India

ABSTRACT

Immunology deals with the physiological activity of organisms to defend against pathogen and toxin invasion. Invertebrates residing in aquatic ecosystems often face toxicological threat arises from habitat pollution. The aquatic habitat of invertebrates is in the precarious risk of pollution caused by diverse groups of environmental toxins. Immunotoxins have been considered as a special group of pollutants capable of affecting the immunological profile of organisms. Invertebrates residing in water bear ecological, economical, medicinal, industrial, nutritional and biotechnological significance. Global aquatic bioresource is largely composed of invertebrates belonging to multiple Phyla. These organisms, including insects, snails, clams, mussels, crabs and sponges, are physiologically dependent on innate immunological response for defense against pathogen and environmental contaminants. External physicochemical barriers of invertebrates act as primary line of defen against toxin entry. Principal barriers have been identified as shell, tunic, test, carapace, mucus, etc., in diverse species. Toxin-induced morphological damage of specialized immunocytes of invertebrates has been reported. Toxin-induced shift in density, surface adhesion efficacy and aggregation of blood cells or haemocytes have been identified as major xenobiotic stress in invertebrates. Various environmental toxins are capable of initiating alteration in the innate phagocytic response and cytotoxicity of blood cells. Lysosomes of invertebrate haemocytes are functionally involved in intracellular destruction of environmental pathogens. Toxins like arsenic, pyrethroid pesticides, azadirachtin and washing soda were reported to increase the relative fragility of lysosomal membranes of immunocytes. This often leads to impairment in the efficacy of invertebrates to destroy pathogen under the exposure of pollutants. Xenobiotics like pyrethroid pesticides have been recorded to affect apoptosis and necrosis of invertebrate immunocytes. Selected toxin-induced morphological damages of heart, gill, digestive gland, mantle and antennae may result in the overall impairment in homeostatic levels of invertebrates inhabiting the polluted environment. Global environment, in recent times, is under the serious threat of contamination by diverse chemical compounds of unknown or less known toxicity. A thorough ecotoxicological analysis at cellular and molecular levels needs to be carried out in invertebrates occupying the different realms of the planet in future.

Keywords: Pesticides, arsenic, washing soda, flow cytometry, respiratory burst, phagocytosis, nitric oxide, molluscs, crab, sponge

INTRODUCTION

Invertebrates constitute more than ninety-five per cent of the existing biodiversity of the world. They present an extreme range of variation in terms of their body architecture, adaptation, food preference, habitat preference, behaviour and physiological response against environmental stressors. Evolutionarily, they are an ancient group of animals which have survived an extreme range of environmental adversities and perturbations from the time of their origin. Invertebrates occupied diverse kinds of habitats including terrestrial, freshwater, marine and estuarine ecosystems. Aquatic invertebrates are usually distributed in the multiple spatial compartments of the water bodies from the surface to the bottom region called the benthic stratum. Aquatic invertebrates are also found in the conspicuous ecological areas of the mud–water interface and subsoil region at the floor of aquifers. Freshwater ponds of India serve as habitat for diverse aquatic organisms including members of Porifera, Mollusca and Arthropoda. Majority of the ecotoxicological research was carried out on *Eunapius carteri* (Porifera: Demospongiae: Spongillidae); *Bellamya bengalensis* (Mollusca: Gastropoda: Prosobranchia); *Pila globosa* (Mollusca: Gastropoda: Ampullariidae) and *Lamellidens marginalis*(Mollusca: Bivalvia: Eulamellibranchiata).

Aquatic invertebrates bear ecological, economical, nutritional and biotechnological significance and demand a special scientific attention. Many of them are the dietary source of nutrition for human and other organisms. Their importance as potential source of bioactive substances and pharmacologically active compounds cannot be denied. Current scientific reports indicate that the rapid and unrestricted contamination of freshwater and marine ecosystem by diverse pollutants poses serious ecotoxicological threat for the existence of invertebrates in their natural habitats. Various anthropogenic activities like habitat destruction and habitat contamination have been identified as the causative factors of dwindling of invertebrate species on earth. Continuous addition of various toxic chemical compounds into the global environment is a major environmental challenge encountered by human and other organisms. Many of these environmental pollutants are of either industrial or agricultural origin and need a thorough toxicological screening in invertebrate models.

The overall evolutionary success and survival efficacy of invertebrates depended on several factors including the development of a highly advanced immunological system. Immunology deals with the typical strategies of biological defence against environmental toxins and pathogens. Components of immune system and their coherent functional attributes enabled the invertebrates to overcome the toxin-induced chemical stresses of the primitive and modern global environment (Figure 1). Many of the environmental pollutants of the current hydrosphere are less researched, with limited toxicological information. Invertebrates over a period of time are assumed to evolve novel and unique modes of immunological reactivities to defend against the toxic insults of environmental pollutants at the cellular and molecular levels. Aquatic invertebrates are relatively a neglected group of organisms with reference to their immunotoxicological status in a biounsafe environment. The principal immunotoxicological responsiveness of invertebrates centre around the reactivities of circulating immunocytes and selected effector organs like gill, digestive gland, intestine, labial palp, etc.

IMMUNOTOXINS AND THEIR IMPACT ON AQUATIC INVERTEBRATES

Immunotoxins are chemical compounds that can modulate the immune-related parameters and can adversely affect the biological and physiological functioning of the immune system of organisms including human. The freshwater ecosystem of India is being contaminated by diverse groups of xenobiotics of known and unknown chemistry. Ray et al. (2011) reported mineral acids, alkalis, detergents, metalloids and pesticides as major environmental contaminants of the Indian freshwater ecosystem [1]. Scientists reported [2] the toxic effects of pyrethroids, cypermethrin and fenvalerate in aquatic molluscan invertebrates and claimed them as immunotoxins for their adverse effect on the immune parameters of host animal. Toxicity of these pyrethroid pesticides on non-target aquatic invertebrates appears to be detrimental and can seriously affect their survival efficacy and reproductive success. Arsenic, another metalloid and a potent immunotoxin, bears toxic effects on aquatic invertebrates and can modulate the histopathology and immune parameters of freshwater mollusc [3] and estuarine mud crab [4]. Biopesticide azadirachtin has been recently identified as an immunotoxin due to its potential to affect the immunofunctioning of mussels. Moreover, during monsoon and flood, pesticide-laden agricultural runoff often contaminates the freshwater ecosystem and poses serious threat to its inhabitants. Washing soda, chemically known as anhydrous sodium carbonate, is an important aquatoxin that can alter the selected physiological parameters of diverse groups of invertebrates. Mukherjee et al. (2015b) reported the toxicity of washing soda in a freshwater sponge of India with reference to its phagocytic and cytotoxic status [5]. Altered functioning of the immune system may lead to opportunistic invasion of environmental pathogens and parasites into the body of host and increase the vulnerability of these biofilter species in polluted environment.

The nature and magnitude of immunotoxicity depend on multiple parameters including the concentration of toxin, type and span of exposure and route of entry to the host. Immunotoxins are difficult to identify as they can cause a wide magnitude of adversity on the immune status of organisms. Immunotoxicology deals with the assessment of toxicological response in an organism by estimating the responsiveness and reactivity of its immunological parameters. Immunotoxicology of invertebrates, in recent times, has been gaining a special scientific attention for its efficacy in monitoring the health of environment – both aquatic and terrestrial. Accuracy and precision of selected toxicological responses of aquatic invertebrates enabled a few species to function as suitable biomonitoring agents of water pollution [6]. Several effective immunomarkers of aquatic pollution have recently been established in model invertebrates [7].

Figure 1: Attributes of cell-mediated immune responses of invertebrate immunocytes exposed to various aquatic pollutants.

EXTERNAL PHYSICOCHEMICAL BARRIERS OF INVERTEBRATES AS FIRST LINE OF IMMUNE DEFENCE AGAINST POLLUTANTS

External physicochemical barriers of the invertebrates act as the first line of defence against the invasion of parasite, pathogen and toxin in a polluted aquatic environment. External physicochemical barriers of the invertebrates include shell, tunic, test, cuticle, carapace, pinacoderm and others. In mollusc, hard calcareous shell or valves provide the primary protection against entry of pathogens and toxins. However, in shell-less molluscs, thick external cuticular sheet known as mantle or pallium are assumed to play a significant physiological role against toxin and pathogen entry. Soft body wall of molluscs consists of cuticle, epidermis and muscles, which are thought to take an active part in the innate immune defence against environmental pathogens and toxins. In a polluted environment, where environmental pathogens and toxins are present in sufficient quantity, the two valves of the mussel are kept closed to minimize the entry of undesired agents. In addition to external shell, mucus secreted from the internal viscera provides another line of defence against the invasion of pathogens and toxins. Mucus acts as a protective barrier preventing the direct contact of toxins to epithelia. Secretion of mucus is reported to be an important detoxification and evasive mechanism of invertebrates. Pathogens are trapped within the mucus secreted by the organism leading to elimination. Calcareous shell made up of calcium carbonate and mucus is considered as external physiochemical barrier and plays an important role in the immune defence of aquatic invertebrates. Carapace, the external physiochemical barrier of crab provides the first line of defence in crustaceans. Corrosive toxins including mineral acids, alkalis, pesticides and detergents appear to be the potential threats of the invertebrate. Contamination of water bodies by these pollutants often results in breaches of the physicochemical barriers of external body surface, which leads to facilitation of invasion of toxic microorganisms and parasites into the viscera of target invertebrates. Components of innate immune response comprise physicochemical barrier against external pathogen and parasite entry. Cuticle covered with waxy material serves as a mechanical barrier of crab against parasitic infection. This chitinous exoskeleton or cuticle helps in the process of wound healing by preventing the fatal loss of haemolymph from body, maintenance of tissue architecture and prevention of opportunistic invasion of pathogen. Whenever this cuticle is damaged by injury or infection, the wound is rapidly sealed by clotting of immunocytes, preventing blood loss and pollutant entry. Once the clot is formed, wound is darkened and accumulation of melanin occurs. Melanin is reported to be involved in sealing of wound and synthesis of new cuticle. On the other hand, pinacoderm made up of flattened pinacocytes forms a continuous layer on the external surface of freshwater sponges and acts as a first line of defence against foreign invaders.

An experiment was performed to investigate the possible role of mucus as a primary barrier in removal of external particles. In this context, the shell of freshwater gastropod, *P. globosa* was drilled to create a micropore and a solution of cultured yeast (*Saccharomyces cerevisiae*) suspended in sterile phosphate buffered saline was injected into the body through a syringe (Figure 2). The mollusc was under scientific surveillance and the activity of the mollusc was photodocumented. After two hours of injection, hypersecretion of mucus was recorded by the experimental *P. globosa* within a time span of approximately thirty minutes. The expelled mucus contained more than ninety-five per cent of yeast particles, indicating the efficacy of mucus to act as external physicochemical barrier.

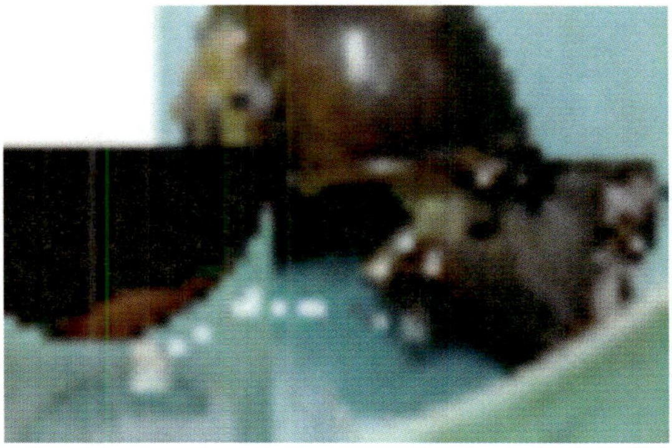

Figure 2: Rapid elimination of foreign particulates entrapped in the mucus of aquatic invertebrate, *P. globosa*.

XENOBIOTIC-INDUCED SHIFT IN HAEMOCYTE DENSITY AND MORPHOLOGICAL DAMAGES OF BLOOD CELLS

Haemocytes are the immunocompetent cells which are functionally responsive to various xenobiotics present in the aquatic environment. Homeostasis of total haemocyte density of aquatic invertebrates within the permissible physiological limit may be considered as an important immunological parameter [8] of cell-mediated immune response of molluscs [2]. Chakraborty et al. (2008) reported suppression in the total count of haemocytes of the freshwater edible mollusc, *L. marginalis* under the sublethal exposure of sodium arsenite, an aquatic pollutant [9]. Mukherjee et al. (2006) reported modulation in the total cell density of the same specimen under the sublethal concentrations of azadirachtin, a neem-based pesticide, a common contaminant of pond water [10]. According to them, toxin-induced alteration in the total cell dynamics may lead to a gradual decline of this species in its natural habitat. The total count of haemocyte of *Villorita cyprinoides* was found to decrease under the exposure of copper [11]. Ray et al. (2013b) reported dynamics of the total haemocyte density of *B. bengalensis* and *L. marginalis*exposed to environmentally realistic sublethal concentrations of cypermethrin and fenvalerate, respectively [2]. Authors reported a significant increase in total haemocyte count following exposure to experimental concentrations of cypermethrin and fenvalerate to *B. bengalensis* and *L. marginalis*, respectively. Russo and Madec (2007) reported effect of pollutant fomesafen pesticide on the haemocyte density in the snail, *Lymnaea stagnalis* [12]. Authors reported the increase in total number of circulating haemocytes under the exposure of fomesafen as "facilitated cell turnover." The differential cell count, on the other hand, is also considered as an important immunotoxicological marker of environmental pollution [13]. Moreover, heterogenous populations of invertebrate blood cells have been reported to perform diverse physiological activities including nonself recognition, encapsulation and generation of cytotoxic molecules. Chakraborty et al. (2008) reported that haemocyte density can be considered as a potential biomarker of toxicity of sodium arsenite, a vital pollutant of the Sunderbans estuary of India [9]. Their experimental findings revealed the variation in the relative densities of granulocytes, agranulocytes, blast-like cells, hyalinocytes and asterocytes under the exposure of sodium arsenite. Ray et al. (2013b) reported a non-linear fluctuation in the differential cell density of *B. bengalensis* and *L. marginalis* following sublethal exposure of cypermethrin and fenvalerate, respectively [2]. According to them, aquatic pesticide-induced alteration in the differential haemocyte density was indicative to stress-induced loss of cell balance, which might alter the physiological homeostasis of these species distributed in pesticide-contaminated habitat. Qubella et al. (1993) proposed that fluctuation in the differential haemocyte density may be a result of reversible migration of haemocytes from tissues to haemolymph or *vice versa* [14].

Mukherjee et al. (2008) reported an increase in relative percentage of pro-haemocytes subpopulation in aquatic mollusc *L. marginalis* under the sublethal exposure of azadirachtin [15], a contaminant of water bodies. According to them, pesticide-induced depletion in the densities of granulocytes, agranulocytes, hyalinocytes and asterocytes was indicative to a possible impairment of haemocytic function of mussels distributed in polluted environment. Ray et al. (2013b) flow cytometrically identified (Figure 5a) five discrete subpopulations of haemocytes, that is, blast-like cells, round hyalinocytes, spindle hyalinocytes, round granulocytes, and granular asterocytes in the freshwater molluscs, *B. bengalensis* and *L. marginalis* [2]. Experimental exposure of 0.5 ppm of cypermethrin for 96 hours resulted in significant increase in the per cent populations of blast-like cells and round hyalinocytes and significant decrease in the subpopulation of granular asterocyte in *B. bengalensis* as compared to control. Exposure of 3 ppm of fenvalerate for 96 hours resulted in significant increase in the relative percentage of round granulocyte and significant decrease in the subpopulation of round hyalinocyte in *L. marginalis* as compared to control. Das et al. (2012) reported alteration of relative densities of haemocyte subpopulations of *L. marginalis* under sublethal exposures of cypermethrin [16]. Exposure to sublethal concentrations of cypermethrin yielded decrease in densities of agranulocytes and asterocytes and increase in densities of blast-like cells and granulocytes. Pyrethroids-induced alterations in the differential cell density of freshwater molluscs may lead to impairment of blood cell homeostasis in the experimental species distributed in polluted environment. Morphological alterations of haemocytes under the exposure of environmental toxins play an important role in cell-mediated immune response of molluscs (Figure 3). Ray et al. (2013b) identified cytoplasmic hypervacuolation, rounding up of cell, alteration in cell shape, hypergranulation, increased cytoplasmic spreading, membrane disintegration and membrane blebbing as principal aberrations in the haemocytes of *B. bengalensis* and *L. marginalis* under the exposure of pollutant like pyrethroid [2]. Cypermethrin- and fenvalerate-induced morphological damages in the circulating haemocytes of freshwater edible molluscs may lead to possible impairment of the functioning of cells and their immunological reactivity. Chakraborty and Ray (2009) reported dose- and time-dependent increase in the frequency of binucleated and micronucleated haemocytes and gill cells in the freshwater bivalve, *L. marginalis*, under the exposure of sodium arsenite [6]. Authors claimed arsenic-induced nuclear anomalies may be used as a biomarker of arsenic toxicity in freshwater ecosystems.

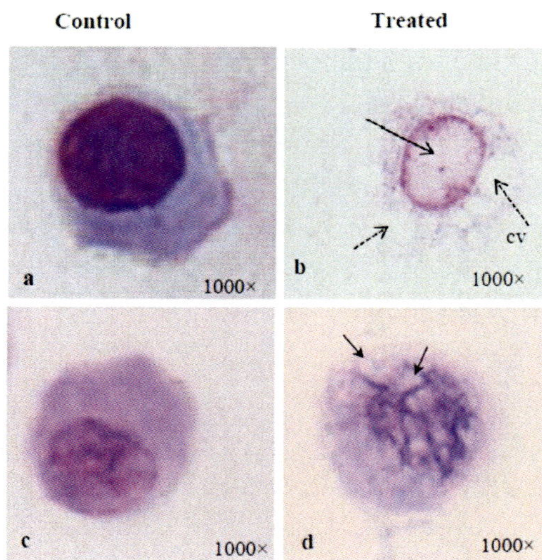

Figure 3: Photomicrographs of normal haemocytes of aquatic molluscs, *P. globosa* (a) and *L. marginalis* (c). Cypermethrin treatment (1.5 ppm/7 days) yielded intense cytoplasmic vacuolation (cv) (b) in the haemocytes of *P. globosa* and nuclear disintegration (d) in the haemocytes of *L. marginalis*.

NONSELF SURFACE ADHESION AND AGGREGATION RESPONSE OF HAEMOCYTES

Nonself surface adhesion is considered as an important immunological mechanism which is fundamentally related to self–nonself discrimination. Cell adhesion and interaction between cell and the substratum play a pivotal role in the development, maintenance and immune recognition in multicellular animals. Mukherjee et al. (2007) studied the glass surface adhesion efficacy of haemocytes of *L. marginalis* under the sublethal concentrations of azadirachtin [17]. Authors reported a decrease in activity of surface adhesion property of haemocyte under prolonged exposure of pesticide. According to them, the data were indicative to impairment of immunological response of *L. marginalis* in its natural habitat leading to a possible dwindling of this biofilter species from aquatic ecosystem. Ghosh et al. (2008) studied the kinetics of nonself surface adhesion in the haemocytes of mud whelk, *Telescopium telescopium* exposed to diesel in the Sunderbans biosphere reserve [18]. Vehicle diesel is reported [19] to be a serious pollutant of Indian estuary. Authors reported diesel-induced inhibition in the nonself surface recognition efficacy and a shift in the kinetics of adhesion. Saha et al. (2008a) reported modulation in the nonself surface adhesion efficacy of haemocytes of edible mud crab, *Scylla serrata*, under the sublethal exposures of sodium arsenite [20]. Ray et al. (2012) reported arsenic-induced glass surface adhesion of haemocytes of juvenile mud crab [21]. According to them, arsenic-induced alteration in the reactivity of haemocyte may render the juvenile mud crab to become immunologically impaired in the parasite- and pathogen-contaminated natural habitat.

In the dynamic freshwater ecosystem, the inhabitants often compete for niche for their better survival and propagation. Overlapping in the niche leads to a state of acute predation and fighting among animals. As a result, the animals may encounter acute competition and subsequent physical damage and loss of body fluid. Cellular aggregation is a functional attribute offered by the haemocytes of invertebrates [22] to prevent the accidental blood loss by formation of biological plug at the site of wound and resist the entry of pathogenic microorganism. Selected pollutants like sodium arsenite, washing soda, pyrethroid, azadirachtin are reported to affect the aggregation response of many aquatic invertebrates (Figure 4). Hence, cell–cell aggregation is considered as an immunological response for host defence. Aggregation of haemocytes around invaded microorganisms is termed as "encapsulation response" and is considered as an important immunological response [23]. When successful encapsulation occurs, a host animal can restrict the proliferative and invasive property of a pathogen. Encapsulation reaction is mediated by specific population of immunoactive blood cells and is effective in cell-mediated immunity of invertebrates. Mukherjee et al. (2011) reported anticoagulant and carbohydrate-induced interference of cellular aggregation of mussel, *L. marginalis*, under experimental exposure of azadirachtin [24]. Persistent exposure of azadirachtin inhibited the cellular aggregation response in freshwater mussel. Workers apprehend such scenario in the natural environment may lead to a decline in the population of freshwater mussel and loss of freshwater biodiversity of India. Furthermore, a drastic increase in the occurrence of free cells was recorded against ethylene diamine tetraacetic acid and manure treatment, which was suggestive to possible role of these chemical agents as inhibitor of cellular aggregation. Ray et al. (2012) reported sodium-arsenite-induced inhibition in aggregation of haemocytes of juvenile mud crab, *S. serrata* [21]. According to them, arsenic-induced alteration of immune status may impart a state of vulnerability in juvenile crab inhabiting the arsenic-polluted environment.

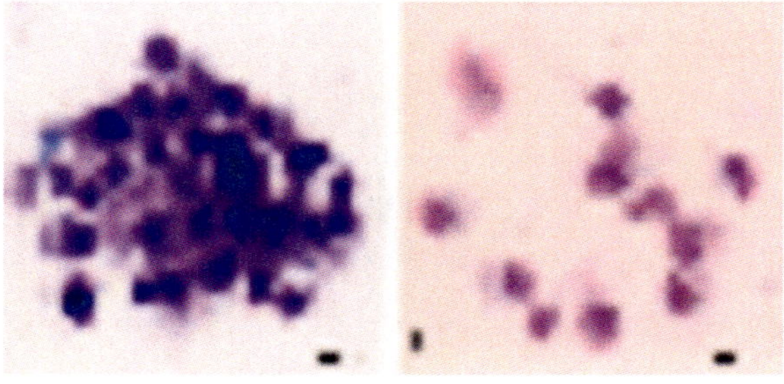

Figure 4: Typical aggregation response of haemocytes of a freshwater gastropod, *Pila globosa* (a). Cypermethrin is a contaminant of the habitat of *P. globosa* which inhibits the degree of haemocyte aggregation (b).

PHAGOCYTIC RESPONSE IN THE FACE OF ENVIRONMENTAL STRESSORS

Phagocytosis, in general, is considered a classical innate immune response reported in the majority of the invertebrate Phyla. It is an established immunological response and is considered as a biomarker of aquatic pollution [13]. Phagocytic response enables invertebrates to combat against invading pathogens and pollutants of known and unknown chemistry. Haemocyte-mediated phagocytosis of nonself particles provides natural immunity in the bivalves [25]. Chakraborty et al. (2009) reported the inhibitory effect of sodium arsenite on the phagocytic response of *L. marginalis* under the challenge of yeast at various sublethal concentrations [7]. Inhibition in phagocytic response was also recorded under similar laboratory condition for the haemocytes of arsenic-treated *L. marginalis* when challenged with human red blood corpuscles [26]. According to them, impairment in the phagocytic potential of the haemocytes of arsenic-treated mussels may lead to compromisation of the immune status of the animals distributed in the contaminated habitat. Mukherjee et al. (2011) reported azadirachtin-induced suppression in the phagocytic potential of haemocytes of *L. marginalis* under the experimental challenge of charcoal particulate [27]. Azadirachtin-induced inhibition in the phagocytic response indicated a state of immune suppression of mussels. Ray et al. (2012) screened the immunotoxicological reactivity of haemocytes of juvenile mud crab, *S. serrata* of the Sunderbans biosphere reserve and reported sodium-arsenite-induced inhibition in the phagocytic potential under the challenge of yeast [21]. Arsenic-induced altered reactivity of haemocytes may affect the propagation and survival of mud crab population by increasing its vulnerability to higher rate of disease and parasite attack. Sponges, on the other hand, are non-selective filter feeders which depend on phagocytosis for the purposes of feeding and digestion (Figure 5b, c). Phagocytosis, in sponge, plays a dual physiological attribute in the form of food procurement and innate immune response [28]. Mukherjee et al. (2015b) reported washing-soda-induced inhibition in the phagocytic potential of cells of freshwater sponge, *E. carteri*, under the challenge of yeast [5]. According to them, decrease in phagocytic potential is suggestive to a possible impairment of both food capture efficiency and innate immune status, leading to suppression of immunological status of *E. carteri* distributed in detergent-contaminated natural habitat.

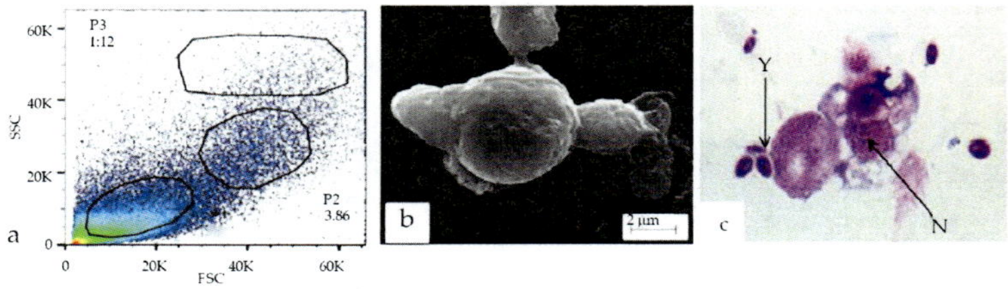

Figure 5: Flow cytometry of isolated cells of sponge, *E. carteri*, depicting their relative size and granularity (a).Phagocytosis of yeasts (Y) by the cells of *E. carteri* (b, c) exposed to washing soda. N = Nucleus.

CYTOTOXICITY OF BLOOD CELLS AS AN EFFECTIVE IMMUNE STRATEGY

Cytotoxic molecules generated by the immunocompetent cells of invertebrates are reported to play an important role in the destruction and deactivation of foreign engulfed pathogens [29]. Authors reported superoxide anion, nitric oxide and phenoloxidase as established cytotoxic molecules of freshwater molluscs affected by various environmental pollutants. Contamination of natural habitat by toxic metals, metalloids, pesticides and washing soda results in a significant alteration in the cytotoxic status of invertebrates. Cytotoxic molecules are considered as an effective component of innate immune defence of invertebrates distributed in polluted environment. Nappi and Ottaviani (2000) reported nitric oxide and superoxide anions as potential "killing agents" of invertebrates [30]. Generation of reactive oxygen intermediate and reactive nitrogen intermediate by the phagocytic cells are mediated by NADPH oxidase and nitric oxide synthase (NOS), respectively. According to them, generation of superoxide anions can be correlated with increased respiratory burst activity in phagocytic cells. Mukherjee et al. (2012) reported a dose-dependent increase in the generation of superoxide anion in the cells of freshwater edible bivalve, *L. marginalis* under the exposure of azadirachtin, a neem-based pesticide [31]. According to them, this response of haemocytes may be considered as cellular stress under the sublethal and environmentally realistic concentrations of azadirachtin. Nitric oxide is considered as a signalling molecule generated in the biological system by the immunocytes of invertebrates as a potent cytotoxic agent for killing of invading microorganisms. It is generated during the conversion of L-arginine to L-citrulline by nitric oxide synthase. Saha et al. (2008b) reported sodium-arsenite-induced increase in the generation of the intracellular nitric oxide in estuarine mud crab, *S. serrata* [32]. According to them, dose-dependent increase in the generation of nitric oxide may be considered as a biomarker of arsenic pollution in the Sunderbans delta of India. Chakraborty et al. (2009) reported inhibition in generation of nitric oxide in *L. marginalis* under prolonged exposure of arsenic [7]. Authors claimed this studied parameter as possible biomarker of arsenic toxicity in aquatic environment. Phenoloxidase, on the other hand, is considered as another cytotoxic molecule and is involved in the process of nonself recognition, phagocytosis and melanisation. The production of toxic quinoid derivative by phenoloxidase is an early step of biosynthesis of melanin for host defence. Chakraborty et al. (2010a) reported depletion in generation of cytotoxic molecule like nitric oxide and activity of phenoloxidase in the gill of freshwater bivalve, *L. marginalis*, under the sublethal exposures of sodium arsenite [33]. According to them, sodium-arsenite-induced alteration in the cytotoxic status in gill indicated a state of immunocompromisation in the animal. Chakraborty et al. (2013) reported inhibition in the generation of superoxide anions and nitric oxide and activity of phenoloxidase in the digestive gland of the same experimental bivalve under prolonged exposure of arsenic [34]. Generation of cytotoxic molecules like superoxide anions and nitric oxide and activity of phenoloxidase are important immunological responses offered by the cells of sponge under the exposure of pathogens and toxins [5]. Authors reported washing-soda-induced alteration in the generation of superoxide anion, nitric oxide and activity of phenoloxidase in the dissociated cells of freshwater sponge, *E. carteri.* Washing-soda-induced alteration in cytotoxic response of *E. carteri* may lead to an undesirable shift in the immune status of the animal distributed in detergent contaminated natural habitat.

LYSOSOMAL MEMBRANE STABILITY AND ACTIVITY OF PHOSPHATASES

Lysosome is an important subcellular organelle involved in the process of degradation of foreign engulfed particulate. After phagocytosis, the phagocytic vacuole with engulfed foreign particles fuses with lysosome to form phagolysosome. As lysosome plays an important role in secretion of various digesting enzymes, maintenance of lysosomal membrane integrity has been gaining special scientific attention from the immunological point of view. Chakraborty and Ray (2009) reported impairment of lysosomal membrane integrity by neutral red retention assay in the haemocytes of *L. marginalis*exposed to sodium arsenite [6]. According to them, arsenic-induced fragility of lysosomal membrane may lead to leakage of lysosomal enzymes into the cytosol and subsequent destruction of the adjoining self cell or tissue. Ray et al. (2013b) reported lysosomal membrane integrity of haemocytes of *B. bengalensis* and *L. marginalis* under the experimental exposures of cypermethrin and fenvalerate, respectively [2]. Authors reported pyrethroid-induced lysosomal membrane fragility of haemocytes suggesting substantial damage and destabilization of lysosomal membrane (Figure 6). Screening of lysosomal membrane integrity of molluscan haemocytes by neutral red retention assay is considered as a possible biological marking of arsenic toxicity [6] and as an early warning tool of environmental pollution [35]. Phosphatases are principal lysosomal enzymes which are involved in pathogen destruction and are reported as markers of environmental stress [36]. Chakraborty et al. (2010a) reported suppression in the activity of hydrolytic enzymes, acid and alkaline phosphatases in the gill of freshwater edible bivalve, *L. marginalis* under the exposure of sodium arsenite [33]. According to them, inhibition in the activity of these enzymes might cripple the immune status and nutrient mobility in the gill of *L. marginalis.* Chakraborty et al. (2013) reported inhibition in the activity of phosphatases in the haemocyte and digestive tissue of edible bivalve under similar toxic insult by environmental arsenic [34]. Workers apprehend a state of immunocompromisation of the organisms under persistent exposure of arsenic. Saha et al. (2009) reported inhibition in the activity of phosphatases in the haemocytes of estuarine mud crab, *S. serrata* under the exposure of arsenic [37]. According to them, this situation in the natural environment might result in impairment of immunological activity and opportunistic invasion of parasite and pathogen into the body of the organism.

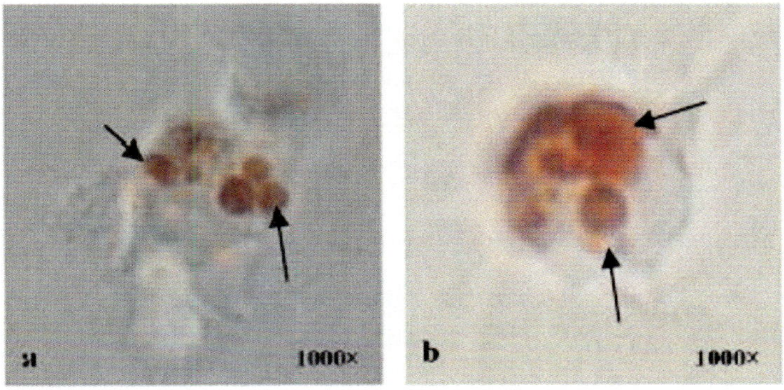

Figure 6: Cypermethrin-induced fragility of lysosome membrane of haemocytes of aquatic mollusc as evident from neutral red dye retention assay. Dye neutral red concentrated in the lysosomal compartments at zero minute of the assay (a). Complete diffusion of neutral red in the cytoplasm of cypermethrin-exposed haemocytes has been designated as "end point" of the assay (b).

POLLUTANT-INDUCED APOPTOSIS AND NECROSIS OF INVERTEBRATE IMMUNOCYTES

Apoptosis, the programmed cell death, is functionally involved in developmental and immunological processes of organisms. Moreover, apoptosis helps in the process of removal of damaged cells and thus is considered an important machinery of survival of host animal in polluted environment. It is involved in specific cell signaling mechanism, which is yet to be studied in invertebrates in detail. However, pesticides and other environmental xenobiotics are reported to affect the apoptotic and necrotic pathways of the immunoactive cells of invertebrates (Figure 7). Current scientific reports suggest apoptosis as an effective biomarker of aquatic pollution. Apoptosis presented changes in the morphological characteristics of cells including membrane blebbing, nuclear condensation, cytoplasmic shrinkage and membrane asymmetry. Kiss (2010) reported the translocation of phosphatidylserine from inner leaflet of the plasma membrane to the outer leaftet, which was considered a hallmark of apoptosis [38]. Ray et al. (2013b) reported the apoptotic and necrotic cell deaths of haemocytes of *B. bengalensis*and *L. marginalis* under the sublethal exposures of cypermethrin and fenvalerate, respectively, employing flow cytometry [2]. According to them, cypermethrin and fenvalerate treatment yielded decrease in the percentage of apoptotic and necrotic haemocyte morphotypes of *B. bengalensis* and *L. marginalis.* Pyrethroid-induced apoptosis of molluscan haemocytes is considered as impairment of immunological status of *B. bengalensis* and *L. marginalis.*

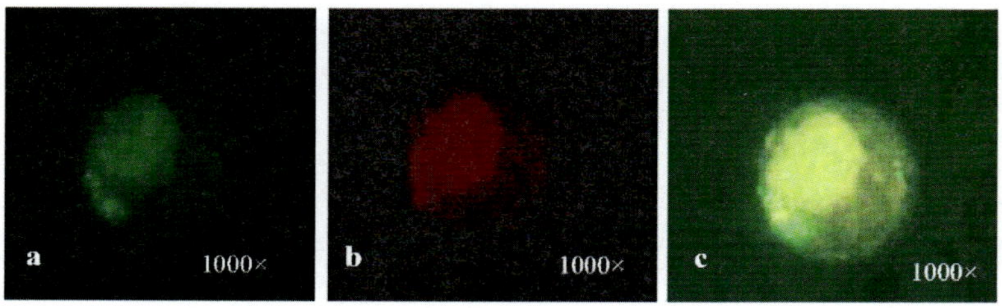

Figure 7: Immunofluorescent detection of apoptosis of haemocytes of pond snail, *B. bengalensis* exposed to sublethal (3000 ppm/7 days) concentrations of washing soda. (a) and (b) represent positive staining by Annexin V-FITC and propidium iodide, respectively; (c) depicts co-localization of fluorescent signals.

TOXIN-INDUCED DAMAGE OF TARGET ORGANS AND TISSUES

For filter feeding aquatic organisms, the gill, heart, digestive gland, mantle and antennae act as major target organs of common environmental xenobiotics (Figure 8). Mollusca bear a well-developed gill which serves as an organ of gaseous exchange, filter feeding and immunosurveillance. Molluscan gill is a thin membranous and vascularized organ that is in continuous contact with the environmental toxicants distributed in water. During the process of respiration and filter feeding, gill is exposed to various toxins of known and unknown chemistry. Exposure of gill to aquatic sodium arsenite, pyrethroid pesticide and washing soda of pond water yields a structural damage of lamellae. Chakraborty et al.

(2010a) reported hyperchromatic anaplastic cells in the gills of *L. marginalis* exposed to sublethal concentrations of arsenic [33]. This is often associated with tissue rupture and formation of dense fibrosis. The heart, in general, acts as an organ of body fluid pumping. Exposure of cypermethrin results in severe histopathological damage of heart of mollusc. Similar kind of morphological damage was recorded due to exposure of aquatic mollusc to washing soda, a common pollutant of pond water. Digestive gland of mollusc bears immense physiological significance. The organ acts as an important site of diverse metabolic activities and biochemical detoxification. Exposure of sodium arsenite yields histopathological damage of the digestive gland of *L. marginalis* [34]. Toxin-induced morphological alteration of digestive gland of mollusc includes hyperinfiltration of haemocytes, vacuolation and inflammatory lysis.

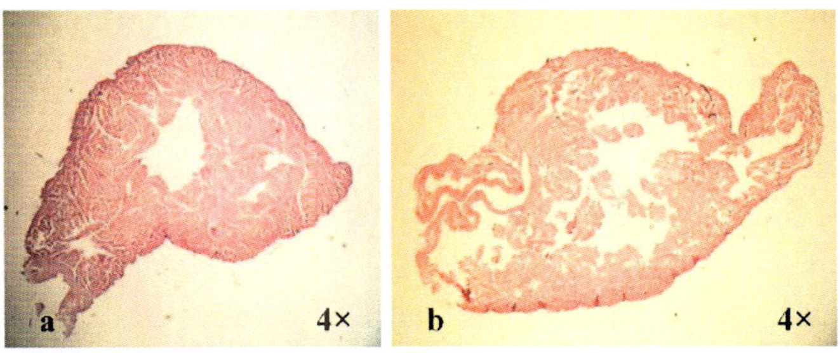

Figure 8: Cypermethrin (1.5 ppm/7 days)-induced histopathological damage of heart of mollusc, *P. globosa*.

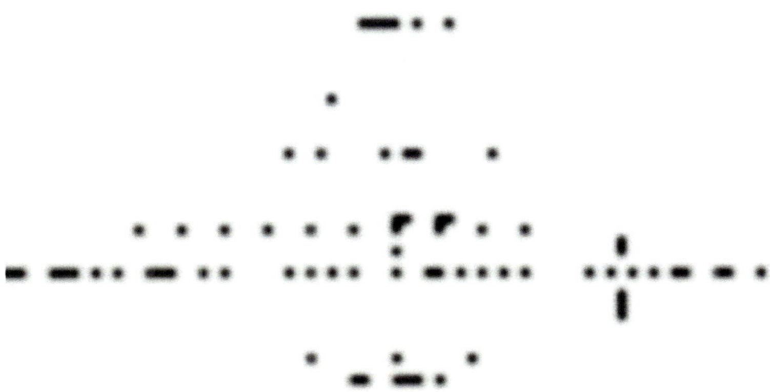

Figure 9: Ecophysiological consequences of environmental immunotoxins on aquatic invertebrates.

DISCUSSION

Global environment in recent times is characterised by the presence of various xenobiotics of known, less known and unknown toxicity and chemistry. Information of immunological attributes of chemical compounds in aquatic invertebrates is limited in the current scientific literature. Limited but significant reports indicate a substantial impairment of the immunological status of invertebrates under the exposure of selected ecotoxins (Figure 9). Acute, subchronic and chronic exposure of common toxins like pesticides, arsenic and alkaline washing soda cause severe damage in the morphological and functional profiles of haemocytes, the chief immunoeffector cells of invertebrates and other organs and tissues. Sponges, in general, are devoid of well-developed organ system. A variety of specialized cells of sponges are functionally involved in various immunological activities. Exposure of immunotoxins like washing soda largely affected the density dynamics of sponge cells as well as the cytotoxic and phagocytic status in freshwater sponge. Invertebrates of freshwater ecosystem act as important economical resource for developed and developing countries. A thorough toxicological analysis of the functional performance of target cells and tissues of invertebrates needs to be carried out for the purpose of conservation and culture of this species in their natural environment.

ACKNOWLEDGEMENTS

Authors gratefully acknowledge the scientific contributions of Suman Mukherjee, Sudipta Chakraborty and Sanjib Saha – the former members of the Aquatic Toxicology Laboratory. CRNN of University of Calcutta, UGC SAP DRS II and DST INSPIRE are acknowledged for financial and analytical support as well.

REFERENCES

1. Ray S, Ray M, Chakraborty S, Mukherjee S. Immunotoxicity of environmental chemicals in the pearl forming mussel of India-a

review. In: McGevin LE (ed.) Mussels: Anatomy, Habitat and Environmental Impact. Nova Science Publishers Inc. NY, USA; 2011. p429-440.

2. Ray M, Bhunia AS, Bhunia NS, Ray S. Density shift, morphological damage, lysosomal fragility and apoptosis of hemocytes of Indian molluscs exposed to pyrethroid pesticides. Fish Shellfish Immunol 2013b; 35:499-512.
3. Chakraborty S, Ray M, Ray S. Arsenic toxicity: A heart breaking saga of a freshwater mollusc. Tissue Cell 2012; 44:151-155.
4. Ray S, Saha S. Arsenic Toxicity in Estuarine Mudcrab. Lambert Academic Press, Germany; 2011 pp. 1-80.
5. Mukherjee S, Ray M, Ray S. Immunotoxicity of washing soda in a freshwater sponge of India. Ecotoxicol Environ Safety 2015b; 113:112-123.
6. Chakraborty S, Ray S. Nuclear morphology and lysosomal stability of molluscan hemocyte as possible biomarkers of arsenic toxicity. Clean 2009; 37(10):769-775.
7. Chakraborty S, Ray M, Ray S. Evaluation of phagocytic activity and nitric oxide generation by molluscan haemocytes as biomarkers of inorganic arsenic exposure. Biomarkers 2009; 14(8):539-546.
8. Mello DF, Proenca LA, Barracco MA. Comparative study of various immune parameters in three bivalve species during a natural bloom of Dinophysis acuminata in Santa Catarina Island, Brazil. Toxins 2010; 2:1166-1178.
9. Chakraborty S, Ray M, Ray S. Sodium arsenite induced alteration of hemocyte density of Lamellidens marginalis – an edible mollusc from India. Clean 2008; 36(2):195-200.
10. Mukherjee S, Ray M, Ray S. Azadirachtin induced modulation of total count of hemocytes of an edible bivalve Lamellidens marginalis. Proc Zool Soc Calcutta 2006; 59(2): 203-207.
11. Suresh K, Mohandas A. Number and types of hemocytes in Sunetta scripta and Villoritacyprinoids var. Cochinensis (Bivalvia) and leucocytosis subsequent to bacterial challenge. J Invert Pathol 1990; 55:312-318.
12. Russo J, Madec L. Hemocyte apoptosis as a general cellular immune

response of the snail, Lymnaea stagnalis, to a toxicant. Cell Tiss Res 2007; 328(2):431-441.

13. Oliver LM, Fisher WS. Appraisal of prospective bivalve immunomarkers. Biomarkers 1999; 4:510-530.
14. Qubella R, Maes P, Paillard C, Auffret M. Experimentally induced variation in hemocyte density for Ruditapes philippinarum and R. decussatus (Mollusca, Bivalvia). Dis Aqua Organ 1993; 15:193-197.
15. Mukherjee S, Ray M, Ray S. Dynamics of hemocyte subpopulation of Lamellidens marginalis exposed to a neem based pesticide. In: Ramakrishna and Chatterjee RN (ed.) Zoological Research in Human Welfare. Zoological Survey of India, Kolkata; 2008. p389-394.
16. Das K, Ray M, Ray S. Cypermethrin induced dynamics of hemocyte density of Indian mollusc Lamellidens marginalis. Anim Biol J 2012; 3(1):1-11.
17. Mukherjee S, Ray M, Ray S. Impairment of immunoadhesion of molluscan hemocytes exposed to azadirachtin. Icfai J Life Sci 2007; 1(1):40-44.
18. Ghosh S, Ray M, Ray S. Kinetics of nonself surface adhesion and viability of immunocytes of blood of economically important mollusc Telescopium telescopium of Sunderbans Estuary. Icfai J Life Sci 2008; 2(3):7-14.
19. Ray S, Ghosh S. Immunotoxicity of Diesel in Estuarine Mud Whelk. Lambert Academic Press, Germany; 2011pp- 1-74.
20. Saha S, Ray M, Ray S. Kinetics of nonself surface adhesion and phagocytic response of hemocyte of Scylla serrata exposed to sodium arsenite. Toxicol Int 2008a; 15(1):15-19.
21. Ray S, Saha S, Ray M. Immunotoxicological reactivity of hemocyte of juvenile mudcrab of Sunderbans biosphere reserve of India. In: Eder DL. (ed.) Aquatic Animals: Biology Habitats and Threats. Nova Science Publishers Inc. NY, USA; 2012. p53-75.
22. Ray S, Ray M, Bhunia NS. Toxicity of smoke in an amphibian mollusc from India. In: Gallo D, Mancini R (ed.) Environmental and Regional Air Pollution. Nova Science Publishers Inc. NY, USA; 2009. p351-366.

23. Nappi AJ, Christensen BM. Melanogenesis and associated cytotoxic reactions: Applications to insect innate immunity. Insect Biochem Mol Biol 2005; 35:443-459.

24. Mukherjee S, Ray M, Ray S. Anticoagulant and carbohydrate induced interference of aggregation of mussel haemocyte under azadirachtin exposure. In: McGevin LE (ed.) Mussels: Anatomy, Habitat and Environmental Impact. Nova Science Publishers Inc. NY, USA; 2011b. p441-451.

25. Lo'pez C, Carballal MJ, Azevedo C, Villalba A. Differential phagocytic ability of the circulating hemocyte types of the carpet shell clam Ruditapes Decussates (Mollusca : Bivalvia). Dis Aqua Organ 1997; 30:209-215.

26. Chakraborty S, Ray M, Ray S. Cytotoxic response and nonself phagocytosis as innate immunity of mollusc under arsenic exposure. Anim Biol J 2010b; 1(3):173-183.

27. Mukherjee S, Ray M, Ray, S. Phagocytosis of charcoal particulate as immunological marker of azadirachtin exposure. In: Gupta VK (ed.) Perspective of Animal Ecology and Reproduction. Daya Pub. New Delhi, India; 2011a. p55-67.

28. Mukherjee S, Ray M, Ray S. Phagocytic efficiency and cytotoxic responses of Indian freshwater sponge (Eunapius carteri) cells isolated by density gradient centrifugation and flow cytometry: a morphofunctional analysis. Zoology 2015a; 118:8-18.

29. Ray M, Bhunia NS, Bhunia AS, Ray S. A comparative analyses of morphological variations, phagocytosis and generation of cytotoxic agents in flow cytometrically isolated hemocytes of Indian molluscs. Fish Shellfish Immunol 2013a; 34:244-253.

30. Nappi AJ, Ottaviani E. Cytotoxicity and cytotoxic molecules in invertebrates. BioEssays 2007; 22:469-480.

31. Mukherjee S, Ray M, Ray S. Oxidative stress in freshwater bivalve of India exposed to azadirachtin based pesticide. Roman J Biol-Zool 2012; 57(1):79-88.

32. Saha S, Ray M, Ray S Nitric oxide generation by immunocytes of mud crab exposed to sodium arsenite. In: Ramakrishna and Chatterjee RN (eds.) Zoological Research in Human Welfare. Zoological Survey of India, Kolkata; 2008b; 425-428.

33. Chakraborty S, Ray M, Ray S. Toxicity of sodium arsenite in the gill of an economically important mollusc of India. Fish Shellfish Immunol 2010a; 29:136-148.

34. Chakraborty S, Ray M, Ray S. Cell to organ: Physiological, immunotoxic and oxidative stress responses of Lamellidens marginalis to inorganic arsenite. Ecotoxicol Environ Safety 2013; 94:153-163.

35. Lee BT, Kim KW. Lysosomal membrane response of earthworm, Eisenia fetida, to arsenic contamination in soils. Environ Toxicol 2009; 24:369-376.

36. Yin X, Li Z, Yang K, Lin H, Guo Z. Effect of guava leaves on growth and nonspecific immune response of Penaus monodon. Fish Shellfish Immunol 2014; 40:190-196.

37. Saha S, Ray M, Ray S. Activity of phosphatase in the hemocytes of estuarine edible mudcrab, Scylla serrata exposed to arsenic. J Environ Biol 2009; 30(5):655-658.

38. Kiss T. Apoptosis and its functional significance in molluscs. Apoptosis 2010; 15:313-321.

11
CHAPTER

LETHAL AND SUBLETHAL EFFECTS OF PESTICIDES ON AQUATIC ORGANISMS: THE CASE OF A FRESHWATER SHRIMP EXPOSURE TO ROUNDUP

Paul K. Mensah[1], Carolyn G. Palmer[1] and Wilhelmine J. Muller1

[1] Institute for Water Research, Rhodes University, Grahamstown, South Africa

INTRODUCTION

In the last few decades, rapid human population growth with its concomitant astronomical increase in urbanisation, industrialisation and technology has had its toll on natural resources of the world. Climate change, acid rain, nutrient enrichment of aquatic environments, pollution by pesticides, metals, and synthesised toxic substances on local, regional and global scales are the result of such anthropogenic disturbances. Recent events, as witnessed the world over such as large scale mortality of wildlife (e.g. sea mammals, birds), increasing menace to human health (e.g. cancerous cells, chronic respiratory disease, damage to organs such as brain, lung, heart, liver, kidneys) and algal bloom in many water bodies are all effects of the anthropogenic perturbations of the biosphere. The biosphere is part of the earth that supports life. It comprises of the lithosphere, hydrosphere and atmosphere. The hydrosphere is the total mass of water on planet Earth, which includes oceans, lakes, streams, groundwaters and glaciers. Saline water account for 97.5 % while freshwater accounts for 2.5 %. The bulk of freshwater, 68.7 %, is stored in ice and permanent snow cover, while 29.9 % exists as groundwater. Only 0.26 % is found in lakes, river systems and reservoirs [1]. However, among all the components of hydrosphere, freshwater ecosystems are the most vulnerable to pollution due to anthropogenic stresses [2-3]. Agricultural, industrial and domestic activities are the major sources of this pollution [4]. These activities use more than one-third of the Earth's accessible freshwater resources and have contaminated water with numerous synthetic and geogenic compounds [4]. For instance, about 300 billion kilograms of synthetic compounds used in industrial, consumer and agricultural products find their way into natural freshwater systems every year [5]. Ten percent of the globally accessible runoff is used, generating a stream of wastewater, which flows or seeps into groundwater, rivers, lakes, or the oceans [5].

The use of agrochemicals is necessary to control pests and increase yields in order to produce adequate food for the global population, estimated at 6.8 billion in 2009 [5], and recently reported to have reached 7 billion [6]. Underdeveloped countries, where 1.02 billion people (15 %) are undernourished and 1.3 billion people (19 %) live on an inadequate diet [5], need an adequate food supply. However, the agricultural sector's annual application of over 140 billion kilograms of fertilizers and large amounts of pesticides creates massive sources of diffuse pollution of freshwater systems [4]. The presence of these toxic chemicals in both aquatic and terrestrial ecosystems has become an important issue globally. Growing research-based evidence shows that pesticides, metals and many industrial chemicals interfere with the health and normal functioning of the endocrine systems of a wide range of organisms, including humans [7-9]. It is believed that effects of these chemicals on the normal functioning of the endocrine system are responsible for a number of developmental anomalies in a wide range of species, from invertebrates to higher mammals [10-13].

PESTICIDES

Pesticide Pollution in Ecosystems

Pesticides are substances or mixture of substances designed to control, repel, mitigate, kill or regulate the growth of undesirable biological organisms. These undesirable biological organisms (pests) do not only compete with humans for food, transmit diseases and destroy property, but are generally nuisance. Pests include insects, plant pathogens, weeds, molluscs, fish, birds, mammals, nematodes and microorganisms such as bacteria and viruses. Pesticides may be classified as being biological or synthetic. Biological pesticides are derived from natural sources such as extracts from plants (e.g. pyrethrin insecticide from chrysanthemum plants and azadriachtin from neem trees). Majority of pesticides are synthetic as they are made through industrial processes. A pesticide may also be classified as broad-spectrum when used to control a wide range of species or as narrow-spectrum when used to control a small group of species. However, the most common classification of pesticides is based on the type of pest they are used to control. These include insecticides (control insects), herbicides (control weeds) and fungicides (control fungi). Pesticides are used in agriculture to maintain high production efficiency and there is a constant demand for stable crop production to support the growing human population. Therefore, use of pesticides is expected to increase in the near future [14]. However, their use is an environmental hazard and can affect non-targeted organisms, other than the targeted pests [15].

Pesticide pollution affects both aquatic and soil ecosystems. Factors that promote pesticide pollution include drainage patterns, properties of the pesticide, rainfall, microbial activity, treatment surface and rate of application. Pesticides are able to move from one ecosystem to another through processes such as transfer (mobility) and transformation (degradation). Transfer may occur through surface runoff, vapourization to atmosphere, sorption (adsorption/desorption), plant uptake or soil water fluxes. Transformation occurs through chemical, microbial and photo-degradation [16]. A risk to a water body by a particular pesticide is dictated by the unique properties of the pesticide. For example, half-life, mobility and solubility are three properties of pesticides which determine their specific effects. Although pesticides are used on a local scale, their effects are ubiquitous and can be felt regionally and globally [17]. They are transported into aquatic systems through processes such as direct applications, surface runoffs, spray drifts, agricultural returns and groundwater intrusions; either as single chemicals or complex mixtures [18]. The transportation of pesticides to their final destination in the aquatic ecosystem may result in adverse health effects on the organisms found there. All members that form the different communities of an ecosystem, from the smallest invertebrates to birds and humans, are affected by pesticides. Most toxic pesticides in urban and agricultural settings are responsible for the deaths of many birds, fish and zooplanktons that fish depend on for food [19]. It has been reported that pesticides contaminate many breeding sites of amphibians and that some of them may persist in the environment for a very long time even at lower concentrations. Some effects of pesticides only become highlighted after long term exposure. For example, the survival patterns for early green frogs and late wood frogs are affected only after 24 days of exposure to atrazine [17].

Herbicides: Weed Control Pesticides

Weeds are plants that grow in places people do not wish them to grow because they compete with "beneficial and desirable" plant species. Until the last century, much of the energy used in farming went into removing weeds to provide suitable conditions for efficient cropping. However, during the industrial revolution, more people moved to work in factories, thus creating a shortage of labour on farms and it became necessary to develop more efficient ways to control weeds [5]. Herbicides are chemical substances used to suppress or kill unwanted vegetation (weeds). They are only one of the many types of pesticides that include insecticides, fungicides, rodenticides and nematocides [5]. Herbicides may be classified based on the time of application: pre-plant herbicides are applied to the soil before the crop is planted; pre-emergence herbicides are applied to the soil after the crop is planted, but before the crop or weeds emerge; post-emergence herbicides are applied to both crop and weeds after they have germinated and emerged from the soil. Herbicides may also be classified by the way they kill or suppress plants. These include hormone inhibitors, cell division inhibitors, photosynthesis inhibitors, pigment synthesis inhibitors, lipid synthesis (cell membrane) inhibitors, or cell metabolism (e.g. amino acid biosynthesis) inhibitors [20]. All herbicide products have chemical properties that influence their ability to suppress growth or kill plants. While some of these properties are inherent in the chemical nature of the herbicides, others are added to enhance their efficacy. The following are some chemical properties of herbicides that influence their use:

Chemical structure: The biologically active portion of a herbicide product is the *active ingredient*. It is the fundamental molecular composition and configuration of the herbicide. The physical and chemical properties of a herbicide can also determine the method of application and use.

Water solubility and polarity: Herbicides that are produced as salts dissolve quite well in water and are usually formulated to be applied in water, while non-polar herbicide sources are not. Water is the main substance used to disperse (spray) herbicides, and hence the water solubility of a herbicide influences the type of product that is formulated, how it is applied and the movement of the herbicide in the soil profile.

Volatility: Herbicides with a high vapour pressure volatilise easily, while those with a low vapour pressure are relatively non-volatile. The volatility of a herbicide can determine the mode of action and the herbicide's fate in the environment.

Formulations: Commercial herbicide products contain an active ingredient and "inert" ingredients. An "inert" ingredient could be a carrier that is used to dilute and disperse the herbicide (e.g. water, oil, certain types of clay, vermiculite, plant residues, starch polymers, certain dry fertilizers) or an adjuvant (e.g. activator, additive, dispersing agent, emulsifier, spreader, sticker, surfactant, thickener, wetting agent) that enhances the herbicide's performance, handling, or application [20]. In recent years, carriers and adjuvants have been implicated in adding to the toxicity of the active ingredients, and in some cases, have been even more toxic than the active ingredient alone [20].

Before herbicide products are registered for use, the registration authorities require experimental information on their toxicology, biology, chemistry, and biochemical degradation in addition to their effect on air and water quality, soil microorganisms, and wildlife. Although commercial herbicide products contain several different ingredients, toxicity tests are usually conducted only on the active ingredient, which is the component of the product believed to actually affect the target organism [20]. The criteria for assessing the possible effects of herbicides on the safety of humans, animals and the environment are the herbicide's toxicity (including carcinogenicity, mutagenicity, endocrine disruption, reproduction and developmental abnormalities), biomagnification, and persistence in the environment ([20].

Given the scarcity of water resources in South Africa, aquatic herbicides are of special interest. The potential of an aquatic herbicide to adversely affect aquatic organisms depends on its inherent toxicity to the specific organism and the organism's exposure to the compound in terms of concentration and duration [21]. The inherent toxicity of the pesticide, which is due to its mode of action, is a specific relationship between the organism and the chemical, whereas factors such as application rates and techniques, chemical and physical properties of the pesticide, and environmental conditions at the time of application can make exposures highly variable. Herbicides now lead all other pesticide groups in terms of amount produced, total acreage treated, and total value from sale. Over the past decades, public awareness of the worldwide increase in the use of herbicides and their adverse effects on aquatic ecosystems has been growing [22]. Herbicides may reach water bodies directly by overhead spray of aquatic weeds, or indirectly through processes such as agricultural runoff, spray drift and leaching. Potential problems associated with herbicide-use include injury to non-target vegetation, injury to crops, residue in soil or water, toxicity to non-target organisms, and concerns for human health and safety [20]. Herbicides can decrease environmental water quality and ecosystem functioning by reducing species diversity, changing community structure, modifying food chains, altering patterns of energy flow and nutrient recycling, and reducing resilience of ecosystems, among others [22].

ROUNDUP

Glyphosate-Based Herbicides

Glyphosate-based herbicides are the world's leading post-emergent, organophosphonate systemic, broad-spectrum and non-selective herbicides for the control of annual and perennial weeds [22]. Roundup® is the major glyphosate-based herbicide in which glyphosate (the active ingredient) is formulated as isopropylamine (IPA) salt, polyoxyethylene amine (POEA) (a surfactant), and water. Other formulations (e.g. Rodeo®) contain the IPA salt of glyphosate without POEA, and in some countries are primarily used for controlling aquatic weeds [23-24]. Other trade names of glyphosate-based herbicides include Roundup®, Roundup Ultra®, Roundup Pro®, Accord®, Honcho®, Pondmaster®, Protocol®, Rascal®, Expedite®, Ranger®, Bronco®, Campain®, Landmaster®, Fallow Master® and Aquamaster® by Monsanto; Glyphomax®, Glypro® and Rodeo® by Dow Agrosciences; Glyphosate herbicide by Du Pont; Silhouette® by Cenex/Land O'Lakes; Rattler® by Helena; MirageR®by Platte; JuryR® by Riverside/Terra; and Touchdown® by Zeneca [25].

Glyphosate is an aminophosphonic analogue of the natural amino acid glycine [22].The International Union of Pure and Applied Chemistry's (IUPAC) name for glyphosate is N-(phosphonomethyl) glycine and the Chemical Abstracts Service (CAS) registry number is 1071-83-6. The glyphosate molecule has several dissociable hydrogens, especially the first hydrogen of the phosphate group (Figure 1). Thus, a typical glyphosate molecule is an acid, and is often referred to as the technical grade glyphosate.

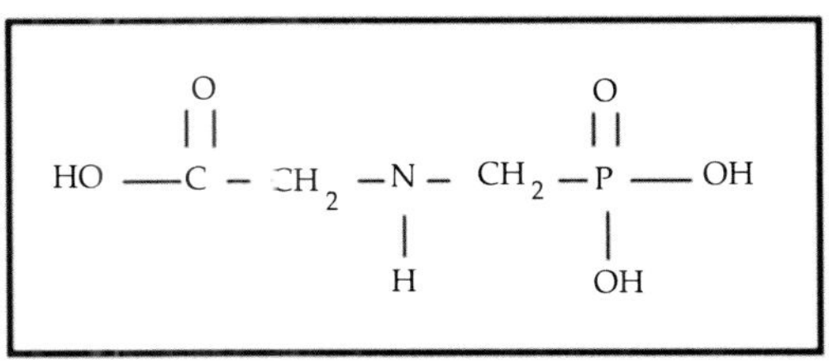

Figure 1: Molecular structure of N-(Phosphonomethyl) glycine

Technical-grade glyphosate has a relatively low solubility in water (12 g/L at 25° C and 60 g/L at 100° C), and is insoluble in other solvents because of strong intermolecular hydrogen bonds that stabilise the crystal lattice [26]. For this reason, commercial herbicide formulations contain glyphosate in the form of salt, which has much higher solubility but still maintains the herbicidal properties of the parent compound [22]. Formulations of glyphosate in salt form include monoammonium salt, diammonium salt, isopropylamine salt, potassium salt, sodium salt, and trimethylsulfonium or trimesium salt. Of these, the isopropylamine, sodium, and monoammonium salt forms are commonly used in formulated herbicide products [27].

The isopropylamine salt is the most commonly used in commercialised formulated products (e.g. Roundup®). The physical and chemical properties of glyphosate acid and two of its salt forms are listed in Table 1. The concentration of glyphosate is commonly expressed as mg a.i./L (active ingredient/Litre) or mg a.e./L (acid equivalents/Litre) [22]. Acid equivalent is the theoretical percent yield of parent acid from a pesticide active ingredient, which has been formulated as a derivative (usually esters, salts or amines) [28].

Table 1: Physical and chemical properties of glyphosate acid, glyphosate isopropylamine salt, and glyphosate ammonium salt [27]

Active ingredient	Form	Vapour pressure	Henry's constant	Molecular weight	Solubility in water	Log $K_{o}W$	$K_{o}C$
Glyphosate acid	Odourless, white solid	1.31×10^{-2} mPa (25° C); 1.84×10^{-7} mmHg (45° C)	4.08×10^{-19} atm· m^3/mol	169.07 g/mol	pH 1.9: 10,500 mg/L; pH 7: 157,000 mg/L	< -3.2	300 - 20,100
Glyphosate Isopropylamine salt	Odourless, white solid	2.1×10^{-3} mPa (25° C); 1.58×10^{-8} mmHg (25° C)	6.27×10^{-27} atm· m^3/mol	228.19 g/mol	pH 4.1: 786 ,000 mg/L	-3.9 or -5.4	300 - 20,100
Glyphosate ammonium salt	Odourless, white solid	9×10^{-3} mPa (25° C); 6.75×10^{-8} mmHg (25° C)	1.5×10^{-13} atm· m^3/mol	186.11 g/mol	pH 3.2: 144,000 mg/L	-3.7 or -5.3	300 - 20,100

Mode of Action of Glyphosate

As a systemic herbicide, glyphosate is readily translocated through the phloem to all parts of the plant. Glyphosate molecules are absorbed from the leaf surface into plant cells where they are symplastically translocated to the meristems of growing plants [22]. Glyphosate's phytotoxic symptoms usually start gradually, becoming visible within two to four days in most annual weeds, but may not occur until after seven days in most perennial weeds. Physical phytotoxic symptoms include progress from gradual wilting and chlorosis, to complete browning, total deterioration and finally, death [22]. The primary mode of action of glyphosate is confined to the shikimate pathway aromatic amino acid biosynthesis, a pathway that links primary and secondary metabolism.

Shikimate (shikimic acid) is an important biochemical intermediary in plants and microorganisms, such as bacteria and fungi. It is a precursor for the aromatic amino acids phenylalanine, tryptophan and tyrosine. Other precursors of the shikimate pathway are indole, indole derivatives (e.g. indole acetic acid), tannins, flavonoids, lignin, many alkaloids, and other aromatic metabolites. The biosynthesis of these essential substances is promoted by the enzyme 5-enolpyruvylshikimate-3-phosphate synthase (EPSPS), the target enzyme of glyphosate (Figure 2). This enzyme is one of the seven enzymes that catalyse a series of reactions, which begins with the reaction between shikimate-3-phosphate (S3P) and phosphoenolpyruvate (PEP). The shikimate pathway accounts for about 35 % of the plant mass in dry weight and therefore any interference in the pathway is highly detrimental to the plant. Glyphosate inhibits the activity of EPSPS, preventing the production of chorismate, the last common precursor in the biosynthesis of numerous aromatic compounds in bacteria, fungi and plants. This causes a deficiency in the production of the essential substances needed by the organisms to survive and propagate [22, 29]. The pathway is absent in animals, which may account for the low toxicity of glyphosate to animals. However, acute effects in animals, following intraperitoneal administration of high glyphosate doses suggest altered mitochondrial activity, possibly due to uncoupling of oxidative phosphorylation during cellular respiration [26]. In summary, glyphosate ultimately interrupts various biochemical processes, including nucleic acid synthesis, protein synthesis, photosynthesis and respiration, which are essential life processes of living things.

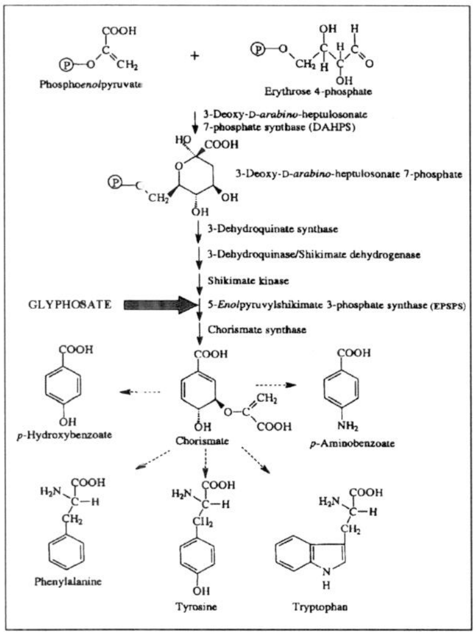

Figure 2: Glyphosate mode of action in plants with red arrow pointing to the target enzyme 5-enolpyruvylshikimate-3-phosphate synthase (modified from [32])

Environmental Fate of Glyphosate

Glyphosate has a strong soil adsorption capacity, which limits its movement in the environment. The average half-life of glyphosate in soil is two months, but can range from weeks to years [25]. The presence of glyphosate in water systems may be due to runoff from vegetation surfaces, spray drift, and intentional or unintentional direct overspray, with an average half-life of two to ten weeks [25]. Glyphosate is susceptible to chemical and photo-degradation, although microbial degradation is the primary dissipation mechanism in soils. The rate of degradation in water is generally slower than in most soils because of fewer microorganisms in water than in soils [30]. When glyphosate degrades, it produces aminomethylphosphonic acid (AMPA) and carbon dioxide [31], both of which reduce pH when dissolved in water. However, pH is known to affect the stability of glyphosate in water. For instance, glyphosate did not undergo hydrolysis in buffered solution with a pH of 3, 6, or 9 at 35° C, while insignificant photodegradation has been recorded under natural light in a pH 5, 7, and 9 buffered solutions [27]. In natural water systems, glyphosate dissipates through degradation, dilution, and adsorption on organic substances, inorganic clays and the sediment (the major sink for glyphosate in water bodies) [25, 30]. With its long half-life and its ability to cause the death of organisms in aquatic systems, it is recommended that glyphosate should be used as an aquatic herbicide to treat only one-third to half a water body at any one time [25].

Toxicology of Glyphosate

Ecotoxicologists are greatly concerned about the exposure of non-target aquatic organisms to glyphosate formulations because of its high water solubility and the extensive use of glyphosate-based herbicides in the environment, especially in shallow water systems [23]. The surfactant polyoxyethylene amine (POEA) is thought to be responsible for the relatively high toxicity of Roundup® to several freshwater invertebrates and fishes, although isopropylamine (IPA) salt of glyphosate also contributes its share [23, 33]. Technical grade glyphosate is slightly to very slightly toxic, with reported LC50 values of greater than 55 mg/L and a 21 d NOEC value of 100 mg/L.

Conversely, formulations of glyphosate are moderately to very slightly toxic with 2 d EC50 values of 5.3-5600 mg/L and 21 d MATC values of 1.4-4.9 mg/L reported [26]. The LC50 values also determine which glyphosate formulation can be applied in aquatic systems. For example, Touchdown 4-LC® and Bronco® have low LC50s for aquatic species (<13 mg/L), and are not registered for aquatic use, while Rodeo® has relatively high LC50s (>900 mg/L) for aquatic species and is permitted for use in aquatic systems. In the same manner, Roundup® is not registered for use in aquatic systems in the United States because its 96-hour LC50 for *Daphnia* is 25.5 mg/L, while that of glyphosate alone is 962 mg/L [25].

Effects of Glyphosate-Based Herbicides on Aquatic Animals

Glyphosate-based herbicides are used globally to control both aquatic and terrestrial weeds. In recent years, its use has increased tremendously and is likely to impact on non-target organisms in the environment. Even though it is generally regarded as having a low potential for contaminating surface waters due to its perceived rapid dissipation and strong adsorption to soils and sediments, it has been detected in surface waters long after being used to kill aquatic weeds [34]. In fact, its mode of action was designed to affect only plants [29], but various studies in recent years have reported adverse impact on non-target animals [23, 33, 35]. These impacts could be lethal or sublethal. Lethal effects are mainly mortality and immobility endpoint measures. However, there are several endpoint measures that can be used to assess sublethal effects. At the 'physical' level, measures of survival, growth, morphological changes, and behavioural changes exposed animals are used as endpoint indicators. Measures of reproductive performance that are often used to assess sublethal response include sexual maturity, time to first brood release, time required for egg development, fecundity, gonad histopathology, and alterations in reproductive characteristics. Biochemical measures used as possible endpoints to assess exposed animals include metabolic disruption, steroid metabolism, vitellogenin induction, lipid peroxidation, acetylcholinesterase activity, cytochrome P450enzymes and blood glucose levels.

EXPOSURE EFFECTS

Classification of Exposure Effects

The effect caused by exposure to chemicals can be classified according to different exposure time (short-term or long-term) and exposure type (lethal or sublethal). Short-term exposure time is usually defined as not more than 96 h, while long-term exposure time defined as being more than 96 h (Table 2). There are different possibilities of effect to expect when animals are exposed to chemicals. Lethal exposure to stress can possibly cause a biological system to respond in short-term or long-term. Similarly, biological systems can experience sublethal responses to a stressor in short-term or long-term (Table 3). Lethal exposure will often result in mortality (i.e. immobility, decolouration and degeneration), whereas sublethal exposure normally results in a cellular, molecular or biochemical level response including growth (length, weight and moulting), reproduction (embryo and gonads) and biochemical (acetylcholinesterase and lipid peroxidation) (Table 3).

Table 2: Exposure-effect classification of chemicals

Exposure classification	**Effect classification**	**Description**
Exposure time	Short-term	≤ 96 h
	Long-term	≥ 96 h
Exposure type	Lethal	Mortality measure as endpoint
	Sublethal	Cellular/molecular/biochemical/physiological level measure as endpoint

Table 3: Different possibilities of effect of animals exposed to chemicals

Effect classification	**Description**
Short-term lethal	≤ 96 h and mortality measure as endpoint
Short-term sublethal	≤ 96 h and cellular/molecular/biochemical/physiological level measure as endpoint
Long-term lethal	≥ 96 h and mortality measure as endpoint
Long-term sublethal	≥ 96 h and cellular/molecular/biochemical/physiological level measure as endpoint

Effects of Lethal Exposure

Mortality is the most common endpoint measure when organisms are exposed to a lethal dose, although immobility is also considered as lethal effect of exposure. In [36] the relevance of using mortality and immobility as endpoints to reflect the toxicity of the organophosphorous insecticide chlorpyrifos in fourteen different freshwater arthropods was evaluated. Using dose response models and species sensitivity distributions (SSDs), they compared the differences in response dynamics during 96 h of exposure with these two endpoints across the different species. Their study suggests that freshwater arthropods vary less in their immobility response than in their mortality response. They suggested immobility as the relevant endpoint for SSDs and ERA (environmental risk assessment) because they found it was a more sensitive endpoint than mortality, with less variability across the tested species. Generally, effect concentrations for immobility and mortality will converge to the same value with time, but this does not occur with the same speed for all species [36]. However, a good match between effective (immobility) and lethal (mortality) concentrations can exist right from the start of a toxicity test where LC50/EC50 ratios equal one, approximately. For some species, the differences between LC50 and EC50 can remain relatively constant within the 96 h of testing. Furthermore, the extent to which LC50 and EC50 values differ for certain time points is species specific [36]. For example, exposure concentrations may not induce any significant incipient mortality in a particular species, but will induce immobility at very low concentrations in another species. This is due to differences in toxicokinetics and/or toxicodynamics between the species. For instance, differences in toxicokinetics may enable one species to decrease or regulate uptake and eliminate the test chemical, or detoxify it quickly, thereby significantly delaying incipient mortality. Toxicodynamic differences, such as differences in the interaction of the stressor and target enzyme, or in the ability to compensate or repair damage, may cause different species to respond differently to the test chemicals [36].

Mortality was also used as an endpoint response measure by [37] when they studied the acute mortality of adults and sub-lethal embryo responses of *Palaemonetes pugio* to endosulfan. Their findings suggest that the insecticide endosulfan may preferentially affect male grass shrimp, and exposed female grass shrimp may produce embryos with delayed hatching times. They suggested that the size difference between male and female grass shrimp might be the cause as mortality decreases by 25 % with a corresponding increase in size of 1 mm. Some studies have reported correlation between lethal and sublethal effects. In [38], the correlation between 96 h mortality and 24 h acetylcholinesterase inhibition in three life stages of the grass shrimp (*Palaemonetes pugio*) after exposure to organophosphate pesticides was investigated. They found a strong positive relationship ($R^2 = 0.962$) between the ratio of the lowest observed effect concentration and 20 % effect concentration (LOEC/EC20). Therefore, they concluded that sublethal endpoints could be used as a predictor of 96 h mortality for the life stages of *P. pugio*.

Growth Measures Used As Sub-Lethal Responses to Exposure

Body weight and length are two direct measures of growth that may be used in the assessment of sub-lethal effects on arthropods. Simple dry weight can be determined by drying sampled animals at an average temperature of 60° C, and a mean drying time of 48 hours to constant weight [39]. However, for many invertebrates, ash-free dry weight (AFDW) is often used as the appropriate weight measurement because the method reduces any inaccuracies that might be introduced by inorganic constituents in the animal's body. Inorganic components may arise from processes such as the development of skeletal components, or from feeding (the ingestion of sediment) [39]. In small-sized crustaceans, such as caridean shrimps and mysids, the removal of ash from the dry weight measurement is unnecessary since it would have a negligible effect on the accuracy of the measurement [40]. Separate determinations should be made for male and female crustaceans because they might be different sizes [39]

Different body length dimensions of shrimp can be measured to determine growth. These may include the distance from the base of the eye-stalks to the tip of the telson or to the tip of the exopod; or from the tip of the carapace to the tip of the exopod along the midline of the body [39]. Sometimes, it is difficult to measure preserved animals because of the body curvature that results from the fixation process. Relaxing the animal and then determining length as the sum of a series of relatively straight-line measurements prior to fixation may reduce inaccuracy. Animals may be anaesthetized in soda water to relax them prior to length measurements [41-42].

Reduced growth may not be a particularly sensitive endpoint, but it is the most common response to sub-lethal exposure to toxicants [39]. Reduced growth is connected to reproductive success since the size of female crustaceans is directly related to fecundity [43]. The age of test animals and the toxicant concentration are related to the effect of toxicant exposure on growth. In general, young crustaceans are more sensitive than adults to toxicant exposure. However, effects of toxicants on juvenile survival do not always lead to reductions in population growth rate since survivors may compensate for the lost individuals by increasing their own reproduction [44]. Similarly, effects at the individual level may sometimes run opposite to the population level effects. This shows complex relationship between toxicant effects on individual performance versus population dynamics. [44] evaluated the effects of nonylphenol on two life-history traits (i.e. juvenile survival and fecundity) of the parthenogenetic springtail, *Folsomia candida,* in relation to population growth rate. They reported that the presence of nonylphenol stimulated fecundity and the body-growth rate of test organisms, but did not affect population growth rate. The authors found that the effect of the test chemical on fecundity was the main contributor of the observed effect on growth rate. However, since relative sensitivity of fecundity (elasticity) was very low, large changes in fecundity resulted in a minimal effect on population growth rate. Conversely, juvenile survival had higher elasticity, but was not affected by nonylphenol, and hence did not contribute to effects on population growth rate. The study by [44] revealed that increase in body size and fecundity after exposure to chemicals does not necessarily translate into increase in population growth rate. Their study also shows that effects of chemicals on individual life-history traits are attenuated at the population level and that population growth rate is an appropriate endpoint for ecotoxicological studies.

Moulting is an important physiological process in arthropods because it allows them to grow [45-46]. It is regulated by the interaction of moult stimulating hormones (MSHs, generally referred to as ecdysteroids), and nervous system secretions produced in the cephalothorax, and with moult-inhibiting hormones (MIHs) produced in the eyestalks [39]. In higher crustaceans such as the Malacostraca, paired cephalic endocrine organs called Y-organs (absent in lower crustaceans such as Entomostraca) secrete three different ecdysteroids, namely ecdysone (E), 25-deoxyecdysone (25dE), and 3-dehydroecdysone (3DE). Usually these organs produce either E + 25dE, or E + 3DE. Activities of the Y-organ are regulated mainly by the MIH, an inhibitory neuropeptide secreted from the X-organ-sinus gland complex [45-46]. Since hormones regulate moulting in crustaceans, moulting is a clear indicator of the adverse effects of endocrine disrupting chemicals, which include many pesticides. Hormonal regulation of moulting in crustaceans makes the process vulnerable to the adverse effect of endocrine disrupting chemicals (EDCs), including many pesticides [39]. Furthermore, since substantial growth in crustaceans can only occur as a result of moulting, any disruption in the moulting process could affect growth. Therefore, estimation of moulting frequency may be a useful endpoint.

Moult stage is a useful technique for measuring growth [39]. If moult stages are classified based on duration of different stages under normal laboratory conditions, then the environmental effects on relative duration of stages can be evaluated, using the moult-stage technique [47]. However, moult-cycle chronology is a prerequisite for the use of moult staging in growth studies. The moult-stage technique was used to determine the main moult stages for juveniles and young adults of *Mysis mixta*and *Neomysis integer* under different temperature conditions and feeding. The technique was also used in the field to determine the moult cycle duration of *Mysis mixta* [47].

Reproductive Measures Used As Sublethal Responses to Exposure

Embryotoxicity and gonad histopathology are two main reproductive measures used as sublethal responses to exposure. Embryo development time (or incubation period) in caridean shrimps is measured as the number of days between the first appearance of embryos in the brood pouch and the first release of neonates. In uncontaminated systems, incubation period is related to environmental temperature, salinity and an interaction between the two factors. However, the effect of most contaminants is to lengthen hatching time. In many embryotoxicity studies, either gravid females are placed in exposure containers, or fertilized eggs are removed from the female and placed in exposure containers where they develop to hatching. In [48], both gravid maternal and isolated embryos of *Daphnia magna* were exposed to the agricultural fungicide fenarimol to evaluate embryo development and susceptibility to the anti-ecdysteroidal properties of the fungicide. They reported that exposure of either gravid maternal animals or isolated embryos to the test toxicant resulted in embryo abnormalities which ranged from early partial developmental arrest to incomplete development of antennae and shell spines. They found that such developmental abnormalities were linked to suppressed ecdysone levels in the embryos and that the abnormalities could be prevented by co-exposure to 20-hydroxyecdysone. The results also showed how environmental anti-ecdysteroids, such as fenarimol, in many agro-chemicals disrupt the normal development of crustacean embryos.

Effective embryotoxicity investigations are based on identification of specific developmental features during embryogenesis and the susceptibility of such features to chemical exposure. Embryonic development of *C. nilotica* under laboratory conditions was investigated by [49] and identified stages in embryonic development which could be used as quantifiable experimental endpoints in toxicity tests. The author identified and described seven potential developmental stages that could be used in toxicity tests to study exposure-response relationships to stressors.

Histopathology is a technique that combines knowledge and experience of fundamental animal anatomy, physiology, endocrinology, pathology, and toxicology. It can enhance relevant biological information in sublethal exposure tests by allowing proper and more specific hazard identification, such as the organs targeted by toxic substances and mechanisms of action in aquatic ecotoxicological studies [50]. Histopathology is relevant to an ecological assessment of toxicants because it can detect critical adverse biological effects (e.g. reproductive abnormality) and is more sensitive than the classical toxicological testing, since histological effects are visible at lower exposure concentrations than they are at toxicological endpoints, such as mortality or behavioural changes [50]. The use of small crustaceans in practical histopathology makes it possible to embed the animals *in situ* for a quick overview of various relevant organs, making screening fast and comprehensive [50].

Biochemical Measures Used As Sub-Lethal Responses to Exposure

Acetylcholinesterase activity (AChE) and lipid peroxidation (LPx) are two biochemical measures often used as to assess sublethal responses to exposure. The main physiological function of the enzyme AChE is to hydrolyze acetylcholine (ACh), a neurotransmitter of cholinergic synapses during transduction of nerve impulses. Inhibition of AChE prevents the hydrolysis of ACh in nerve synapses and neuromuscular junctions, causing accumulation of excess ACh at these sites. This results in over-excitation of the synaptic and muscular tissues, which may lead to abnormal behaviours such as hyperactivity, asphyxia and death. AChE activity is therefore regarded as a good biomarker to detect a range of toxic compounds in aquatic animals, including insecticides, herbicides, surfactants and metals [51-52]. In a study to evaluate AChE activity in the oyster *Crassostrea corteziensis*, [53] exposed the organisms to the pesticide dichlorvos. The results of their study revealed that AChE activity was 65 % lower in oysters exposed to the pesticides than in control animals. Based on this outcome, they suggested using AChE activity in oysters as early biomarkers of effects and exposure to pesticides in aquatic environments. Similar observations and suggestions were made when the mosquito fish *Gambusia affinis* was exposed to the pesticide chlorpyrifos [54]. Although AChE is used as a classical biomarker in biomonitoring studies with regard to the exposure of a number of organophosphate and carbamate pesticides, recent studies have shown the existence of sublethal effects of glyphosate-based compounds on biomarkers of neurotoxicity including AChE [33, 55-56]

Lipid peroxidation is a recognised mechanism of cellular injury in plants and animals, and is used as an indicator of oxidative stress in cells and tissues. Lipid peroxides are unstable and decompose to form a complex series of compounds which include reactive carbonyl compounds. Polyunsaturated fatty acid peroxides decompose to produce malondialdehyde (MDA) and 4-hydroxyalkenals (HAE), and the measurement of MDA and HAE is used as an indicator of lipid peroxidation. Whether cells and tissues are susceptible to oxidative stress when exposed to pesticides reflects the balance between oxidative stress and the anti-oxidant defence capability. Since free radicals and hydroperoxides are potentially harmful, toxicants that stimulate lipid peroxidation and/or weaken anti-oxidant defence capability may cause or increase cellular susceptibility to oxidative damage. Animals exposed to pesticides may have their anti-oxidant defence capabilities directly or indirectly modified, rendering them susceptible to oxidative stress. Oxidative damage of cells and tissues of animals exposed to pesticides may be the result of insufficient anti-oxidant potential [57]. Developing biomarkers of oxidative stress as a pollution-mediated mechanism of toxicity requires knowledge of how anti-oxidant biochemical systems and target molecules are influenced by test toxicants [58].

Different toxicants may produce different anti-oxidant/pro-oxidant responses in organisms, depending on whether the organism can produce reactive oxygen species and anti-oxidant enzymes to detoxify them. Changes in juveniles of the freshwater crustacea *Daphnia magna* anti-oxidative processes in were assessed by [58] after exposure to paraquat and endosulfan in a 48 h sublethal toxicity test. They evaluated lipid peroxidation and activities of key anti-oxidant enzymes including catalase, superoxide dismutase, glutathione peroxidase and glutathione S-transferases. They found that increased lipid peroxidation produced low anti-oxidant enzyme activity for endosulfan, while decreased lipid peroxidation enhanced levels of anti-oxidant enzyme activities for paraquat. In [59], the authors suggested that glyphosate exposure and metabolism in the liver of animals can lead to excessive production of MDA and oxidative stress through unregulated generation of reactive oxygen species (ROS) such as superoxide anion, hydrogen peroxide, hydroxyl radical, peroxyl radicals and singlet oxygen. Excessive ROS in turn can be detrimental to cell structure through oxidative damage of lipids, proteins or DNA, and altered regulation of gene functions critical for development, differentiation, and aging.

LETHAL AND SUBLETHAL EFFECTS OF *CARIDINA NILOTICA* EXPOSURE TO ROUNDUP®

Caridina Nilotica: A Decapod Freshwater Shrimp

The exposure of non-target aquatic organisms to glyphosate-based herbicides is of great concern because of the high water solubility of glyphosate and its extensive use in the environment. Thus, it is important to investigate the effects of these bioactive chemicals on aquatic organisms. *Caridina nilotica* (Decapoda: Atyidae) (Figure 3) is the most common of four indigenous freshwater caridean species found in South Africa. It has been used in ecotoxicological studies in South Africa since the early 1990s. Roundups® was selected as a representative of glyphosate-based herbicides by the virtue of it being the most popular and widely used herbicide in South Africa and most parts of the world [60-61]. In this section of this chapter, summary of findings of some exposure tests are given to demonstrate lethal and sublethal effects of Roundup® to *C. nilotica* at different biological scales. Mortality was the lethal effect investigated, whereas the sublethal effects studied were growth, acetylcholinesterase activity and lipid peroxidation. The tests were all aimed at demonstrating the use of *C. nilotica* as an early detection sensor system of pesticides pollution in South African aquatic ecosystems. Comprehensive reports of these tests are reported in [62-65].

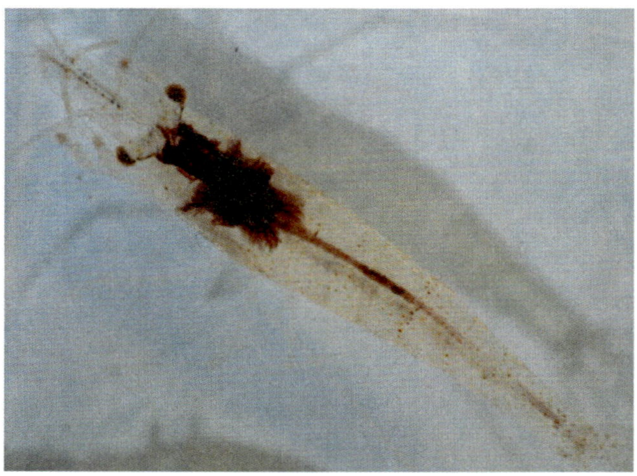

Figure 3: Adult *Caridina nilotica*

Short-Term Lethal Tests — Mortality

The toxicity of the herbicide Roundup® was assessed using three different life stages of *C. nilotica*. Neonate (<7 days post hatching (dph)), juvenile (>7 dph and <20 dph) and adult (>40 dph) shrimps were exposed to varying concentrations of the herbicide in 48 and 96 h short-term lethal tests in order to determine the most sensitive life-stage. Mortality was calculated at the end of each test period. Based on this, Roundup® 48 h and 96 h LC50 (median lethal concentration) values and the associated 95 % confidence limits were calculated for *C. nilotica* using the probit method. The results showed neonates to be more sensitive to Roundup® than both juveniles and adults. The estimated 96 h LC50 of neonates is much lower than the field application rate, though the application's impact will depend on the dilution rate of the applied concentration in the environment. This study shows that low levels of the Roundup®may adversely affect *C. nilotica* health and survival. Thus, the herbicide should be carefully managed to minimize any negative impact on non-target freshwater organisms.

Long-Term Sublethal Tests — Growth

The possible use of growth measures in *C. nilotica* as biomarkers of Roundup® pollution as part of aquatic life in South Africa were evaluated. Using static-renewal methods in a 25-d growth toxicity test, 40 dph shrimps were exposed to different sublethal Roundup® concentrations. Shrimp total lengths and wet weights were measured every fifth day. These data were used to determine the shrimp's growth performance and feed utilization in terms of percent weight gain (PWG), percent length gain (PLG), specific growth rate (SGR), condition factor (CF), feed intake (FI), feed conversion ratio (FCR) and feed conversion efficiency (FCE). Moulting was observed for 14 d and the data used to determine the daily moult rate for each concentration. Results of growth performance and food utilization indices showed that growth was significantly impaired in all exposed groups compared to control. Moulting frequency was also higher in all exposed groups than in control. Although all the tested growth measures proved to be possible biomarkers of Roundup® pollution, moulting frequency gives a clearer indication of the sublethal effects of Roundup® toxicity to *C nilotica*.

Short-Term and Long-Term Sublethal Tests — Biochemical

The use of *C. nilotica* whole-body acetylcholinesterase (AChE) activity and lipid peroxidation as potential biomarkers of Roundup® pollution of aquatic ecosystems was investigated. Forty days post hatch (dph) shrimps were exposed to different concentrations of Roundup® in a 96 h short-term sublethal test and a 21 d long-term sublethal test. Shrimp whole-body AChE activities were determined at the end of the exposure periods by spectrophotometric assay of sample extract. Final AChE activities were expressed as nmol/min/mg proteins. Shrimp whole-body LPx was estimated by thiobarbituric acid reactive species (TBARS) assay, performed by a malondialdehyde (MDA) reaction with 2-thiobarbituric acid (TBA) measured spectrophotometrically. Final MDA concentrations were expressed as nmol MDA produced/mg protein. The results showed that AChE activity was concentration-dependent, with percent activity levels decreasing monotonically from control to the highest concentration. Conversely, LPx was significantly lower in control than in shrimps exposed to different Roundup® concentrations, increasing monotonically. The study provides ecotoxicological basis for the possible use of AChE activity and LPx in *C. nilotica* as possible biomarkers for monitoring effects of Roundup® pollution in freshwater systems.

CONCLUDING REMARKS

In this chapter, the effects of rapid human population growth on aquatic ecosystems have been discussed. These effects are seen in such phenomena as climate change, nutrient enrichment of aquatic environments, and pollution by all types of chemicals including pesticides on local, regional and global scales. These anthropogenic disturbances adversely impact the normal functioning of organisms and are responsible for a number of developmental anomalies in a wide range of species; from invertebrates to higher mammals. It is expected that the use of pesticides, especially herbicides, will continue to increase and eventually becoming environmental hazard to non-target organisms at different biological scale levels unless proactive measures are taken. The case study, i.e. lethal and sublethal exposures of *C. nilotica* to varying environmentally relevant concentrations of Roundup®, showed that *C. nilotica* can be used as early detection system to assess glyphosate-based herbicides pollution effects on aquatic ecosystems.

REFERENCES

1. Shiklomanov, I.A. (1998). World Water Resources: A new Appraisal and Assessment for the 21st Century. An IHP report, UNESCO, Paris, France.
2. Moiseenko, T.I. (2008). Aquatic ecotoxicology: theoretical

principles and practical application. Water Resources, 35(5), 530-541.

3. Vorosmarty, C.J., McIntyre, P.B., Gessner, M.O., Dudgeon, D., Prusevich, A., Green, P., Glidden, S., Bunn, S.E., Sullivan, C.A., Liermann, C.R. and Davies, P.M. (2010). Global threats to human water security and river biodiversity. Nature, 467, 555-561.
4. Schwarenbach, R.P., Escher, B.I., Fenner, K., Hoffstetter, T.B., Johnson, C.A., von Gunten, U. and Wehrli, B. (2006). The challenge of micropollutants in aquatic systems. Science, 313, 1072-1077.
5. Jurado, A.S., Fernandes, M.A.S., Videira, R.A., Peixoto, F.P. and Vicente, J.A.F. (2011). Herbicides: The Face and the Reverse of the Coin. An in vitro Approach to the Toxicity of Herbicides in Non-Target Organisms, Herbicides and Environment, Andreas Kortekamp (Ed.), ISBN: 978-953-307-476-4, InTech. http://www.intechopen.com/articles/show/title/herbicides-the-face-and-the-reverse-of-the-coin-an-in-vitro-approach-to-the-toxicity-of-herbicides-i. (Accessed September, 2013).
6. PRB (Population Reference Bureau) (2012). 2012 World population data sheet, published by the Population Reference Bureau, 1875 Connecticut Ave., NW, Suite 520, Washington, DC 20009 USA, http://www.prb.org/pdf12/2012-population-data-sheet_eng.pdf, (accessed December 2012).
7. LeBlanc, G.A. (2007). Crustacean endocrine toxicology: a review. Ecotoxicology, 16, 61-81.
8. Correia, T.G., Narcizo, A.M., Bianchini, A., and Moreira, R.G. (2010). Aluminum as an endocrine disruptor in female Nile tilapia (Oreochromis niloticus). Comparative Biochemistry and Physiology, Part C, 151, 461-466.
9. Pedersen, M., Halldorsson, T.I., Mathiesen, L., Mose, T., Brouwer, A., Hedegaard, M., Steffen, L., Kleinjans, J.C.S., Besselink, H. and Knudsen, L.E. (2010). Dioxin-like exposures and effects on estrogenic and androgenic exposures and micronuclei frequency in mother–newborn pairs. Environment International, 36, 344-351.
10. Abel P.D. (2002). Water Pollution Biology (2nd edn.), Taylor and Francis Ltd, London.
11. London, L., Dalvie, M.A. and Cairncross, E. (2005). Approaches

for regulating water in South Africa for the presence of pesticides. Water SA, 31(1), 53-59

12. Mihaich, E.M., Friederich, U., Caspers, N., Hall, A.T., Klecka, G.M., Dimond, S.S., Staples, C.A., Ortego, L.S. and Hentges, S.G. (2009). Acute and chronic toxicity testing of bisphenol A with aquatic invertebrates and plants. Ecotoxicology and Environmental Safety, 72, 1392-1399.
13. Benstead, R.S., Baynes, A., Casey, D., Routledge, E.J., Jobling, S. (2011). 17 β-Oestradiol may prolong reproduction in seasonally breeding freshwater gastropod molluscs. Aquatic Toxicology, 101, 326-334.
14. Lipika, P. and Patra, A. K. (2006). Haemoatopoietic alterations induced by carbaryl in Clarias batrachus (LINN), Journal of applied Sciences and Environmental Management, 10 (3), 5-7.
15. Boran, M., Altinok, I., Capkin, E., Karacam, H. and Bcer, V. (2007). Acute toxicity of carbaryl, methiocarb and carbosulfan to the rainbow trout (Oncorhynchus mykiss) and guppy (Poecilia reticulate), Turk. J. Vet. Anim. Sci., 31.
16. Maharaj S. (2005). Modelling the Behaviour and Fate of Priority Pesticides in South Africa. MSc Thesis, Department of Earth Sciences, University of Western Cape, South Africa.
17. Storrs, S. I. and Kiesiecker, (2004). Survivorship patterns of larval amphibians exposed to low concentrations of atrazine, Environmental Health Perspectives, 112 (10), 1054-105.
18. Scholz, N.L., Incardona, J.P., Baldwin, D.H., Berejikan, B.A., Dittman, A.H., Feist, B.E. and Jordan, C. (2003). Evaluating the sublethal impacts of current use pesticides on the environmental health of salmonids in Columbia River Basin. Bonneville Power Administration FY 2003 Provincial Project Review, 1-41.
19. Khan, M.Z., Tabassum, R., Naqvi, S.N.H., Shah, E.Z., Tabassum, F., Ahmad, I., Fatima, F. and Khan, M.F. (2003). Effect of cypermethrin and permethrin on cholinesterase activity and protein contents in Rana tigrina (Amphibia), Turkey Journal of Zoology, 27, 243-246.
20. Radosevich, S.R., Holt J.S. and Ghersa C.M. (2007). Ecology of Weeds and Invasive Plants: Relationship to Agriculture and Natural Resource Management (3rd edn.), Wiley-Interscience, Hoboken,

USA.

21. Wilson, C. (2009). Aquatic toxicology notes: predicting the fate and effects of aquatic and ditchbank herbicides. Soil and Water Science Department, Institute of Food and Agricultural Sciences, University of Florida, Gainesville, USA.
22. Pérez, G.L., Vera, M.S. and Miranda, L. (2011). Effects of Herbicide Glyphosate and Glyphosate-Based Formulations on Aquatic Ecosystems, Herbicides and Environment, Andreas Kortekamp (Ed.), ISBN: 978-953-307-476-4, InTech. http://www.intechopen.com/articles/show/title/effects-of-herbicide-glyphosate-and-glyphosate-based-formulations-on-aquatic-ecosystems (Accessed November, 2011).
23. Tsui, M.T.K. and Chu, L.M. (2003). Aquatic toxicity of glyphosate-based formulations: comparison between different organisms and the effects of environmental factors. Chemosphere, 52, 1189-1197.
24. Zhou, D-M., Wang, Y-J., Cang, L., Hao, X-Z., and Luo, X-S. (2004). Adsorption and cosorption of cadmium and glyphosate on two soils with different characteristics. Chemosphere, 57, 1237–1244.
25. Tu, M., Hurd, C. and Randall, J.M. (2001).Weed Control Methods Handbook: Tools and Techniques for Use in Natural Areas, The Nature Conservancy. http://tncweeds.ucdavis.edu (Accessed July, 2009).
26. WHO (World Health Organization) (1994). Environmental Health Criteria 159 – Glyphosate, International Programme on Chemical Safety. World Health Organization, Geneva, Switzerland.
27. Miller, A., Gervais, J.A., Luukinen, B., Buhl, K. and Stone, D. (2010). Glyphosate Technical Fact Sheet; National Pesticide Information Center, Oregon State University Extension Services. http://npic.orst.edu/factsheets/glyphotech.html (accessed January 2012).
28. Nordby, D.E., Hager, A.G. (2011). Herbicide formulations and calculations: active ingredient or acid equivalent, a Weed Fact sheet. Integrated Pest Management Handbook, University of Illinois, USA. http://ipm.illinois.edu/weeds/aeai.pdf. Accessed 4 November 2011.
29. Stenersen, J. (2004). Chemical Pesticides: Mode of Action and Toxicology. CRC Press, Boca Raton, Florida, USA, 296 pp.

30. Schuette, J. (1998). Environmental fate of glyphosate, Environmental Monitoring & Pest Management, Department of Pesticide Regulation, Sacramento, CA 95824-5624, USA.
31. Meyer, M.T., Loftin, K.A., Lee, E.A., Hinshaw, G.H., Dietze, J.E. and Scribner, E.A. (2009). Determination of Glyphosate, its Degradation Product Aminomethylphosphonic Acid, and Glufosinate, in Water by Isotope Dilution and Online Solid-Phase Extraction and Liquid Chromatography/Tandem Mass Spectrometry: U.S. Geological Survey Techniques and Methods, book 5, chap. A10, 32p.
32. Herrmann, K.M and Weaver, L.M. (1999). The shikimate pathway. Annual Review of Plant Physiology and Plant Molecular Biology, 50, 473-503.
33. Giesy, J.P., Dobson, S. and Solomon, K.R. (2000). Ecotoxicological risk assessment for Roundup herbicide. Review of Environmental Contamination and Toxicology, 167, 35-120.
34. Glusczak L., Miron D. S., Moraes B. S., Simões R. R., Schetinger M. R. C., Morsch V. M. and Loro V. L. (2007). Acute effects of glyphosate herbicide on metabolic and enzymatic parameters of silver catfish (Rhamdia quelen). Comparative Biochemistry and Physiology-Part C, 146, 519-524.
35. El-Shebly A. A. and El-Kady M. A. H. (2008). Effects of glyphosate herbicide on serum growth hormone (GH) levels and muscle protein content in Nile Tilapia (Oreochromis Niloticus L.). Research Journal of Fisheries and Hydrobiology, 3(2), 84-88.
36. Rubach, M.N., Crum, S.J.H. and Van den Brink, P.J. (2011). Variability in the dynamics of mortality and immobility responses of freshwater arthropods exposed to Chlorpyrifos. Archives of Environmental Contamination and Toxicology, 60, 708-721.
37. Wirth, E.F., Lund, S.A. and Fulton, M.H. (2001). Determination of acute mortality in adults and sublethal embryo responses of Palaemonetes pugio to endosulfan and methoprene exposure. Aquatic Toxicology, 53, 9-18.
38. Key, P.B and Fulton, M.H. (2006). Correlation between 96-h mortality and 24 h acetylcholinesterase inhibition in three grass shrimp larval life stages. Ecotoxicology and Environmental Safety, 63, 389-392.

39. OECD (Organisation for Economic Co-operation and Development) (2005). Detailed review paper on aquatic arthropods in life-cycle and two-generation toxicity tests. An OECD Environment Health and Safety Publications Series on Testing and Assessment, Number 50. Environmental Directorate of Organization for Economic Co-operation and Development, Paris, France.
40. USEPA (United States Environmental Protection Agency) (2000). Methods for Measuring the Toxicity and Bioaccumulation of Sediment-Associated Contaminants with Freshwater Invertebrates. EPA/600/R-99/064. United States Environmental Protection Agency, Office of Prevention, Pesticides and Toxic Substances. Washington, USA.
41. Langdon, C.J., Harmon, V.L., Vance, M.M., Kreeger, K.E., Kreeger, D.A. and Chapman, G.A. (1996). A 7-d Toxicity test for marine pollutants using the Pacific mysid Mysidopsis intii. 1. Culture and protocol development. Environmental Toxicology and Chemistry, 15(10), 1815-1823.
42. Winkler, G. and Greve, W. (2002). Laboratory studies of the effect of temperature on growth, moulting and reproduction in the co-occurring mysids Neomysis integer and Praunus flexuosus. Marine Ecology Progress Series, 235, 177-188.
43. Verslycke, T.A., Fockedey, N., McKenney Jr, C.L., Roast, S.D., Jones, M.B., Mees, J. and Janssen, C.R. (2004). Mysid crustaceans as potential test organisms for evaluation of environmental endocrine disruption: a review. Environmental Toxicology and Chemistry, 23(5), 1219-1234.
44. Widarto, T.H., Krogh, P.H. and Forbes, V.E. (2007). Nonylphenol stimulates fecundity but not population growth rate (λ) of Folsomia candida. Ecotoxicology and Environmental Safety, 67, 369-377.
45. Lachaise, F., le Roux, A., Hubert, M. and Lafont, R. (1993). The moulting gland of crustaceans: localization, activity, and endocrine control (A Review). Journal of Crustacean Biology, 13(2), 198-234.
46. Spaziani, E., Mattson, M.P., Wang, W.L. and McDougall, H.E. (1999). Signaling pathways for ecdysteroid hormone synthesis in crustacean Y-organs. American Zoology, 39, 496-512.
47. Gorokhova, E. (2002). Moult cycle and its chronology in Mysis

mixta and Neomysis integer (Crustacea, Mysidacea): implications for growth assessment. Journal of Experimental Marine Biology and Ecology, 278(2), 179-194.

48. Mu, X. and LeBlanc, G.A. (2002). Environmental antiecdysteroids alter embryo development in the crustacean Daphnia magna. Journal of Experimental Zoology, 292, 287-292.
49. Ketse, N. (2006). The effects of selected reference toxicants on embryonic development of the freshwater shrimp Caridina nilotica (Decapoda: Atyidae), MSc. Thesis, Unilever Centre for Environmental Water Quality (UCEWQ), Rhodes University, Grahamstown, South Africa.
50. Wester, P.W., van der Ven, L.T.M., Vethaak, A.D., Grinwis, G.C.M. and Vos, J.G. (2002). Aquatic toxicology: opportunities for enhancement through histopathology. Environmental Toxicology and Pharmacology, 11, 289-295.
51. Parvez, S. and Raisuddin, S. (2005). Protein carbonyls: novel biomarkers of exposure to oxidative stress-inducing pesticides in freshwater fish Channa punctata (Bloch). Environmental Toxicology and Pharmacology, 20, 112-117.
52. Richardson N., Gordon A.K., Muller W.J., Pletschke B.I. and Whitfield, A.K. (2010). The use of liver histopathology, lipid peroxidation and acetylcholinesterase assays as biomarkers of contaminant induced stress in the Cape stumpnose, Rhabdosargus holubi (Teleostei: Sparidae), from selected South African estuaries. Water SA, 36(4), 407-415.
53. Bernal-Hernandez, Y.Y., Medina-Dıaz, I.M., Robledo-Marenco, M.L., Velazquez-Fernandez, J.B., Giron-Perez, M.I., Ortega-Cervantes, L., Maldonado-Vazquez, W.A. and Rojas-Garcıa, A.E. (2010). Acetylcholinesterase and metallothionein in oysters (Crassostrea corteziensis) from a subtropical Mexican Pacific estuary. Ecotoxicology, 19, 819-825.
54. Kavitha, P.J. and Venkateswara, R. (2008). Toxic effects of chlorpyrifos on antioxidant enzymes and target enzyme acetylcholinesterase interaction in mosquito fish, Gambusia affinis. Environmental Toxicology and Pharmacology, 26, 192-198.
55. Modesto, K.A. and Martinez, C.B.R. (2010). Effects of Roundup

Transorb on fish: hematology, antioxidant defenses and acetylcholinesterase activity. Chemosphere, 81, 781-787.

56. Menendez-Helman, R.A., Ferreyroa, G.V., dos Santos Afonso, M. and Salibian, A. (2012). Glyphosate as an Acetylcholinesterase Inhibitor in Cnesterodon decemmaculatus. Bulletin of Environmental Contamination and Toxicology, 88, 6-9.

57. Banerjee, B.D., Seth, V., Bhattacharya, A., Pasha, S.T. and Chakraborty, A.K. (1999). Biochemical effects of some pesticides on lipid peroxidation and free-radical scavengers. Toxicology Letters, 107, 33-47.

58. Barata, C., Varo, I., Navarro, J.C., Arun, S. and Porte, C. (2005). Antioxidant enzyme activities and lipid peroxidation in the freshwater cladoceran Daphnia magna exposed to redox cycling compounds. Comparative Biochemistry and Physiology, Part C, 140, 175-186.

59. Beuret, C.J., Zirulnik, F., and Gimenez, M.S. (2005). Effect of the herbicide glyphosate on liver lipoperoxidation in pregnant rats and their fetuses. Reproductive Toxicology, 19, 501–504.

60. Bold, T. (2007). Working for Water management treatment summary guide for aquatic and terrestrial alien plants, updated in May 2007, http://www.dwaf.gov.za/wfw/Control (Accessed July, 2009).

61. Romero D.M., Molina M.C.R. and Juarez A.B. (2011). Oxidative stress induced by a commercial glyphosate formulation in a tolerant strain of Chlorella kessleri. Ecotoxicology and Environmental Safety, 74, 741-747.

62. Mensah, P.K., Muller, W.J., Palmer, C.G., 2011. Acute toxicity of Roundup® herbicide to three life stages of the freshwater shrimp Caridina nilotica (Decapoda: Atyidae). Physics and Chemistry of the Earth, 36, 905-909.

63. Mensah, P.K., Muller, W.J., Palmer, C.G., 2012. Using growth measures in the freshwater shrimp Caridina nilotica as biomarkers of Roundup® pollution of South African freshwater systems. Physics and Chemistry of the Earth, 50-52, 262-268.

64. Mensah, P.K., Muller, W.J., Palmer, C.G., 2012. Acetylcholinesterase activity in the freshwater shrimp Caridina nilotica (Decapoda: Atyidae) as a biomarker of the herbicide Roundup® pollution of

freshwater systems in South Africa. Water Science and Technology, 66(2), 402-408.

65. Mensah, P.K., Palmer, C.G., Muller, W.J., 2012. Lipid peroxidation in the freshwater shrimp Caridina nilotica as a biomarker of Roundup® herbicide pollution of freshwater systems in South Africa. Water Science and Technology, 65(9), 1660-1666.

REFERENCES

1. Albuquerque, A. F. *et al.* Pesticides in Brazilian freshwaters: a critical review. *Environ. Sci. Process. Impacts* **18,** 779–787 (2016).
2. Antwi, F. B. & Reddy, G. V. P. Toxicological effects of pyrethroids on non-target aquatic insects. *Environ. Toxicol. Pharmacol.* **40,** 915–923 (2015).
3. Ashton, F. & Craft, A. *Mode of Action of Herbicides*. (John Wiley and Sons, Inc, 1973).
4. Ashton, F. & Crafts, C. *Mode of Action of Herbicides*. (John Wiley and Sons, Inc, 1981).
5. Audus, L. in *The physiology and biochemistry of herbicides* (ed. Audus, L.) (Academic Press, 1964).
6. Bergek, S., Ma, Q., Vetemaa, M., Franzén, F. & Appelberg, M. From individuals to populations: Impacts of environmental pollution on natural eelpout populations. *Ecotoxicology and Environmental Safety* (2012). doi:10.1016/j.ecoenv.2012.01.019
7. Bevenue, A. The bioconcentration aspects of DDT in the

environment. *Residue Rev.* 37–112 (1976).

8. Bondarev, V. S., Spiridonov, Y. Y. & Shestakov, V. G. Effect of sorption on rate of degradation of picloram in soil. *Sov. Soil Sci.* (1986).
9. Boudou, A. & Ribeyre, F. Aquatic ecotoxicology: From the ecosystem to the cellular and molecular levels. *Environmental Health Perspectives* (1997). doi:10.1289/ehp.97105s121
10. Burton, J. in *Herbicide Activity: Toxicology, Biochemistry and Molecular Biology1* (ed. Roe, R. et al) 187–205 (997).
11. Cairns, J., Heath, A. G. & Parker, B. C. The effects of temperature upon the toxicity of chemicals to aquatic organisms. *Hydrobiologia* (1975). doi:10.1007/BF00036747
12. Calderbank, A. & Slade, P. in *Herbicides: Chemistry, degradation and mode of action* (eds. Kearney, P. & Kaufman, D.) (Marcel Dekker Inc, 1976).
13. Canton, J. H. The toxicity of benomyl, thiophanate-methyl, and BCM to four freshwater organisms. *Bull. Environ. Contam. Toxicol.* **16,** 214–218 (1976).
14. Cao, Y., Yi, L., Huang, L., Hou, Y. & Lu, Y. Mechanism and pathways of chlorfenapyr photocatalytic degradation in aqueous suspension of TiO2. *Environ. Sci. Technol.* **40,** 3373–3377 (2006).
15. Christoffoleti, P. J., de Figueiredo, M. R. A., Peres, L. E. P., Nissen, S. & Gaines, T. Auxinic herbicides, mechanisms of action, and weed resistance: A look into recent plant science advances. *Sci. Agric.* **72,** 356–362 (2015).
16. Cope, O. in *Research in Pesticides* (ed. Chichester, F.) (Academic Press, 1965).
17. Copplestone, J. F. The development of the WHO Recommended Classification of Pesticides by Hazard. *Bull. World Health Organ.* **66,** 545–51 (1988).
18. Costa, L. in *Casarett and Doull's Toxicology: The Basic Science of Poisons* (ed. Klassen, C.) 883 (McGraw-Hill, 2008).
19. Cripe, C. R. & Pritchard, P. H. Aquatic test systems for studying the fate of xenobiotic compounds. *ECOL. RES. SER. U.S. ENVIRON. PROT. AGENCY.* (1989).

20. Culture, P. T. & Application, I. B. Uptake and Metabolism of Aldrin in Plant Tissue Cultures. (1977).

21. Davies, P. J. in (ed. Davies, P.) (Martinus Nijhoff Publishers, 1987).

22. DeLorenzo, M. E., Scott, G. I. & Ross, P. E. Toxicity of pesticides to aquatic microorganisms: a review. *Environ. Toxicol. Chem.* **20,** 84–98 (2001).

23. Devine, M., Duke, S. & Fedtke, C. *Physiology of herbicide action.* (Prentice Hall Inc, 1993).

24. Dion, S., Gorse, I. & Tanguay, S. *Guide de classement des ingrédients actifs par groupes chimiques. Fortune* (2007).

25. Dowson, R. J. An introduction to the principles of neurophysiology. *Pestic. Sci.* **8,** 651–660 (1977).

26. Edwards, C. in *Pesticides in Aquatic Environments* (ed. Khan, M.) (Plenum Press, 1977).

27. Eldefrawi, A. in *Comprehensive Insect Physiology, Biochemistry and Pharmacology* (eds. GA, K. & Gilbert, L.) 115 (Pergamon Press, 1985).

28. Epstein, E., Chen, K. H. & Cohen, J. D. Identification of indole-3-butyric acid as an endogenous constituent of maize kernels and leaves. *Plant Growth Regul.* **8,** 215–223 (1989).

29. Food And Agriculture Organization of The United Nations (FAO). Chapter 4 : Pesticides as water pollutants. 1–16 (2014).

30. Foy, C. in *Herbicidides: Chemistry, Degradation, and Mode of Action vol. 2* (eds. Kearney, P. & Kaufman, D.) (Marcel Dekker Inc, 1975).

31. Funari, E., Bottoni, P. & Giuliano, G. in *Chemistry, Agriculture and the Environment* (ed. Richardson, M.) (The Royal Society of Chemistry, 1991).

32. Ghisi, N. D. C. Relationship Between Biomarkers and Pesticide Exposure in Fishes : A Review. *Pestic. – Adv. Chem. Bot. Pestic.* 357–382 (2012). doi:37966

33. Giesy, J. P. & Hoke, R. A. Freshwater sediment toxicity bioassessment: rationale for species selection and test design. *J. Great Lakes Res.* (1989).

34. Graczyk, T. K. & Conn, D. B. Molecular markers and sentinel

organisms for environmental monitoring. *Parasite* (2008). doi:10.1051/parasite/2008153458

35. Gronwald, J. in *Herbicide Resistance in Plants: Biology and Biochemistry* (eds. Holtum, J. & Powles, S.) (Lewis Publishers, 1994).
36. Harris, C. R. Influence of soil type and soil moisture on the toxicity of insecticides in soils to insects. *Nature* **202,** 724 (1964).
37. Hart, L. in *Pesticides in Aquatic Environments* (ed. Khan, M.) (Plenum Press, 1977).
38. Hartley, G. & Graham-Bryce, I. *Physical Principles of Pesticide Behaviour - The dynamics of applied pesticides in the local environment in relation to biological response.* (Academic Press, 1980).
39. Hladik, M. & McWayne, M. USGS Techniques and Methods 5-C3: Methods of Analysis—Determination of Pesticides in Sediment By Using Gas Chromatography/ Mass Spectrometry. (2012).
40. Holt, J. S., Powles, S. B. & Holtum, J. A. M. Mechanisms and agronomic aspects of herbicide resistance. *Annu. Rev. Plant Biol.* **44,** 203–229 (1993).
41. Hurlbert, S. Secondary effects of pesticides on aquatic systems. *Residue Rev.* 82–148 (975).
42. IUPAC. *ompendium of Chemical Terminology 2nd ed ('The Gold Book'). Compiled by A. D. McNaught and A. Wilkinson.* (Blackwell Scientific Publications, 1997). doi:http://dx.doi.org/10.1351/goldbook.I03352
43. Kanazawa, J. Uptake and excretion of organophosphorus and carbamate insecticides by fresh water fish, motsugo, Pseudorasbora parva. *Bull. Environ. Contam. Toxicol.* **14,** 346–352 (1975).
44. Katagi, T. Bioconcentration, bioaccumulation, and metabolism of pesticides in aquatic organisms. *Rev. Environ. Contam. Toxicol.* (2010). doi:10.1007/978-1-4419-1440-8_1
45. Khan, M. in *Pesticides in Aquatic Environments* (ed. Khan, M.) (Plenum Press, 1977).
46. Khan, M., Forte, F. & Payne, J. in *Pesticides in Aquatic Environments* (ed. Khan, M.) (Plenum Press, 1977).

47. Klämbt, D. A view about the function of auxin-binding proteins at plasma membranes. *Plant Mol. Biol.* **14,** 1045–1050 (1990).

48. Kogan, M., Metz, A. & Ortega, R. Adsorption of glyphosate in Chilean soils and its relationship with unoccupied phosphate binding sites. *Pesqui. Agropecu. Bras.* **38,** 513–519 (2003).

49. Kuhr, R. & Dorough, H. *Carbamate insecticides: Chemistry, Biochemistry and Toxicology*. (CRC Press, 1976).

50. Leuba, V. & Letourneau, D. " Plant G wth gulation Auxin Activity of Phenylacetic Acid in Tissue Culture. 71–76 (1990).

51. Lichtenthaler, H. K. Mode of Action of Herbicides Affecting Acetyl-CoA Carboxylase and Fatty Acid Biosynthesis. *Zeitschrift fur Naturforsch. - Sect. C J. Biosci.* (1990). doi:10.1515/znc-1990-0538

52. Lissalde, S. *et al.* Aquatic Toxicology. *Aquat. Toxicol.* (2010). doi:10.1016/j.aquatox.2010.06.006

53. Lydy, M. J., Lohner, T. W. & Fisher, S. W. Influence of pH, temperature and sediment type on the toxicity, accumulation and degradation of parathion in aquatic systems. *Aquat. Toxicol.* (1990). doi:10.1016/0166-445X(90)90010-M

54. Macek, K. in *The biological impact of pesticides in the environment* (ed. Gillet, J.) (Oregon State University, 1970).

55. Matsumura, F. *Toxicology of Pesticides*. (Plenum Press, 1985).

56. Matthews, G. *Pesticides - Health, Safety and the Environment*. (Blackwell Publishing Ltd, 2006).

57. McEwen, L. & Stephenson, G. *The use and significance of pesticides in the environment*. (John Wiley and Sons, Inc, 1979).

58. Monica, S., Comput, I. & Bull, C. Model Ecosystem for the Evaluation of Pesticide Biodegradability and Ecological Magnification. *Experimental* **5,** 709–713 (1971).

59. Morley, H. in *Pesticides in Aquatic Environments* (ed. Khan, M.) (Plenum Press, 1977).

60. Nikinmaa, M. & Nikinmaa, M. in *An Introduction to Aquatic Toxicology* (2014). doi:10.1016/B978-0-12-411574-3.00001-3

61. Nikinmaa, M. & Nikinmaa, M. in *An Introduction to Aquatic Toxicology* (2014). doi:10.1016/B978-0-12-411574-3.00012-8

62. Nikinmaa, M. & Nikinmaa, M. in *An Introduction to Aquatic Toxicology* (2014). doi:10.1016/B978-0-12-411574-3.00016-5

63. Nikinmaa, M. & Nikinmaa, M. in *An Introduction to Aquatic Toxicology* (2014). doi:10.1016/B978-0-12-411574-3.00014-1

64. Nikinmaa, M. & Nikinmaa, M. in *An Introduction to Aquatic Toxicology* (2014). doi:10.1016/B978-0-12-411574-3.00006-2

65. Nikinmaa, M. & Nikinmaa, M. in *An Introduction to Aquatic Toxicology* (2014). doi:10.1016/B978-0-12-411574-3.00017-7

66. Nobel Media. The Nobel Prize in Physiology or Medicine 1948. *Nobelprize.org* (2013). Available at: http://www.nobelprize.org/nobel_prizes/medicine/laureates/1948/. (Accessed: 3rd May 2017)

67. Nowell, L. H. & Resek, E. A. National standards and guidelines for pesticides in water, sediment, and aquatic organisms: application to water-quality assessments. *Rev. Environ. Contam. Toxicol.* **140,** 1–164 (1994).

68. O'Brien, R. in *Biochemistry of Insects* (ed. Wilkinson, C.) 271 (Plenum Press, 1976).

69. Pandey, R. M. Cytotoxic Effects of Pesticides in Somatic Cells of Vicia faba L . 1. **42,** 373–377 (2008).

70. Perspectives, C. Aquatic Toxicology, Molecular, Biochemical and Cellular Perspectives,. *Aquat. Toxicol.* (1995). doi:10.1577/1548-8659-124.5.786

71. Peterson, G. *The Discovery and Development of 2 , 4-D.* **41,** (1967).

72. Pimentel, D. Environmental and economic costs of the application of pesticides primarily in the United States In Integrated Pest Management: Innovation-Development Process. *Environ. Dev. Sustain.* **7,** 229–252 (2005).

73. Porte, C. *et al.* Endocrine disruptors in marine organisms: Approaches and perspectives. *Comparative Biochemistry and Physiology - C Toxicology and Pharmacology* (2006). doi:10.1016/j.cbpc.2006.03.004

74. Reinecke, D. M. & Bandurski, R. S. Auxin Biosynthesis and Metabolism. *Plant Horm. their Role Plant Growth Dev.* **3,** 24–42 (1987).

75. *Herbicide Activity: Toxicology, Biochemistry and Molecular*

Biology. (IOS Press, 1997).

76. Rotchell, J. M. & Ostrander, G. K. Molecular markers of endocrine disruption in aquatic organisms. *J. Toxicol. Environ. Health. B. Crit. Rev.* (2009). doi:10.1080/10937400306476
77. Rubery, P. Auxin receptors. *Annu. Rev. Plant Physiol.* (1981).
78. Sappington, K. Case Study : USEPA Benthic Invertebrate Risk Assessment for Endosulfan. (2013).
79. Schnellmann, R. G. & Manning, R. O. Perfluorooctane sulfonamide: A structurally novel uncoupler of oxidative phosphorylation. *BBA - Bioenerg.* **1016,** 344–348 (1990).
80. Section, E. Mechanism and Pathways of Chlorfenapyr Photocatalytic. 3373–3377 (2006). doi:10.1021/es052073u
81. Shankland, D. in *Insecticide Biochemistry and Physiology* (ed. Wilkinson, C.) 229 (Plenum Press, 1976).
82. Singh, B. K., Walker, A. & Wright, D. J. Persistence of chlorpyrifos, fenamiphos, chlorothalonil, and pendimethalin in soil and their effects on soil microbial characteristics. *Bull. Environ. Contam. Toxicol.* **69,** 181–188 (2002).
83. Solomon, K. R. *et al.* Ecological risk assessment of atrazine in North American surface waters. *Environ. Toxicol. Chem.* (1996). doi:10.1002/etc.5620150105
84. Solomon, K. R. *et al.* Ecological risk assessment of atrazine in North American surface waters. *Environ. Toxicol. Chem.* (1996). doi:10.1897/1551-5028(1996)015<0031:ERAOAI>2.3.CO;2
85. Sterling, T. M. & Hall, J. C. in *Herbicide Activity: Toxicology, Biochemistry and Molecular Biology* (ed. Roe, R. et al) 111–141 (IOS Press, 1997).
86. Stinson, E. R. H. L. A. W. D. L. H. P. Pesticides and Aquatic Animals : A Guide to Reducing Impacts on Aquatic Systems Acknowledgments : *Virginia Polytech. Inst. State Univ.* 24 (2009).
87. Sundaram, K. *Degradation dynamics of fenitrotion in aqueous systems*. (1973).
88. Taylor, W. S., Hixon, M., Chi, H., Marsilii, E. & Rendina Stine-Haskell Research Center, Newark, DE 19714 (USA)), A. R. (E. I. P. de N. and C. Inhibition of acetyl-coenzyme A carboxylase by

coenzyme A conjugates of grass-selective herbicides. *Pesticide Science (United Kingdom)* **v. 43,** (1995).

89. Trinci, A., Ryley, J. F. & others. Mode of action of antifungal agents. in *Mode of action of antifungal agents. symposium of the British Mycological Society held at the University of Manchester, Sept. 1983* (1984).
90. United States Department of Agriculture. Agronomical Technical Note No. 4 - The Use of Chemicals for Weed Control. (1999).
91. US EPA. Overview of Wood Preservative Chemicals - Ingredients Used in Pesticide Products. Available at: https://www.epa.gov/ingredients-used-pesticide-products/chemically-related-groups-active-ingredients. (Accessed: 3rd May 2017)
92. USEPA. *Methods for Measuring the Acute Toxicity of Effluents and Receiving Waters to Freshwater and Marine Organisms. United States Environmental Protection Agency* (2002). doi:Fifth Edition October 2002
93. Walczak, M. & Reichert, M. Characteristics of selected bioaccumulative substances and their impact on fish health. *J. Vet. Res.* **60,** 473–480 (2016).
94. Weaver, L., Gunnerson, C. G., Breidenbach, A. & Lichtenberg, J. J. Chlorinated Hydrocarbon Pesticides in Major U.S. River Basins. *Public Health Rep.* **80,** 481–494 (1965).
95. *Pesticides in Agriculture and the Environment.* (Marcel Dekker Inc, 2002).
96. Whitford, F., Brichford, S. & Turco, R. Pesticides and Water Quality: Principles Policies and Programs. 1–60 (2001).
97. World Health Organization. *Environmental Health Criteria 9: DDT and its Derivatives. International Programme on Chemical Safety* (1979).
98. World Health Organization. The Who Recommended Classification of Pesticides By Hazard and Guidelines To Classification 2009. *World Heal. Organ.* 1–60 (2010). doi:ISBN 978 92 4 154796 3
99. Yu, S. J. *The Toxicology and Biochemistry of Insecticides.* (CRC PRess, 2015).
100. Zelikoff, J. T. *et al.* Biomarkers of immunotoxicity in fish: From the

lab to the ocean. in *Toxicology Letters* (2000). doi:10.1016/S0378-4274(99)00202-7

101. Pesticides and environment, an overview. *Pesticide Action Network* (2010).

102. House of Commons Committees - ENVI (36-2) - Pesticides: Making the Right Choice for the Protection of Health and the Environment - Tax Incentives for Organic Agriculture.

CITATION

CHAPTER 8

Sánchez-Bayo F, Goka K and Hayasaka D (2016) Contamination of the Aquatic Environment with Neonicotinoids and its Implication for Ecosystems. *Front. Environ. Sci.* 4:71. doi: 10.3389/fenvs.2016.00071

CHAPTER 9

Francisco Antonio Guardiola , Alberto Cuesta , José Meseguer and Maria Angeles Esteban, Risks of Using Antifouling Biocides in Aquaculture, doi:10.3390/ijms13021541.

CHAPTER 10

Sajal Ray, Soumalya Mukherjee, Niladri Sekhar Bhunia, Anindya Sundar Bhunia and Mitali Ray (2015). Immunotoxicological Threats of Pollutants in Aquatic Invertebrates, Emerging Pollutants in the Environment - Current and Further Implications, Dr. Marcelo Larramendy (Ed.), InTech, DOI: 10.5772/60216.

CHAPTER 11

Paul K. Mensah, Carolyn G. Palmer and Wilhelmine J. Muller (2014). Lethal and Sublethal Effects of Pesticides on Aquatic Organisms: The Case of a Freshwater Shrimp Exposure to Roundup®, Pesticides - Toxic Aspects, Dr. Sonia Soloneski (Ed.), InTech, DOI: 10.5772/57166.

INDEX

P

R

S

T

U

W